Nanotechnology in Cancer Treatment

Nanotechnology in Cancer Treatment

Special Issue Editors

Steven Fiering
Robert Ivkov

MDPI • Basel • Beijing • Wuhan • Barcelona • Belgrade • Manchester • Tokyo • Cluj • Tianjin

Special Issue Editors
Steven Fiering
Geisel School of Medicine at Dartmouth
USA

Robert Ivkov
Johns Hopkins University School of Medicine
USA

Editorial Office
MDPI
St. Alban-Anlage 66
4052 Basel, Switzerland

This is a reprint of articles from the Special Issue published online in the open access journal *International Journal of Molecular Sciences* (ISSN 1422-0067) (available at: https://www.mdpi.com/journal/ijms/special_issues/nanotechnology_cancer_treatment).

For citation purposes, cite each article independently as indicated on the article page online and as indicated below:

LastName, A.A.; LastName, B.B.; LastName, C.C. Article Title. *Journal Name* **Year**, *Article Number*, Page Range.

ISBN 978-3-03936-374-2 (Pbk)
ISBN 978-3-03936-375-9 (PDF)

Cover image courtesy of Michael Hausmann and Martin Falk.

Contents

About the Special Issue Editors

Steven Fiering, Ph.D., is Professor of Microbiology and Immunology at the Geisel School of Medicine at Dartmouth. His lab's interest is in novel approaches to cancer therapy and has included a variety of approaches utilizing nanoparticles. The approach in his lab is based on developing immunotherapy to treat cancer. A general strategy of interest, in situ vaccination, introduces immunostimulatory reagents into tumors in order to counteract tumor-mediated immunosuppression, generate a strong local antitumor immune response, and then build on that to generate a systemic antitumor immune response to treat metastatic tumors that were not directly treated.

Robert Ivkov, Ph.D., is an Associate Professor in the Department of Radiation Oncology and Molecular Radiation Sciences at The Johns Hopkins University School of Medicine, Baltimore. His lab's research focus includes developing and characterizing magnetic nanoparticles and alternating magnetic field devices for magnetic fluid hyperthermia as an effective heat-based radiation sensitizing cancer therapy. This approach to cancer therapy depends upon cross-disciplinary activities that include developing and characterizing nanoparticle formulations and their magnetic properties, designing and optimizing magnetic coil components to produce therapeutic heat in animal models of human cancer, and understanding the immune-biological effects of iron oxide nanoparticles and the heat they generate to enhance treatment effectiveness.

Preface to "Nanotechnology in Cancer Treatment"

We are pleased to offer this compendium on the theme of "Nanotechnology for Cancer Treatment". The collection is diverse in the topics covered as well as the contributors, who come from across the globe. The overall topic is inherently a complex one since it integrates the exceptional complexity of cancer biology and treatment with the large and rapidly growing number of potential nanostructures and modifications that could be applied to treatment. In both cancer and nanotechnology, a very small change of a relevant variable can generate exceptional variability in biological outcome, creating inherent challenges to clinical adoption. As with many high-potential fields in medicine, the recognition of exceptional potential for nanotechnology in cancer therapy, with associated expectations of rapid clinical impact, has been followed by disappointingly slow progress that highlights not failure but rather unrealistic expectations for rapid progress. We have passed that period of expected rapid clinical impact for nanoparticle treatment of cancer and are now well-entrenched in the more realistic phase of exploiting the easiest applications, in this case packaging of established chemotherapy drugs, while carefully studying the exceptional numbers of relevant variables that must be evaluated for efficacy and safety of innovative strategies. Although never as rapid as hoped, the progress is clear and clinical impact is growing.

The nanoparticles (NPs) that are already in clinical use are all delivering established chemotherapeutics with lower toxicity and equal or higher efficacy. This has been the area of rapid contribution of NPs to cancer treatment. This remains a major area of effort, and there are undoubtedly many more important contributions that will further improve results with known chemotherapeutics when packaged in NPs. Six of the articles provide results of careful studies intended to understand how nanoparticles interact with various cell types within mammals and how that influences delivery of cancer chemotherapy. Lee et al. evaluate how modifications of hydrophobicity of silicon NPs modifies interactions with various cell types in the liver. Folate targeting of NPs carrying arsenic trioxide is studied as treatment for cervical cancer by Ahktar et al. Vu et al. investigate delivery of carboplatin by dendrimer NPs. Shih et al. develop and study 264 nm pectin-type B gelatin NPs carrying curcumin, a high-potential chemotherapeutic that has been limited by poor solubility and bioavailability. Their studies model oral administration of the NPs for colorectal cancer and evaluate performance in the extremes of pH found in the GI tract. Pantshwa et al. describe a drug depot system that releases antibody-functionalized methotrexate-delivering nanomicelles embedded within and released from a complex chitosan-composite hydrogel. The system is designed to be implanted intraperitoneally to slow release the micelles, in this case targeted to the MUC16 surface molecule often found on ovarian cancer. Tested in vivo in mice, the system supported increased survival in ovarian-cancer-challenged mice as compared to IV drug treatments or placebo. Mahmud et al. combine liposomes carrying doxorubicin with targeting using a chlorotoxin peptide fused to a human IgG Fc region in glioblastoma models. The chlorotoxin peptide has affinity for MMP2, which is often found on glioblastoma cells. In vivo mouse studies showed that the chlorotoxin-Fc carrying doxorubicin liposomes increased animal survival significantly better than either free doxorubicin or non-targeted doxorubicin liposomes, and both liposome formulations had less toxicity than free doxorubicin.

The broad array of materials available for NP formulation supports testing of many different particles clinically. However, as noted by Wang et al., the inclusion of novel materials carrying established chemotherapeutic agents slows clinical adoption because the carrier materials must be carefully tested clinically for toxicity. The article reviews a totally different concept, formulating the drugs themselves into nanomaterials without excipients. The review covers the development of this field, delineating how the various approaches are categorized by nanoparticle formation strategy and outlining the potential of these strategies and their associated challenges.

Metals NPs have a variety of important properties and potential usages for cancer therapy. Bhat et al. analyzed interactions of 10 nm gold or silver NPs with artificial membrane vesicles doped with various concentrations of cholesterol. The interactions were quite different between these NPs, with gold NPs concentrating between the lipid bilayers and silver NPs accumulating on the surface of the vesicle in small clusters. Pagacova et al. investigated the potential for very small (2–10 nm) gold or platinum NPs for sensitizing cancer cells to radiation. The study used quite sophisticated assay approaches including single molecule localization microscopy to assay the effects on double-strand breakage and repair. The complex data generated by these careful studies indicated that there are multiple effects of these NPs on cellular response to radiation and they are mediated by both nuclear and cytoplasmic changes.

In addition to the impact of nanotechnology on established chemotherapy and radiation approaches, there is a therapeutic revolution that is harnessing the patient's own immune system to treat cancer, and nanotechnology will contribute. Gallego-Lleyda et al. developed lipid NPs that carry the TRAIL ligand and tested them as treatment for sarcomas. Some tumors carry death receptors, TRAIL R-1 or R-2, that mediate apoptosis when engaged by TRAIL ligand that is often expressed on leukocytes. The study tested treatment of sarcoma cells using lipid nanoparticles with surface tethered TRAIL that mimics the physiological secretion of TRAIL on the membranes of exosomes. The study showed that the nanoparticles are more effective against sarcoma cells than soluble TRAIL.

Historically, nanoparticle therapeutics for cancer tried to avoid ingestion of the NP by phagocytic leukocytes like macrophages. However, as interest in immunotherapy for cancer has increased and the role of immunosuppressive innate immune cells has become clear, the ability to change the phenotype of immunosuppressive phagocytes using NPs has become a hot topic. The review by Ponzoni et al. considers multiple aspects of how this is being approached and highlights results in preclinical models.

The process of translating new lab discoveries to the clinic is never as rapid as hoped; despite some rapid repackaging of established chemotherapeutics, nanotechnology is not an exception to the slow pace of translation. Notwithstanding this reality, it is clear that nanotechnology was not wrongly claimed to hold considerable promise for cancer therapy, and the slow, careful work that translation requires is continued with reports such as those in this Special Issue. While the eventual high-impact therapies are still impossible to recognize, the promising approaches are coming into focus, and many seeds have been planted and are growing toward eventual harvest.

Steven Fiering, Robert Ivkov
Special Issue Editors

International Journal of
Molecular Sciences

MDPI

Article

Hepatic Cellular Distribution of Silica Nanoparticles by Surface Energy Modification

A-Rang Lee [1,†], Kibeom Nam [2,†], Byeong Jun Lee [2], Seoung-Woo Lee [1], Su-Min Baek [1], Jun-Sun Bang [1], Seong-Kyoon Choi [3], Sang-Joon Park [4], Tae-Hwan Kim [1], Kyu-Shik Jeong [1,5], Dong Yun Lee [2,*] and Jin-Kyu Park [1,*]

1 Department of Veterinary Pathology, College of Veterinary Medicine, Kyungpook National University, Daegu 41566, Korea
2 Department of Polymer Science and Engineering, Kyungpook National University, Daegu 41566, Korea
3 Core Protein Resources Center, Daegu Gyeongbuk Institute of Science and Technology (DGIST), Daegu 42988, Korea
4 Laboratory of Veterinary Histology, College of Veterinary Medicine, Kyungpook National University, Daegu 41566, Korea
5 Stem Cell Therapeutic Research Institute, Kyungpook National University, Daegu 41566, Korea
* Correspondence: dongyunlee@knu.ac.kr (D.Y.L.); jinkyu820@knu.ac.kr (J.-K.P.)
† These authors contributed equally to this study.

Received: 15 June 2019; Accepted: 31 July 2019; Published: 5 August 2019

Abstract: The cellular distribution of silica nanoparticles (NPs) in the liver is not well understood. Targeting specific cells is one of the most important issues in NP-based drug delivery to improve delivery efficacy. In this context, the present study analyzed the relative cellular distribution pattern of silica NPs in the liver, and the effect of surface energy modification on NPs. Hydrophobic NP surface modification enhanced NP delivery to the liver and liver sinusoid fFendothelial cells (LSECs). Conversely, hydrophilic NP surface modification was commensurate with targeting hepatic stellate cells (HSCs) rather than other cell types. There was no notable difference in NP delivery to Kupffer cells or hepatocytes, regardless of hydrophilic or hydrophobic NP surface modification, suggesting that both the targeting of hepatocytes and evasion of phagocytosis by Kupffer cells are not associated with surface energy modification of silica NPs. This study provides useful information to target specific cell types using silica NPs, as well as to understand the relationship between NP surface energy and the NP distribution pattern in the liver, thereby helping to establish strategies for cell targeting using various NPs.

Keywords: silica nanoparticles; surface energy modification; NP-based drug delivery

1. Introduction

Nanoparticle (NP)-based drug delivery has emerged as a new alternative therapeutic strategy for many diseases [1]. Since the NPs allow the drug to reach whole liver cells such as liver sinusoidal endothelial cells (LSECs), hepatocytes, and Kupffer cells, the cell fate of the NPs is important [2,3]. Drug delivery effects of NPs can vary depending on targeted cell types as well as the locations of targeted cells in the liver. Targeting the right cells can significantly improve the efficiency of NP-based drug delivery and also enhance stability and safety [4]. Conversely, when NPs are mainly engulfed by Kupffer cells, this can lower the level of NPs acting on other cells [2]. Therefore, the selective delivery of NPs to the targeted specific cells is one of the most important issues in NP-based drug delivery. Modification of the NP surface is a universal approach to improving the efficacy of NP targeting to specific cell types [4]. Information on the relative NP uptake quantity per liver cell is essential to ensuring accurate cell targeting.

An NP is a small object with a diameter of 1–500 nm that can be composed of various materials, such as silica, polylactic-co-glycolic acid, and silver [2,5]. NPs are widely applied in medicine and pharmaceutical fields. Particularly, low-toxicity silica NPs are used in many biomedical applications, due to their biocompatibility, low toxicity, and expandable composition [6–8]. Precision control of silica particle size, porosity, crystallinity, and shape allows nanostructures to be tuned for a variety of applications [6]. In addition, many possible surface modifications of silica NPs enable the precise control of surface chemistry to regulate drug or chemical loading, and the dispersion, blood circulation, and site-specific targeting of the NPs [9]. The ability to combine these properties makes silica NPs a viable platform for biomedical imaging, analysis, delivery, monitoring, and resection therapies. For instance, silica nanoparticles can be functionalized as contrast agents to enable bio-imaging [10]. NP-based drug delivery systems can also selectively target the tumor area due to improved permeability and retention effects [10]. NPs also have the potential to monitor and treat the disease of interest. Imaging agents are already integrated into nanomaterials and can be easily functionalized to include therapeutic functions [10]. Thus, it is possible to fabricate multimodal silica NPs for therapeutic applications, such as diagnostic imaging components or gene modifying components.

Cancer is the second most common cause of death worldwide, and statistics show that the number of deaths by cancer was 9.6 million (almost 1 in 6 deaths) in 2018 [11]. Existing cancer therapies include radiation, hormones, and chemotherapy to reduce cancer size, and surgical removal of cancer tissue. Among these methods, radiation therapy and surgical removal are commonly used and considered as the most effective treatment for various types of cancer [12]. However, there are still various limits such as the difficulty for early detection of cancer cells and inadequate anticancer drug concentration in target cancer cells, and the inability to monitor treatment response [13]. It is essential to deliver anticancer drugs selectively to the appropriate cells [14]. Thus, several research studies using NP-based drug delivery have emerged over the past years for targeting cancer cells. However, the stability and safety of NPs must be ensured to be applied to humans for therapeutic purposes. In recent years, several studies about NPs being undertaken have completed the toxicity test and have been approved by the Food and Drug Administration (FDA) [2].

Therefore, reducing the toxicity of nanomaterials and improving their efficiency are important in order to develop highly potential therapies in cancer research.

Hepatocellular carcinoma (HCC), a common primary malignant tumor, is the leading cause of cancer-related deaths worldwide [15]. HCC usually occurs in patients with chronic liver disease. Among these, viral hepatitis, non-alcoholic fatty liver disease and alcoholic liver disease are the main causes of HCC [16]. Hepatocytes are not only targets for the treatment of hepatocellular carcinoma, but also have stem cell-like abilities for almost infinite regeneration, which can be a cause of hepatocellular carcinoma [16]. Successful targeting of hepatocytes may enable liver cancer to be treated more efficiently. Therefore, it is important to target hepatocytes specifically for the successful treatment of hepatocellular carcinoma.

Surface charge or hydrophobicity of NPs is one of the most important factors for cell-targeted NP delivery [4,17]. Previous studies have demonstrated that the binding of plasma proteins to NPs is mainly reliant on the hydrophobicity, surface charge, composition, and size of the NP [17,18]. Specifically, the affinity of NPs to albumin, immunoglobulin, fibrinogen, and lipoprotein can affect and determine the NP biodistribution pattern via the bloodstream, as well as the cellular distribution of NPs among various parenchymal and non-parenchymal cells, including immune cells, endothelial cells, and mesenchymal cells [4,19,20]. Moreover, the cellular distribution pattern of the NPs can vary depending on the quantity and properties (e.g., size, surface energy) of the proteins binding to the NPs. Therefore, an evaluation of the cellular distribution pattern of NPs is very important in optimizing cell-targeted NP delivery.

In the liver, the size of the fenestrae in LSECs is estimated to be about 280 nm [21,22]. Thus, NP size should be under 280 nm for NP delivery to hepatocytes [23]. The present study analyzed the relative cellular distribution patterns of silica NPs in the liver, according to NP surface energy modification. We modified the surface polarity of silica NPs to hydrophilic or hydrophobic.

2. Results

2.1. Silica NP Synthesis, Surface Modification, and Characterization

NPs were prepared through the Stöber method, using rhodamine B isothiocyanate (RBITC) as a core structure, to enable NP observation or tracking by fluorescence microscopy (Figure 1A). Since prepared silica NPs have a hydroxyl group on their surfaces, they have a hydrophilic nature. To make hydrophobic silica NPs, aliphatic hydrocarbon groups (i.e., *n*-octadecyltrichlorosilane, ODTS) were attached to the surface of NPs through hydrolysis and condensation reactions (Figure 1A). This surface modification of the silica NPs was conducted to determine the cellular distribution property according to the surface energy of the NPs. Silica NPs have an average uniform diameter of approximately 260 nm (Figure 1B). To identify the exact size of hydrophilic and hydrophobic NPs, we measured particle size in solution environments by dynamic light scattering (DLS) (Figure 1C). When hydrophilic NPs and hydrophobic NPs are dispersed in olive oil and PBS solution, hydrophilic NPs in PBS and hydrophobic NPs in olive oil had similar sizes, whereas hydrophobic NPs dispersed in PBS showed different sizes even though we used the same particles (Figure 1C). This result shows that silica NPs are aggregated by the solvent property and have a different distribution in solution. In the modified silica NPs, C-H_x stretching modes at around 2800 and 2900 cm^{-1} are notable compared with those of pure silica, which means that the silane moieties have combined with silica via the silanization reaction (Figure 1D). The stretching vibration of Si-OH was identified near 800 cm^{-1}, the Si-O-Si stretching vibration was identified at 1000–1300 cm^{-1}, and the peak around 3500 cm^{-1} is the characteristic mode of $Si(O\text{-}H)_x$ (Figure 1D). RBITC encapsulated in NPs emitted red light at around 590 nm when excited by light generated from the mercury lamp. When the solvent containing the silica NPs was visualized under a fluorescence microscope, only RBITC-containing NPs could be observed (Figure 1E). Interestingly, the dispersion pattern of silica NPs varied, based on the characteristics of the NP surface and the solvents. Untreated silica NPs (hydrophilic-NP-phosphate-buffered-saline (PBS)) dispersed well, without aggregation, due to their hydrophilic property and hydrogen-bonding interactions with the surrounding medium (Figure 1E). However, hydrophobic ODTS-modified silica NPs formed a large aggregated mass when PBS was employed as a solvent (hydrophobic-NP-PBS). It is thought that the ionic PBS buffer solution generated repulsive forces against hydrophobic silica NPs and prevented them from dispersing in the solvent as single particles. When olive oil was adopted as the solvent, ODTS-modified silica NPs (hydrophobic-NP-olive oil) showed much better dispersion than those dispersed in PBS solution. Hence, the polarity of solvents should be considered to disperse silica NPs homogeneously into solutions, depending on the characteristics of the materials.

2.2. Hydrophobic Surface Modification of Silica NPs Resulted in Increased NP Delivery to the Liver

Silica NPs were intraperitoneally injected into mice once, and the mice were sacrificed after 24 h. In the liver sections, NP distribution was detected at 555 nm under a fluorescence microscope. Interestingly, hydrophobic NP-injected mice (hydrophobic-NP-PBS, hydrophobic-NP-olive oil) exhibited a much increased delivery of silica NPs to the liver compared with hydrophilic-NP-PBS (Figure 2A). There was no difference in the NP delivery amount between hydrophobic-NP-PBS treated liver and hydrophobic-NP-olive oil treated liver (Figure 2A). The NP fluorescence intensity value was significantly higher for the hydrophobic NP-treated liver than the hydrophilic-NP-PBS treated liver (Figure 2B). However, the number of NP-positive cells was similar among all groups (Figure 2C), suggesting that the quantity of NPs taken up by each cell might be much higher in hydrophobic NP-treated livers compared with hydrophilic-NP-PBS treated liver. As expected, the NP fluorescence intensity level per cell was significantly higher in hydrophobic-NP-PBS treated liver and hydrophobic-NP-olive oil treated liver relative to that in the hydrophilic-NP-PBS treated liver

(Figure 2C). There was no significant difference in the NP fluorescence intensity value between hydrophobic-NP-PBS treated liver and hydrophobic-NP-olive oil treated liver (Figure 2B,D). These results indicate that the delivery of silica NPs to the liver can be improved by preparing silica NPs with hydrophobic surface characteristics.

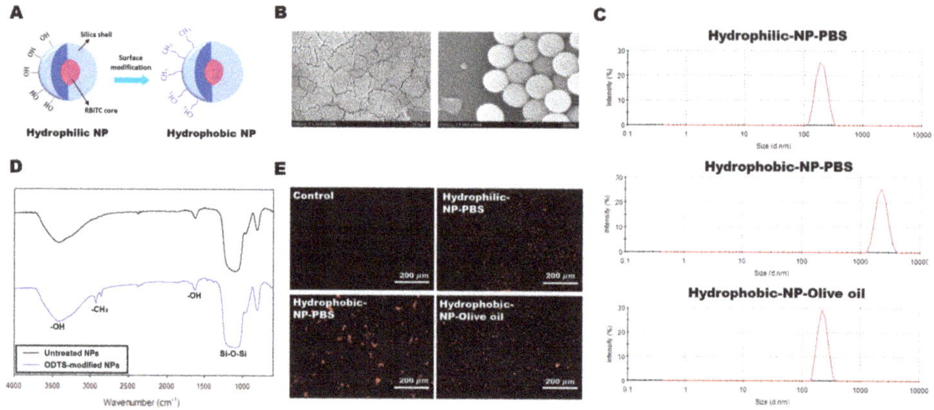

Figure 1. Silica nanoparticle (NP) synthesis, surface modification, and characterization. (**A**) Schematic illustration of rhodamine B isothiocyanate (RBITC)-loaded silica NPs and surface modification with *n*-octadecyltrichlorosilane (ODTS). (**B**) SEM images at (left) low and (right) high magnification of silica NPs. (**C**) Particle size determination by dynamic light scattering (DLS). (**D**) FT-IR spectra of pure silica (black line) and ODTS-modified silica (blue line). (**E**) Fluorescence images of untreated and ODTS-modified silica NPs in different solutions. Dispersion pattern of silica NPs (red) varied based on the characteristics of the NP surface and solvents.

2.3. No Difference Existed in the Ratio of NPs Absorbed by Kupffer Cells among Hydrophilic-NP-PBSTreated Liver, Hydrophobic-NP-PBS Treated Liver, and Hydrophobic-NP-Olive Oil Treated Liver

To determine the cellular distribution of silica NPs, depending on surface characteristics, the NP distribution per each cell type, including Kupffer cells, LSECs, hepatic stellate cells (HSCs), and hepatocytes, was analyzed by immunofluorescence. First, the NP distribution taken up by Kupffer cells was assessed. Accordingly, immunofluorescence with CD68 antibody was used to identify both NP-positive and CD68-positive Kupffer cells (Figure 3A). The NP-positive and CD68-positive Kupffer cells were quantitatively similar among all types of NP-treated livers, without any significant differences (Figure 3A,B). The proportion of NP-positive Kupffer cells among the entire NP-positive liver cell population was constituted by 37 ± 3.9% hydrophilic-NP-PBS, 36 ± 3.7% hydrophobic-NP-PBS, and 32 ± 5.7% hydrophobic-NP-olive oil (Figure 3D). Despite the lack of significant differences in the values among the distinct NP types (Figure 3C), the data suggested that the amount of NPs consumed per Kupffer cell might be higher in the hydrophobic NP-treated liver than in the hydrophilic-NP-PBS treated liver. As expected, the NP fluorescence intensity value per CD68-positive Kupffer cell was significantly higher in hydrophobic-NP-PBS treated liver and hydrophobic-NP-olive oil treated liver in comparison to the hydrophilic-NP-PBS treated liver (Figure 3D). There was no significant difference in the NP fluorescence intensity value between hydrophobic-NP-PBS treated liver and hydrophobic-NP-olive oil treated liver. It inferred that the surface characteristic (hydrophilic or hydrophobic) of silica NPs did not affect their cellular distribution in the liver, although the amount of NPs reaching the liver was greater in the hydrophobic NP-treated liver relative to that of the hydrophilic-NP-PBS treated liver.

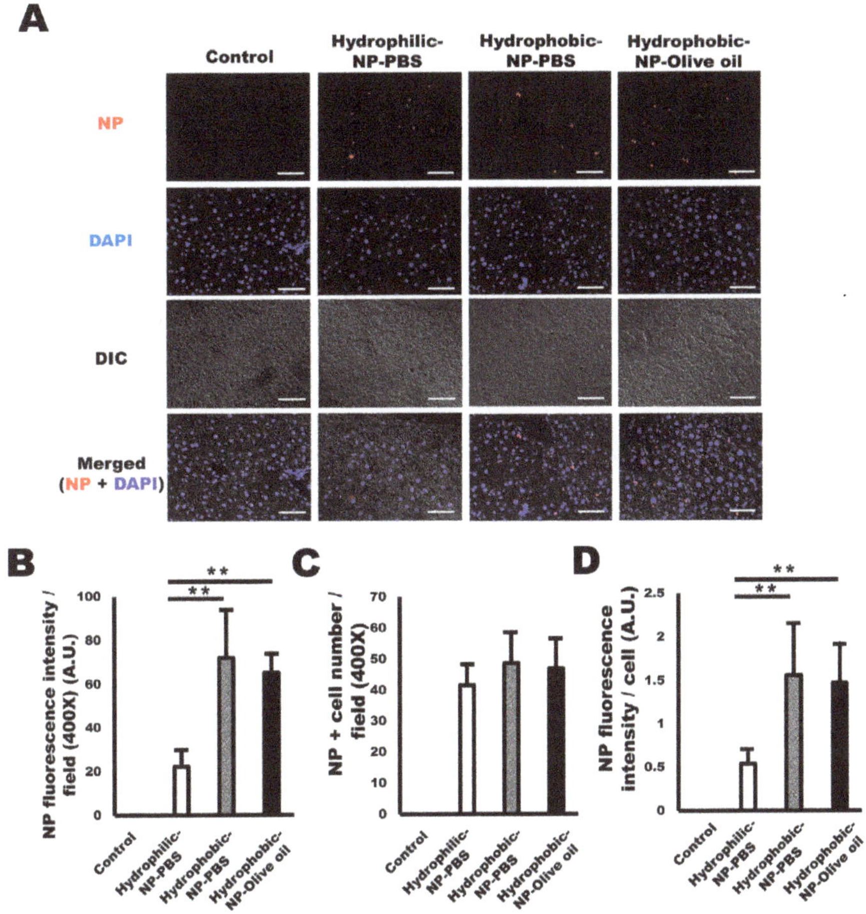

Figure 2. Enhanced NP delivery to the liver by hydrophobic surface modification of silica NPs. (**A**) Representative immunofluorescence micrographs of NPs (red) delivered to the liver. Scale bar = 50 µm. Blue = DAPI. (**B**) NP fluorescence intensity values (arbitrary unit) of NP-treated liver. Hydrophobic NP-treated livers show significantly higher values compared with that of hydrophilic-NP-PBS treated liver. (**C**) The number of NP-positive cells per field based on each surface modification of silica NPs and solvent condition. (**D**) NP fluorescence intensity (arbitrary unit) per cell in NP-treated liver. All data were quantified from 10 fields (400×) per tissue and are shown as mean ± SD. $^{**}p < 0.01$.

2.4. NP Delivery to LSECs was Enhanced by Hydrophobic Surface Modification

Next, we attempted to analyze NP uptake by LSECs in all types of silica NP-treated liver. Immunofluorescence was performed using CD34 antibody to visualize LSECs taking up the NPs (Figure 4A). There was a significantly higher number of both NP-positive and CD34-positive LSECs in hydrophobic-NP-PBS treated liver and hydrophobic-NP-olive oil treated liver when compared with the hydrophilic-NP-PBS treated liver (Figure 4A,B), possibly implying that silica NPs with a hydrophobic surface might have a higher affinity for LSECs than their hydrophilic counterparts. Surprisingly, the percentage of NP-positive LSECs contributing to the entire NP-positive liver cell population was constituted by 29 ± 4.0% hydrophilic-NP-PBS, 42 ± 4.1 % hydrophobic-NP-PBS, and 39 ± 6.9% hydrophobic-NP-olive oil (Figure 4C). Moreover, the NP-positive LSEC ratio was significantly higher in hydrophobic NP-treated liver than in hydrophilic-NP-PBS treated liver, suggesting that silica NPs with a hydrophobic surface have a greater tendency to be taken up by LSECs than those with a hydrophilic surface.

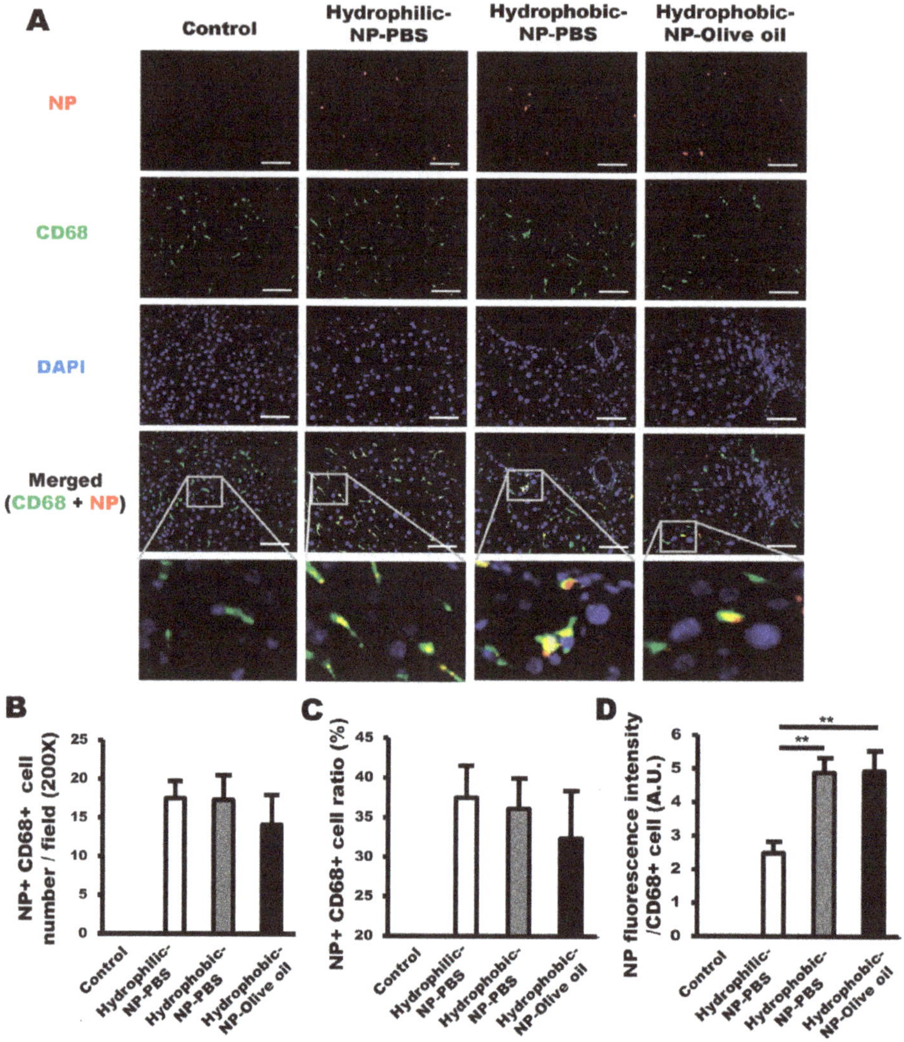

Figure 3. NP uptake by Kupffer cells among hydrophilic-NP-PBS treated liver, hydrophobic-NP-PBS treated liver, and hydrophobic-NP-olive oil treated liver. (**A**) Representative immunofluorescence micrographs of NPs (red) and CD68-positive Kupffer cells (green). Kupffer cells retaining NPs are shown in yellow in the merged images. Scale bar = 100 μm. Blue = DAPI. (**B**) Number of both NP-positive and CD68-positive Kupffer cells per field (200×). (**C**) Ratios of NP-positive Kupffer cells among entire NP-positive cell population. (**D**) Values of NP fluorescence intensity per CD68-positive Kupffer cell in NP-treated liver. All data were quantified from 10 fields (200×) per tissue and are shown as mean ± SD. ** $p < 0.01$.

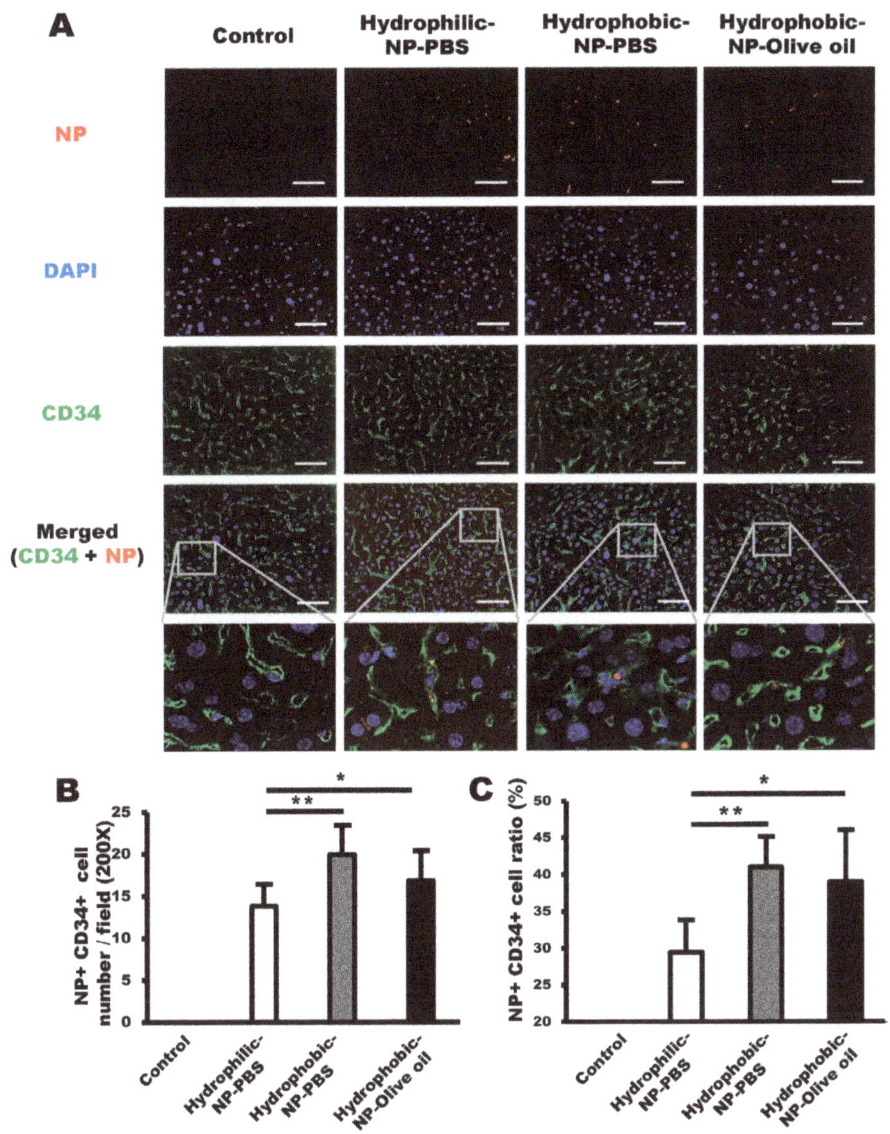

Figure 4. Increased NP delivery to LSECs (liver sinusoidal endothelial cells) by hydrophobic surface modification of silica NPs. (**A**) Representative immunofluorescence micrographs of NPs (red) and CD34-positive LSECs (green). LSECs retaining NPs are shown in yellow in the merged image. Scale bar = 100 μm. Blue = DAPI. (**B**) Number of both NP-positive and CD34-positive LSECs per field (200×). Number of LSECs retaining NPs in hydrophobic-NP-PBS treated liver and hydrophobic-NP-olive oil treated liver was significantly higher compared with that in hydrophilic-NP-PBS treated liver. (**C**) Ratios of NP-positive LSECs among entire NP-positive cell population. All data were quantified from 10 fields (200×) per tissue and are shown as mean ± SD. *$p < 0.05$. **$p < 0.01$.

2.5. Hydrophilic Surface Modification of Silica NPs Resulted in Elevated NP Delivery to HSCs

Desmin antibody was used to observe NP distribution in HSCs. The overall NP population taken up by HSCs was very low. The number of NP-positive HSCs was significantly higher in the hydrophilic-NP-PBS treated liver (Figure 5A,B), meaning silica NPs with a hydrophilic surface might reach more HSCs compared with hydrophobic silica NPs. As expected, there were significant differences in the percentage of NP-positive HSCs contributing to the entire NP-positive cell population among the three different types of silica NP-treated livers, being 29 ± 3.3% in hydrophilic-NP-PBS treated liver, 20 ± 4.4% in hydrophobic-NP-PBS treated liver, and 25 ± 4.3% in hydrophobic-NP-olive oil treated liver (Figure 5C). The ratio of NP-positive HSCs of hydrophilic-NP-PBS treated liver was significantly higher compared with that of hydrophobic-NP-PBS treated liver and, also, almost significantly ($p = 0.05$) higher compared with that of hydrophobic-NP-olive oil treated liver (Figure 5C). Interestingly, there was also a significant difference in the ratio of NP-positive HSCs between hydrophobic-NP-PBS treated liver and hydrophobic-NP-olive oil treated liver (Figure 5C). Since the ratio of NP-positive HSCs was significantly higher in the hydrophobic-NP-olive oil treated liver compared with that of the hydrophobic-NP-PBS treated liver (Figure 5C), it could be surmised that the differences between the two hydrophobic NP-treated livers could be induced by the different characteristics of the solvent used to dissolve the silica NPs.

2.6. Surface Modification of Silica NPs Did Not Affect NP Delivery to Hepatocytes

Differential interference contrast microscopy and fluorescence imaging (excitation filter, 555 nm) were used to evaluate the NP distribution in hepatocytes. NPs were generally observed around the nuclei of hepatocytes, and the NP population in hepatocytes was very few in all NP-treated livers (Figure 6A). The numbers of NP-positive hepatocytes were almost the same, without any significant difference among all types of NP-treated livers (Figure 6B). The percentage of NP-positive hepatocytes as a proportion of the entire NP-positive cell population was about 4 ± 2.3% in hydrophilic-NP-PBS treated liver, 2 ± 1.9% in hydrophobic-NP-PBS treated liver, and 3 ± 3.3% in hydrophobic-NP-olive oil treated liver, respectively (Figure 6C). Namely, the ratio of NP-positive hepatocytes was less than 4% in all types of NP-treated liver regardless of NP surface modification and the solvent used to disperse NPs.

2.7. Hydrophobic-NP-PBS Induced Infiltrations of Inflammatory Cells in the Liver

Next, we performed a histopathological analysis of silica NP-injected liver tissue. Silica NPs generally induced a slight infiltration of inflammatory cells into the liver (Figure 7A). To quantify the inflammation, the number of inflammatory foci was counted per 100× field. For the hydrophilic-NP-PBS treated liver and hydrophobic-NP-olive oil treated liver, one or two inflammatory foci were shown per 100× field (Figure 7A). In comparison, hydrophobic-NP-PBS treated liver showed a significantly increased level of inflammatory foci (about 4 or 5 foci per 100× field) (Figure 7A,B). These data suggest that an increased infiltration of inflammatory cells was largely induced by the condensed NP cluster formation of hydrophobic-NP-PBS.

2.8. Relative Ratio of NP Uptake by Kupffer Cells, LSECs, HSCs, and Hepatocytes Based on Surface Modification of Silica NPs

Finally, we analyzed the relative ratio of NP uptake based on each cell type and based on the surface modification condition of the silica NPs. In hydrophilic-NP-PBS treated liver, Kupffer cells mainly took up NPs (38%), followed, equally, by HSCs (29%) and LSECs (29%), and then hepatocytes (4%). Interestingly, in hydrophobic-NP-PBS treated liver, LSECs were the main cells taking up NPs (41%), followed by Kupffer cells (36%), HSCs (21%), and hepatocytes (2%). The hydrophobic-NP-olive oil treated liver showed that the hierarchy of NP uptake decreased as follows: LSECs (39%) > Kupffer cells (32%) > HSCs (26%) > hepatocytes (3%). These trends prove that hydrophobic surface modification of NPs may be helpful to target LSECs, whereas hydrophilic surface modification of NPs may enhance their affinity to HSCs. Moreover, there was no notable change in NP uptake ratios between hepatocytes and Kupffer cells by the surface modification of silica NPs, indicating that targeting hepatocytes and evasion of phagocytosis by Kupffer cells are not associated with hydrophilic or hydrophobic surface modification of silica NPs.

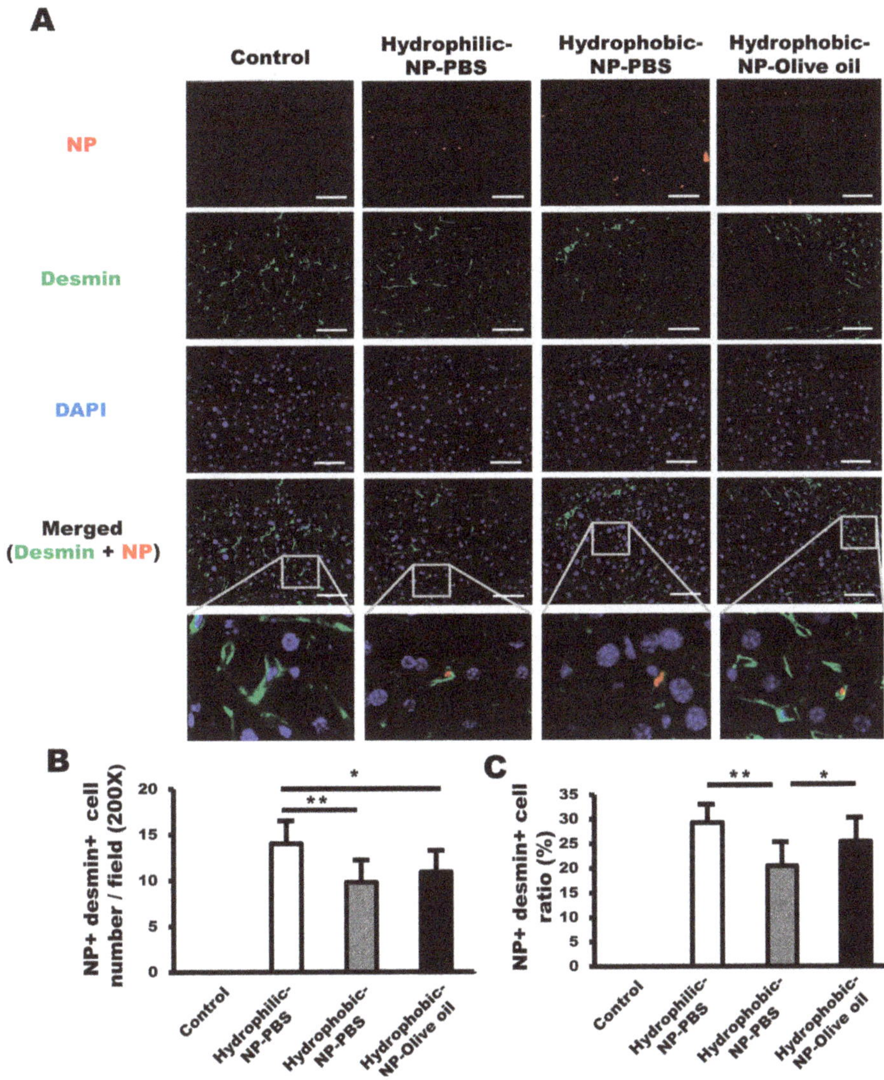

Figure 5. Elevated NP delivery to HSCs (hepatic stellate cells) by hydrophilic surface modification of silica NPs. (**A**) Immunofluorescence image of NPs (red) and desmin-positive HSCs (green). HSCs retaining NPs are shown in yellow in the merged image. Scale bar = 100 µm. Blue = DAPI. (**B**) Number of both NP-positive and desmin-positive HSCs per field (200×). The hydrophilic-NP-PBS treated liver showed the highest number of HSCs retaining NPs. (**C**) Ratios of NP-positive HSCs among entire NP-positive cell population. All data were quantified from 10 fields (200×) per tissue and are shown as mean ± SD. *$p < 0.05$. **$p < 0.01$.

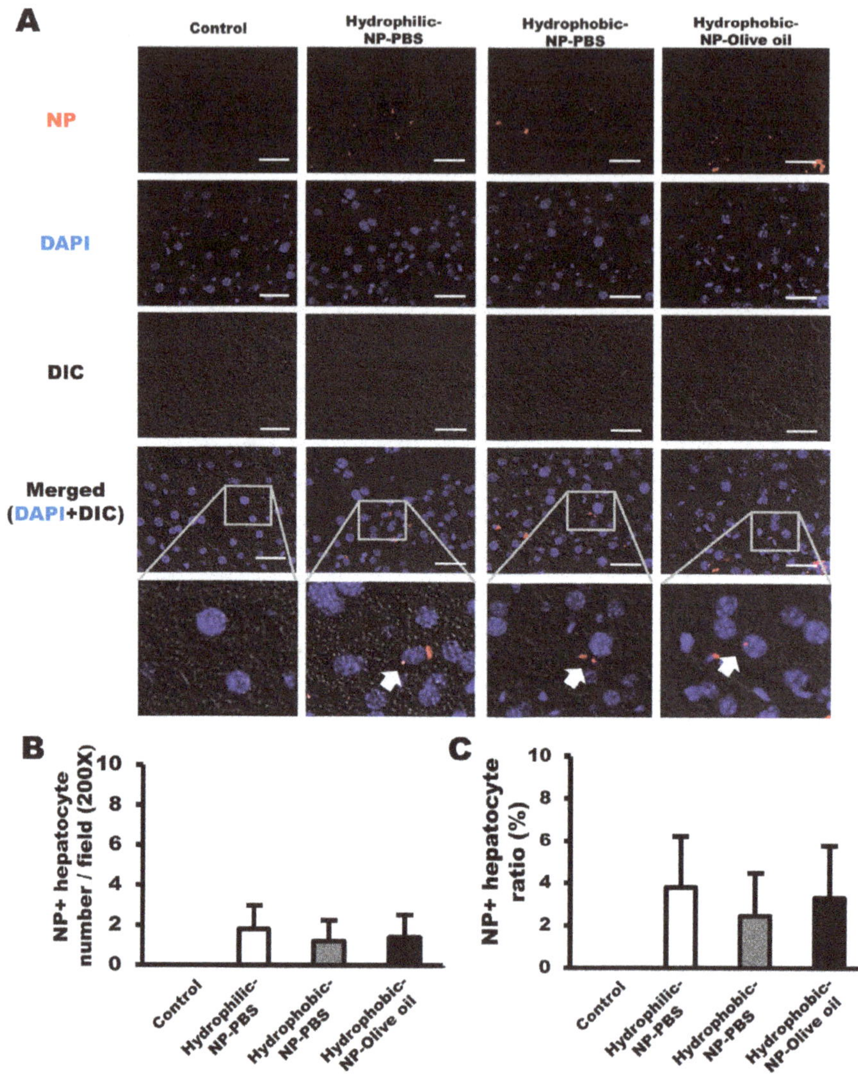

Figure 6. NP delivery to hepatocytes based on surface modification of silica NPs. (**A**) Representative immunofluorescence micrographs of NPs (red) delivered to hepatocytes in the liver. NPs were generally observed in the cytoplasm of hepatocytes around nuclei (blue, DAPI). Blue = DAPI. Gray = Differential Interference Contrast (DIC). (**B**) Number of hepatocytes retaining NPs per field (200×). (**C**) Ratios of NP-positive hepatocytes among entire NP-positive cell population. The ratios were generally less than 4% in all types of NP-treated liver, regardless of NP surface modification and the solvent used to disperse the NPs. All data were quantified from 10 fields (200×) per tissue and are shown as mean ± SD.

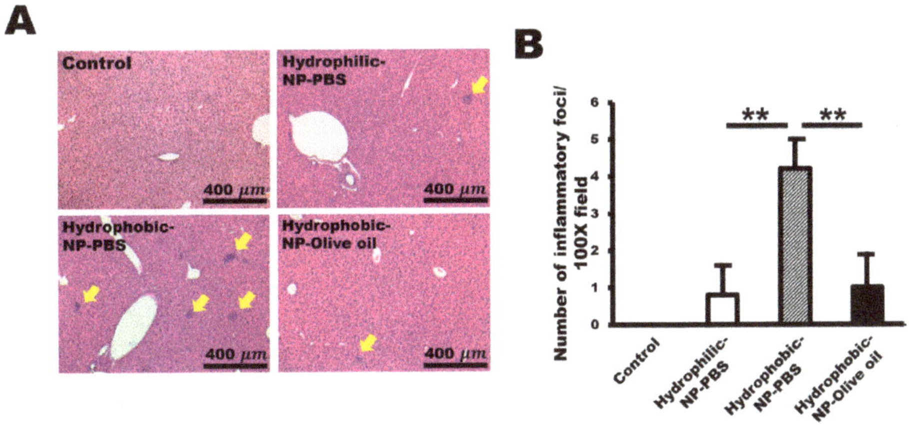

Figure 7. Infiltrations of inflammatory cells in hydrophobic-NP-PBS treated liver. (**A**) Representative histopathologic micrographs of silica NP-injected liver tissue (H&E stain). Hydrophobic-NP-PBS treated liver showed a significantly increased level of inflammatory foci (about 4 or 5 foci per field) compared with those of hydrophilic-NP-PBS treated liver and hydrophobic-NP-olive oil treated liver (arrows). (**B**) Number of inflammatory foci per field (100×) in NP-treated livers. The number of inflammatory foci was counted from 10 fields (100×) per tissue and are shown as mean ± SD. ** $p < 0.01$.

3. Discussion

The present study determined hepatic cellular distribution of surface-modified silica NPs in mouse liver. Hydrophilic surface-modified NPs dispersed as small-sized particles in PBS (Figure 1B). NPs with hydrophobic surface modification dispersed in olive oil but tended to aggregate in PBS (Figure 1B). When those NPs were intraperitoneally injected, considerably more hydrophobic surface-modified NPs were delivered to the liver compared with hydrophilic surface-modified NPs. Hence, NP delivery to the liver can be improved by hydrophobic surface modification of silica NPs. It is believed that the hydrophobic surface condition may facilitate their absorption by the peritoneal tissues, due to the hydrophobicity of the parietal, visceral peritoneum, mesentery, and omentum covering the abdominal visceral organs [24].

In mice, Kupffer cells have been deemed pivotal in removing NPs from the liver [25]. Consistent with this statement, we also observed that NP fluorescence intensity per cell was significantly higher in Kupffer cells than the other types of liver cells, such as LSECs, HSCs, and hepatocytes (Figure S1, Supplementary Materials). Those results can be explained by the cellular location and active phagocytic behavior of Kupffer cells [2]. Interestingly, hydrophilic or hydrophobic surface modification of silica NPs was not associated with NP distribution in Kupffer cells (Figure 3), suggesting that both hydrophobic and hydrophilic surface modification of NPs are not helpful to target and evade Kupffer cells for drug delivery. However, Cheng et al. [26] demonstrated that hydrophobic mesoporous silica NPs were more rapidly removed from liver circulation by Kupffer cells compared with hydrophilic NPs, suggesting hydrophobic NPs rather than hydrophilic NPs were more easily taken up by Kupffer cells. In this study, we synthesized core–shell-type silica NPs and found no significant difference in the relative NP uptake ratio by Kupffer cells among all three types of NP-treated livers (Figure 8). It infers that the affinity of silica NPs for Kupffer cells can be determined by their structure, as well as surface hydrophobicity.

In terms of relative NP uptake ratio by each cell type, LSECs preferentially accumulated the NPs in hydrophobic NP-treated livers (Figure 8B), indicating a hierarchy characterized by the supremacy of LSECs, followed by Kupffer cells, HSCs, and hepatocytes, although the fluorescence intensity per Kupffer cell was much higher when compared with that of the LSECs. Contrary to the hydrophobic

NP-treated liver, the hierarchy of relative NP uptake ratio by each cell type in the hydrophilic NP-treated liver was Kupffer cells > HSCs ≅ LSECs > hepatocytes (Figure 8A). Moreover, the ratio of NP-positive LSECs was significantly lower in hydrophilic-NP-PBS treated liver relative to that in hydrophobic NP-treated liver (Figure 4B,C). The fact that hydrophobic silica NPs have a high affinity for LSECs, whereas hydrophilic surface-modified NPs have a low affinity for LSECs means that hydrophobic surface modification of silica NPs can be a pivotal factor in targeting LSECs, specifically in the liver.

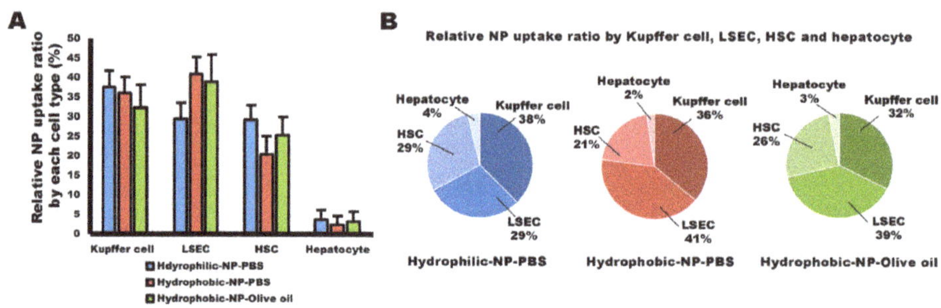

Figure 8. Relative ratio of NP uptake by Kupffer cells, LSECs, HSCs, and hepatocytes based on surface modification of silica NPs. (**A**) Relative NP uptake ratio by Kupffer cells, LSECs, HSCs, and hepatocytes. The ratio values are shown as mean ± SD. (**B**) Hierarchy of NP uptake in all types of NP-treated liver. Kupffer cells mainly took up NPs (38%), followed by HSCs (29%) ≅ LSECs (29%), and hepatocytes (4%) in hydrophilic-NP-PBS treated liver. LSECs were the primary cells taking up NPs (41%), followed by Kupffer cells (36%), HSCs (21%), and hepatocytes (2%) in hydrophobic-NP-PBS treated liver. The hierarchy of NP uptake in hydrophobic-NP-olive oil treated liver was LSECs (39%) > Kupffer cells (32%) > HSCs (26%) > hepatocytes (3%).

LSECs are highly specialized endothelial cells compared with normal continuous endothelial cells in the vessels [27]. LSECs form a permeable barrier, at the interface between the sinusoidal vessels and the space of Disse [28,29]. Therefore, LSECs were regarded as an important barrier to overcoming NP delivery to HSCs or hepatocytes. NP delivery to HSCs and hepatocytes can be impaired by strong endocytosis of LSECs, as well as phagocytosis of Kupffer cells, as evidenced in a previous literature review [30]. This implies a possible convenience of Kupffer cells and LSECs for targeting by NPs, due to their location and strong phagocytic and endocytic activity [31,32]. Based on our results, hydrophobic NP surface modification would be appropriate for targeting LSECs.

Hydrophilic silica NPs seem to be more appropriate to target HSCs rather than LSECs. The number of NP-positive HSCs in the hydrophilic-NP-PBS treated liver was significantly highest among all NP-treated livers (Figure 5B), and so hydrophilic surface-modified NPs might be more likely to reach HSCs when compared with the hydrophobic surface-modified NPs. Besides the surface characteristic of NPs, their size is also a highly important factor affecting NP cellular distribution in the liver [32]. In the liver, NPs should penetrate LSECs to reach HSCs and hepatocytes. NPs mainly pass the LSECs through the fenestrae [32]. Generally, the size of LSEC fenestrae in mice varies between 50 and 280 nm [33], and therefore the NPs we used in the present study could be delivered to HSCs, as well as hepatocytes. However, it is thought that comparatively smaller-sized NPs would be more effective for delivery to HSCs and hepatocytes. Importantly, NPs greater than 280 nm may be highly specific to targeting Kupffer cells. Interestingly, there was a significant difference in the ratio of NP-positive HSCs between hydrophobic-NP-PBS treated liver and hydrophobic-NP-olive oil treated liver. Thus, the NPs' size depends on the presence or absence of silica NP aggregation. In the absence of NP aggregation, NPs seem to pass the LSEC barrier easily via the fenestrae, due to the smaller size of the NPs relative to the fenestrae diameters.

NP delivery to the HSCs has been one of the most important issues for the treatment of chronic liver fibrosis [34–36], a condition induced by the activation of HSCs and their transformation to a myofibroblast-producing collagen-rich extracellular matrix, resulting in severe disruption of the liver architecture. Kupffer cells, LSECs, and a collagenous extracellular matrix have been insurmountable obstacles for NP or drug delivery to HSCs to treat liver fibrosis, despite several studies on this issue [33]. Although some recent research has demonstrated the successful delivery of vitamin A-scaffold lipid or polymer NPs to HSCs by targeting mannose 6-phosphate/insulin-like growth factor II receptor, the collagen type VI receptor, and platelet-derived growth factor β receptor [36–38], several issues remain unresolved. For instance, a means of passing through the LSECs or evading phagocytosis of NPs by Kupffer cells is still considered essential for the successful and efficient NP delivery to HSCs in liver fibrosis.

All types of NPs were not delivered properly to the hepatocytes (Figure 6), suggesting that both hydrophobic and hydrophilic surface modifications are not successful strategies for specific targeting of NPs to the hepatocytes. Considerable emphasis has been placed on NP or drug delivery to hepatocytes in treating liver cancers, such as hepatocellular carcinoma [32,39]. To date, several attempts have been made to design NPs for delivery to the hepatocytes, such as galactosylated chitosan NPs [40], galactose derivative-modified NPs [41], and dye-functionalized theranostic NPs [42]. Most of the NPs are actively targeting specific surface receptors of hepatocellular carcinoma cancer cells, for example, the asialoglycoprotein receptor, ganglioside GM1 receptor, epidermal growth factor receptor, and integrin receptor [39]. However, considering the presence of these receptors in non-targeted normal cells, it is still very important to improve the specificity of NP targeting to hepatocellular carcinoma cells and hepatocytes. The effects of hydrophobic or hydrophilic surface modification in NP delivery to hepatocytes have not yet been fully studied. The size of the silica NPs we used was approximately 250–270 nm, which is smaller than the size of LSEC fenestrae, so it is feasible that the NPs can pass through the fenestrae. Therefore, based on the results of this study, we assumed that the surface polarity of silica NPs was not associated with NP delivery to the hepatocytes.

Although it is important to target hepatocytes to treat HCC effectively, various types of nanoparticles with the modified surface charge used in the experiments were not appropriate for targeting hepatocytes. Modification of the surface charge alone did not directly affect the targeting of hepatocytes. In order to deliver nanoparticles to hepatocytes, it is preferable to apply various conditions in combination. In our study, there was a limit that hydrophilic-NP-PBS had less liver reach than hydrophobic surface modification NPs. However, since hydrophilic NPs have overcome the barrier and reached relatively large numbers of HSCs and hepatocytes, it is expected that more NPs will reach the hepatocyte by decreasing the size of the hydrophilic NPs and administering the higher concentration of nanoparticles.

Silica NPs, which have low toxicity, are usually used as a drug delivery system due to the convenience of modifying or functionalizing the NP surfaces [43]. However, it is reported that silica NPs can induce oxidative stress and inflammation [28,44]. Agglomerated silica NPs caused inflammation, in this study, notably associated with the hydrophobic-NP-PBS rather than the hydrophilic-NP-PBS and hydrophobic-NP-olive oil (Figure 7B). Therefore, inflammation does not seem to be associated with the surface modification of NPs. As mentioned above, the size ranges of LSEC fenestrae in mice are between 50 and 280 nm, and the aggregated NPs must be larger than 280 nm. Therefore, it is speculated that the clustered NPs could not pass the LSEC fenestrae properly, due to their relatively substantial size. In sinusoidal vessels, the aggregated NPs can provide more opportunities to stimulate adjacent Kupffer cells to recruit various secondary inflammatory cells around the clumped NPs. Although silica NPs have been known to have low toxicity, aggregation of NPs seems to cause inflammation and toxicity in the liver.

The modification of the NP surface with a high affinity ligand binding to a cell-specific receptor is one of the most frequently used methods to improve cell targeting efficiency [4]. For example, coating of NP with vitamin A has been tried to target HSC in a previous study [45]. However, it was uncertain whether vitamin A-coated NP is beneficial in targeting HSCs, since the distributions of NPs in other types of liver cells were not determined in the previous study. In this study, we investigated the distribution patterns of NPs in liver by modifying surface energy. Interestingly, the size of the NPs along with the surface charge affected cellular distribution in the liver. Thus, the size of the NPs seems to be another important factor in determining specific NP delivery to target cells.

The present study describes, for the first time, the cellular distribution of silica NPs in the liver, based on the hydrophobic and hydrophilic surface modification of the NPs. The distribution of the NPs is influenced by their size and surface characteristics. Modifying the NP surface hydrophobicity is essential to improving NP delivery to the liver Kupffer cells and LSECs, whereas a hydrophilic surface modification is appropriate to deliver NPs to HSCs. Targeting hepatocytes using NPs is more complicated than targeting Kupffer cells, LSECs, or HSCs because of the location and the presence of reticuloendothelial barriers to protect hepatocytes. This study provides information imperative to targeting the specific types of liver cells using silica NPs and understanding the association between NP surface energy and the distribution pattern in the liver. Regardless of NP formulation, it seems the surface polarity of NPs is a crucial factor in determining their cellular distribution pattern in the liver. Thus, our results would apply to other types of NPs, in terms of cellular distribution in the liver, and thereby provides valuable evidence to establish strategies for cell targeting using various NPs.

4. Materials and Methods

4.1. Chemicals

Tetraethyl orthosilicate (TEOS), NH_4OH (aqueous solution, 28%), RBITC, and 3-aminopropyltriethoxysilane (APTES) were all purchased from Sigma-Aldrich Co. (St. Louis, MO, USA). Ethanol (EtOH) anhydrous was purchased from Duksan, Inc. (Ansan, Korea).

4.2. Synthesis of RBITC-Loaded Core

RBITC-loaded cores were synthesized by a two-step reaction [46]. First, RBITC was dissolved in EtOH at 5 mg/mL, and then 0.4 mL of RBITC solution was mixed with diluted APTES (8 vol % ethanolic APTES). The isothiocyanate groups in RBITC combined with amines of APTES, forming thiourea groups after reaction at room temperature for 24 h. Afterward, 0.1 mL NH_4OH was added dropwise to the reacted solution to induce hydrolysis and condensation reactions of ethoxy groups in APTES. After 24 h, RBITC–APTES molecules chemically aggregated together, forming an RBITC-loaded core. Fluorescent cores were separated by centrifugation at 13,000 rpm and re-dispersed in EtOH. The washing steps were conducted repeatedly.

4.3. Fabrication, Surface Modification, and Characterization of RBITC-Loaded Core–Shell Silica NPs

RBITC-loaded core–shell silica NPs were synthesized via the Stöber process [47]. First, 0.5 mL of TEOS was dissolved in 5 mL EtOH, with stirring for 10 min (TEOS solution); then, 0.65 mL each NH4OH and RBITC-loaded core–shell silica NPs were mixed with 4 mL EtOH for 10 min (core solution). The core solution was poured into the TEOS solution and stirred for 5 h. As a result, the RBITC-loaded cores were surrounded by TEOS via the hydrolysis reaction between TEOS and APTES. To remove impurities from the RBITC-loaded core–shell silica NPs, a rinsing process (×2) was conducted with water and EtOH. Particles were dispersed after 3 h of sonication. Afterward, 3 wt % ODTS was added to modify the surface of the silica NPs with hydrocarbon groups. Prepared ODTS-attached and untreated silica NPs in EtOH underwent solvent exchange process with PBS and olive oil depending on their uses. Three types of solutions were prepared by dispersing ODTS-attached and untreated silica NPs in PBS and ODTS-attached silica NPs in olive oil. For the solvent exchange, silica NPs dispersed in EtOH were separated from EtOH by centrifugation at 13,000 rpm for 5 min and then EtOH was exchanged with PBS and also EtOH was replaced by olive oil. After an hour of re-dispersion into the exchanged solution, centrifugation and dispersion cycle were conducted two more times. To analyze the functional groups of the silica NPs, the untreated and ODTS-modified silica NPs were examined using a Fourier transform infrared spectrometer (Jasco FT/IR-4100, Easton, MD, USA).

4.4. Animals

Thirteen-week-old male C57BL/6 mice weighing 25 g each were housed in individually ventilated cages, maintained at 22–24 °C and 40–50% humidity under 12 h light/dark cycle. Water and normal chow diet were provided ad libitum. For this study, 15 mice were divided into 4 groups (control, hydrophilic-NP-PBS, hydrophobic-NP-PBS, hydrophobic-NP-olive oil) and each group contained 3 or 4 mice. The animals were acclimated under this environment for 7 days before experiments. All animal experiments were performed in accordance with the National Institutes of Health (NIH) guidelines for the care and use of laboratory animals and approved by the Kyungpook National University Institutional Animal Care and Use Committee (IACUC, approval number 2017-0119, 5 May 2017).

4.5. NP Preparation and Administration

Three distinct silica NPs were prepared—hydrophilic-NP-PBS, hydrophobic-NP-PBS, and hydrophobic-NP-olive oil. Hydrophilic silica NPs were dispersed in Dulbecco's PBS, and hydrophobic silica NPs were dispersed in PBS or olive oil. The silica NPs were, respectively, injected intraperitoneally into the mice at a dose of 1.2 mg/25 g. A control group was administered an equivalent volume of Dulbecco's PBS without NP. All mice were sacrificed for collection of liver tissues at 24 h after a single NP injection.

4.6. Immunofluorescence Analysis of Silica NP Distributions in the Liver

For the immunofluorescence observation of NP distribution in the liver, collected liver tissues were fixed in 4% paraformaldehyde at 4 °C for 2 days in the dark and then incubated in 30% sucrose overnight. The tissues sufficiently incubated in sucrose were embedded in OCT compound (Sakura FineTek, Torrance, CA, USA) and rapidly frozen on dry ice. The frozen OCT blocks were cryo-sectioned into 5 μm thick slices. The sections were immunostained with primary antibodies of rat anti-CD68 (Bio-Rad, Hercules, CA, USA), mouse anti-CD34 (Santa Cruz Biotechnology, Santa Cruz, CA, USA), and mouse anti-desmin (Santa Cruz Biotechnology). The sections were subsequently incubated with the secondary antibodies Alexa Fluor®647 donkey anti-rat IgG and Alexa Fluor®647 donkey anti-mouse IgG (Abcam, Cambridge, UK). For nuclear staining, ProLong®Gold Antifade Reagent with DAPI (Cell Signaling, Danvers, MA, USA) was used. Silica NP distributions were observed under an Olympus BX53 fluorescence microscope (Olympus, Tokyo, Japan) at 555 nm, and CD68, CD34, and desmin expressions were detected at 647 nm wavelength. The number of NP-positive cells was counted at 200× magnification. For the evaluation of NP intensity per cell, the total NP fluorescence intensity value per field was divided by the number of NP-positive cells or NP-positive specific cell type per field. For example, total NP fluorescence intensity values from CD68-positive cells per field were divided by the number of both NP- and CD68-positive Kupffer cells (CD68+ NP+ cells). Total NP fluorescence intensity was measured under an Olympus BX53 fluorescence microscope using ToupView software (version x86, 3.7.7817, Hangzhou ToupTek Photonics Co., Zhejiang, China).

4.7. Histological Analysis of Inflammatory Foci

For histological analysis, isolated liver tissues were fixed in 10% neutral-buffered formalin, routinely processed, and embedded in paraffin wax. The blocks were sectioned to 3 μm thickness. The liver sections were deparaffinized in toluene, rehydrated through graded ethanol solutions, washed in distilled water, and stained with hematoxylin and eosin. Stained tissue slides were observed, and representative images were captured using a Leica microscope (Leica Microsystems, Heerbrugg, Switzerland). The number of inflammatory foci was counted at 100× magnification.

4.8. Statistical Analysis

Data were presented as mean ± standard error of the mean. Two-way analysis of variance or the Student's *t*-test was used to determine the statistical significance among multiple experimental groups, by using GraphPad InStat (GraphPad Software, Inc., San Diego, CA, USA). Statistical significance was set at $p < 0.05$.

Supplementary Materials: Supplementary materials can be found at http://www.mdpi.com/1422-0067/20/15/3812/s1.

Author Contributions: A.-R.L. and K.N. designed and performed the experiments, analyzed the data, and wrote the paper; B.J.L., S.-W.L., S.-M.B., and J.-S.B. performed the experiments; S.-K.C., S.-J.P., T.-H.K., and K.-S.J. analyzed the data; D.Y.L. and J.-K.P. designed the study, organized the experimental methodology and data, and wrote and revised the paper critically.

Funding: This work was partly supported by the National Research Foundation of Korea (NRF) grants funded by the Korea government, the Ministry of Education (grant number 2019R1I1A2A01060031) and the Ministry of Science and ICT (grant number 2019R1A2C1009874).

Conflicts of Interest: The authors declare no conflict of interest.

Abbreviations

APTES	3-aminopropyltriethoxysilane
HSC	hepatic stellate cell
LSEC	liver sinusoid endothelial cell
NP	nanoparticle
ODTS	*n*-octadecyltrichlorosilane
RBITC	rhodamine B isothiocyanate
TEOS	tetraethyl orthosilicate

References

1. Liu, J.; Huang, Y.; Kumar, A.; Tan, A.; Jin, S.; Mozhi, A.; Liang, X.J. pH-Sensitive nano-systems for drug delivery in cancer therapy. *Biotechnol. Adv.* **2014**, *32*, 693–710. [CrossRef] [PubMed]
2. Park, J.K.; Utsumi, T.; Seo, Y.E.; Deng, Y.; Satoh, A.; Saltzman, W.M.; Iwakiri, Y. Cellular distribution of injected PLGA-nanoparticles in the liver. *Nanomed. Nanotechnol. Biol. Med.* **2016**, *12*, 1365–1374. [CrossRef] [PubMed]
3. Bartneck, M.; Warzecha, K.T.; Tacke, F. Therapeutic targeting of liver inflammation and fibrosis by nanomedicine. *Hepatobiliary Surg. Nutr.* **2014**, *3*, 364–376. [PubMed]
4. Kang, J.H.; Toita, R.; Murata, M. Liver cell-targeted delivery of therapeutic molecules. *Crit. Rev. Biotechnol.* **2016**, *36*, 132–143. [CrossRef] [PubMed]
5. Kim, J.W.; Kim, L.U.; Kim, C.K. Size control of silica nanoparticles and their surface treatment for fabrication of dental nanocomposites. *Biomacromolecules* **2007**, *8*, 215–222. [CrossRef] [PubMed]
6. Liberman, A.; Mendez, N.; Trogler, W.C.; Kummel, A.C. Synthesis and surface functionalization of silica nanoparticles for nanomedicine. *Surf. Sci. Rep.* **2014**, *69*, 132–158. [CrossRef]
7. Cheng, C.J.; Tietjen, G.T.; Saucier-Sawyer, J.K.; Saltzman, W.M. A holistic approach to targeting disease with polymeric nanoparticles. *Nat. Rev. Drug Discov.* **2015**, *14*, 239–247. [CrossRef]
8. Lehman, S.E.; Morris, A.S.; Mueller, P.S.; Salem, A.K.; Grassian, V.H.; Larsen, S.C. Silica nanoparticle-generated ROS as a predictor of cellular toxicity: Mechanistic insights and safety by design. *Environ. Sci. Nano.* **2016**, *3*, 56–66. [CrossRef]
9. Albanese, A.; Tang, P.S.; Chan, W.C.W. The effect of nanoparticle size, shape, and surface chemistry on biological systems. *Annu. Rev. Biomed. Eng.* **2012**, *14*, 1–16. [CrossRef]
10. Kramer, S.A.; Lin, W. Silica-based Nanoparticles for Biomedical Imaging and Drug Delivery Applications. *Handb. Nanobiomedical Res.* **2014**, 403–437.
11. Siegel, R.L.; Miller, K.D.; Jemal, A. Cancer statistic, 2018. *CA Cancer J. Clin.* **2018**, *68*, 7–30. [CrossRef]
12. Bayne, C.E.; Williams, S.B.; Cooperberg, M.R.; Gleave, M.E.; Graefen, M.; Montorsi, F.; Novara, G.; Smaldone, M.C.; Sooriakumaran, P.; Wiklund, P.N.; et al. Treatment of the Primary Tumor in Metastatic Prostate Cancer: Current Concepts and Future Perspectives. *Eur. Urol.* **2016**, *69*, 775–787. [CrossRef] [PubMed]

13. Ahmad, K.; Rabbani, G.; Baig, M.H.; Lim, J.H.; Khan, M.E.; Lee, E.J.; Ashraf, G.M.; Choi, I. Nanoparticle-based Drugs: A potential Armamentarium of Effective Anti-Cancer Therapies. *Curr. Drug Metab.* **2018**, *19*, 839–846. [CrossRef]
14. Egusquiaguirre, S.P.; Igartua, M.; Hernández, R.M.; Pedraz, J.L. Nanoparticle delivery systems for cancer therapy: Advances in clinical and preclinical research. *Clin. Transl. Oncol.* **2012**, *14*, 83–93. [CrossRef]
15. Grandhi, M.S.; Kim, A.K.; Ronnekleiv-Kelly, S.M.; Kamel, I.R.; Ghasebeh, M.A.; Pawlik, T.M. Hepatocellular carcinoma: From diagnosis to treatment. *Surg. Oncol.* **2016**, *25*, 74–85. [CrossRef]
16. Mu, X.; Español-Suñer, R.; Mederacke, I.; Affò, S.; Manco, R.; Sempoux, C.; Lemaigre, F.P.; Adili, A.; Yuan, D.; Weber, A.; et al. Hepatocellular carcinoma originates from hepatocytes and not from the progenitor/biliary compartment. *J. Clin. Invest.* **2015**, *125*, 3891–3903. [CrossRef]
17. Aggarwal, P.; Hall, J.B.; McLeland, C.B.; Dobrovolskaia, M.A.; McNeil, S.E. Nanoparticle interaction with plasma proteins as it relates to particle biodistribution, biocompatibility and therapeutic efficacy. *Adv. Drug Deliv. Rev.* **2009**, *61*, 428–437. [CrossRef]
18. Saptarshi, S.R.; Duschl, A.; Lopata, A.L. Interaction of nanoparticles with proteins: Relation to bio-reactivity of the nanoparticle. *J. Nanobiotechnol.* **2013**, *11*, 26. [CrossRef]
19. Gao, H.; He, Q. The interaction of nanoparticles with plasma proteins and the consequent influence on nanoparticles behavior. *Expert Opin. Drug Deliv.* **2014**, *11*, 409–420. [CrossRef]
20. Zhang, X.; Zhang, J.; Zhang, F.; Yu, S. Probing the binding affinity of plasma proteins adsorbed on Au nanoparticles. *Nanoscale* **2017**, *9*, 4787–4792. [CrossRef]
21. Sarin, H. Physiologic upper limits of pore size of different blood capillary types and another perspective on the dual pore theory of microvascular permeability. *J. Angiogenes. Res.* **2010**, *2*, 14. [CrossRef] [PubMed]
22. Wisse, E.; Jacobs, F.; Topal, B.; Frederik, P.; De Geest, B. The size of endothelial fenestrae in human liver sinusoids: Implications for hepatocyte-directed gene transfer. *Gene Ther.* **2008**, *15*, 1193–1199. [CrossRef] [PubMed]
23. Rensen, P.C.N.; Sliendregt, L.A.J.M.; Fems, M.; Kieviet, E.; van Rossenberg, S.M.W.; van Leeuwen, S.H.; van Berkel, T.J.C.; Biessen, E.A.L. Determination of the upper size limit for uptake and processing of ligands by the asialoglycoprotein receptor on hepatocytes *in vitro* and *in vivo*. *J. Biol. Chem.* **2001**, *276*, 37577–37584. [CrossRef] [PubMed]
24. Gómez-Suárez, C.; Bruinsma, G.M.; Rakhorst, G.; van der Mei, H.C.; Busscher, H.J. Hydrophobicity of peritoneal tissues in the rat. *J. Colloid Interface Sci.* **2002**, *253*, 470–471. [CrossRef] [PubMed]
25. Sadauskas, E.; Wallin, H.; Stoltenberg, M.; Vogel, U.; Doering, P.; Larsen, A.; Danscher, G. Kupffer cells are central in the removal of nanoparticles from the organism. *Part. Fibre. Toxicol.* **2007**, *4*, 10. [CrossRef] [PubMed]
26. Cheng, S.H.; Li, F.C.; Souris, J.S.; Yang, C.S.; Tseng, F.G.; Lee, H.S.; Chen, C.T.; Dong, C.Y.; Lo, L.W. Visualizing dynamics of sub-hepatic distribution of nanoparticles using intravital multiphoton fluorescence microscopy. *ACS Nano* **2012**, *6*, 4122–4131. [CrossRef]
27. Poisson, J.; Lemoinne, S.; Boulanger, C.; Durand, F.; Moreau, R.; Valla, D.; Rautou, P.E. Liver sinusoidal endothelial cells: Physiology and role in liver diseases. *J. Hepatol.* **2017**, *66*, 212–227. [CrossRef] [PubMed]
28. Braet, F.; Wisse, E. Structural and functional aspects of liver sinusoidal endothelial cell fenestrae: A review. *Comp. Hepatol.* **2002**, *1*, 1. [CrossRef]
29. DeLeve, L.D. Liver sinusoidal endothelial cells in hepatic fibrosis. *Hepatology* **2015**, *61*, 1740–1746. [CrossRef]
30. Smedsrød, B.; Le Couteur, D.; Ikejima, K.; Jaeschke, H.; Kawada, N.; Naito, M.; Knolle, P.; Nagy, L.; Senoo, H.; Vidal-Vanaclocha, F.; et al. Hepatic sinusoidal cells in health and disease: Update from the 14th International Symposium. *Liver Int.* **2009**, *4*, 490–501. [CrossRef]
31. Schroeder, A.; Levins, C.G.; Cortez, C.; Langer, R.; Anderson, D.G. Lipid-based nanotherapeutics for siRNA delivery. *J. Intern. Med.* **2010**, *267*, 9–21. [CrossRef]
32. Poelstra, K.; Prakash, J.; Beljaars, L. Drug targeting to the diseased liver. *J. Control. Release* **2012**, *161*, 188–197. [CrossRef]
33. Snoeys, J.; Lievens, J.; Wisse, E.; Jacobs, F.; Duimel, H.; Collen, D.; Frederik, P.; De Geest, B. Species differences in transgene DNA uptake in hepatocytes after adenoviral transfer correlate with the size of endothelial fenestrae. *Gene Ther.* **2007**, *14*, 604–612. [CrossRef]

34. Toriyabe, N.; Sakurai, Y.; Kato, A.; Yamamoto, S.; Tange, K.; Nakai, Y.; Akita, H.; Harahsima, H. The delivery of small interfering RNA to hepatic stellate cells using a lipid nanoparticle composed of a vitamin A-scaffold lipid-like material. *J. Pharm. Sci.* **2017**, *106*, 2046–2052. [CrossRef]
35. Duong, H.T.T.; Dong, Z.; Su, L.; Boyer, C.; George, J.; Davis, T.P.; Wang, J. The use of nanoparticles to deliver nitric oxide to hepatic stellate cells for treating liver fibrosis and portal hypertension. *Small* **2015**, *11*, 2291–2304. [CrossRef]
36. Beljaars, L.; Molema, G.; Weert, B.; Bonnema, H.; Olinga, P.; Groothuis, G.M.M.; Meijer, D.K.F.; Poelstra, K. Albumin modified with mannose 6-phosphate: A potential carrier for selective delivery of antifibrotic drugs to rat and human hepatic stellate cells. *Hepatology* **1999**, *29*, 1486–1493. [CrossRef]
37. Beljaars, L.; Molema, G.; Schuppan, D.; Geerts, A.; De Bleser, P.J.; Weert, B.; Meijer, D.K.F.; Poelstra, K. Successful targeting to rat hepatic stellate cells using albumin modified with cyclic peptides that recognize the collagen type VI receptor. *J. Biol. Chem.* **2000**, *275*, 12743–12751. [CrossRef]
38. Beljaars, L.; Weert, B.; Geerts, A.; Meijer, D.K.F.; Poelstra, K. The preferential homing of a platelet derived growth factor receptor-recognizing macromolecule to fibroblast-like cells in fibrotic tissue. *Biochem. Pharmacol.* **2003**, *66*, 1307–1317. [CrossRef]
39. Varshosaz, J.; Farzan, M. Nanoparticles for targeted delivery of therapeutics and small interfering RNAs in hepatocellular carcinoma. *World J. Gastroenterol.* **2015**, *21*, 12022–12041. [CrossRef]
40. Zhou, N.; Zan, X.; Wang, Z.; Wu, H.; Yin, D.; Liao, C.; Wan, Y. Galactosylated chitosan–polycaprolactone nanoparticles for hepatocyte-targeted delivery of curcumin. *Carbohydr. Polym.* **2013**, *94*, 420–429. [CrossRef]
41. Huang, K.W.; Lai, Y.T.; Chern, G.J.; Huang, S.F.; Tsai, C.L.; Sung, Y.C.; Chiang, C.C.; Hwang, P.B.; Ho, T.L.; Huang, R.L.; et al. Galactose derivative-modified nanoparticles for efficient siRNA delivery to hepatocellular carcinoma. *Biomacromolecules* **2018**, *19*, 2330–2339. [CrossRef]
42. Press, A.T.; Traeger, A.; Pietsch, C.; Mosig, A.; Wagner, M.; Clemens, M.G.; Jbeily, N.; Koch, N.; Gottschaldt, M.; Bézière, N.; et al. Cell type-specific delivery of short interfering RNAs by dye-functionalised theranostic nanoparticles. *Nat. Commun.* **2014**, *5*, 5565. [CrossRef]
43. Lee, J.A.; Kim, M.K.; Paek, H.J.; Kim, Y.R.; Kim, M.K.; Lee, J.K.; Jeong, J.; Choi, S.J. Tissue distribution and excretion kinetics of orally administered silica nanoparticles in rats. *Int. J. Nanomed.* **2014**, *9*, 251–260.
44. Guo, C.; Xia, Y.; Niu, P.; Jiang, L.; Duan, J.; Yu, Y.; Zhou, X.; Li, Y.; Sun, Z. Silica nanoparticles induce oxidative stress, inflammation, and endothelial dysfunction in vitro via activation of the MAPK/Nrf2 pathway and nuclear factor-κB signaling. *Int. J. Nanomed.* **2015**, *10*, 1463–1477. [CrossRef]
45. Blomhoff, R.; Green, M.H.; Green, J.B.; Berg, T.; Norum, K.R. Vitamin A metabolism: New perspectives on absorption, transport, and storage. *Physiol Rev.* **1991**, *71*, 951–990. [CrossRef]
46. Ow, H.; Larson, D.R.; Srivastava, M.; Baird, B.A.; Webb, W.W.; Wiesner, U. Bright and stable core−shell fluorescent silica nanoparticles. *Nano. Lett.* **2005**, *5*, 113–117. [CrossRef]
47. Rossi, L.M.; Shi, L.; Quina, F.H.; Rosenzweig, Z. Stöber synthesis of monodispersed luminescent silica nanoparticles for bioanalytical assays. *Langmuir* **2005**, *21*, 4277–4280. [CrossRef]

International Journal of
Molecular Sciences

MDPI

Article

Optimisation of Folate-Mediated Liposomal Encapsulated Arsenic Trioxide for Treating HPV-Positive Cervical Cancer Cells In Vitro

Anam Akhtar, Lucy Ghali, Scarlet Xiaoyan Wang, Celia Bell, Dong Li and Xuesong Wen *

Department of Natural Sciences, Middlesex University, The Burroughs, Hendon, London NW4 4BT, UK; a.akhtar@mdx.ac.uk (A.A.); l.ghali@mdx.ac.uk (L.G.); scarlet.wang@kcl.ac.uk (S.X.W.); c.bell@mdx.ac.uk (C.B.); d.li@mdx.ac.uk (D.L.)

* Correspondence: x.wen@mdx.ac.uk; Tel.: +44-208-411-4931

Received: 10 April 2019; Accepted: 29 April 2019; Published: 30 April 2019

Abstract: High-risk human papilloma virus (HPV) infection is directly associated with cervical cancer development. Arsenic trioxide (ATO), despite inducing apoptosis in HPV-infected cervical cancer cells in vitro, has been compromised by toxicity and poor pharmacokinetics in clinical trials. Therefore, to improve ATO's therapeutic profile for HPV-related cancers, this study aims to explore the effects of length of ligand spacers of folate-targeted liposomes on the efficiency of ATO delivery to HPV-infected cells. Fluorescent ATO encapsulated liposomes with folic acid (FA) conjugated to two different PEG lengths (2000 Da and 5000 Da) were synthesised, and their cellular uptake was examined for HPV-positive HeLa and KB and HPV-negative HT-3 cells using confocal microscopy, flow cytometry, and spectrophotometer readings. Cellular arsenic quantification and anti-tumour efficacy was evaluated through inductively coupled plasma-mass spectrometry (ICP-MS) and cytotoxicity studies, respectively. Results showed that liposomes with a longer folic acid-polyethylene glycol (FA-PEG) spacer (5000 Da) displayed a higher efficiency in targeting folate receptor (FR) + HPV-infected cells without increasing any inherent cytotoxicity. Targeted liposomally delivered ATO also displayed superior selectivity and efficiency in inducing higher cell apoptosis in HPV-positive cells per unit of arsenic taken up than free ATO, in contrast to HT-3. These findings may hold promise in improving the management of HPV-associated cancers.

Keywords: arsenic trioxide (ATO); liposome; targeted drug delivery; cervical cancer; human papilloma virus (HPV); folate conjugation

1. Introduction

Cervical cancer is the second most common cancer in women worldwide and imposes a disproportionately high burden (>80%) on the developing world [1,2]. Since the 1990s, HPV infection has been established as the causal agent of cervical cancer with undisputed epidemiological studies, augmented by molecular technology [3,4]. Further studies have indicated that HPV infection might also be responsible for a subset of anal, vulvar, vaginal, penile, upper respiratory-digestive tract, and head and neck cancers [5–8]. Therefore, targeting HPV infection has become a priority in managing HPV-associated cancers, and an anti-HPV agent that can specifically be taken up by HPV-infected cells is ideally required to enhance treatment and minimise off-target toxicity to surrounding non-HPV-infected cells and tissues [9].

Our research team, along with others, have shown that ATO, as a therapeutic drug, downregulated HPV oncogene expression and had an inhibitory effect on HPV-infected HeLa cells [10,11]. ATO is a potent clinical agent for the treatment of acute promyelocytic leukaemia and has demonstrated significant clinical success for other haematological malignancies [12,13]. The mechanisms of ATO's anti-cancer activity has been studied extensively, and various modes of action have been proposed, including the induction of apoptosis [14], mitochondrial toxicity [15,16], and the generation and accumulation of ROS [14,17]. In addition, ATO has been reported to interfere in other cellular events, for instance tubulin polymerisation, DNA repair, and cell cycle progression [16].

Clinical response of solid tumours following ATO treatment, however, has been compromised by its severe side effects due to a requirement of much higher dosages for the treatment [18,19]. In vitro studies on HPV-associated cervical cancers have also clearly indicated that ATO is selective in its action only at low concentrations (~2 μM). When the dose was increased to 5 μM, most of the cells were killed from drug-induced toxicity, irrespective of HPV status [11].

Therefore, an effective delivery system that can reduce ATO's off-target toxicity and expand its clinical utility by increasing its accumulation at tumour sites, extending its circulation time in blood and increasing its specificity in targeting HPV-infected cancer cells, is needed. Encapsulating ATO in liposomes is one such strategy that has the potential to improve the drug's therapeutic index by reducing its side effects and increasing drug concentration in tumours through an enhanced permeability and retention (EPR) effect [20]. In addition, if active targeting strategies are employed, the encapsulated drug can be directed to the target sites with minimal disturbance to the surrounding cells and tissues [21].

Active targeting, also referred to as ligand-mediated targeting, employs the specific affinity of ligands on the surface of the nanoparticles towards specific receptors or surface molecules overexpressed in target cells, tissues, or organs [22–26]. Aided by the EPR effect, which allows the nanoparticle to carry the drug payload to remain in the vicinity of the target cells, active targeting facilitates its cellular internalisation via the specific ligand–receptor interaction [27]. As a result, this approach is increasingly being recognised as an effective strategy to enhance the therapeutic indices of anti-cancer drugs. From our previous findings, FA was selected as the targeting ligand of choice for our target cells, as its corresponding FR was highly expressed on HPV-positive cells while minimally expressed for HPV-negative cell lines. FA conjugation to ATO liposomes thus has the promise to offer the desired selectivity for targeting HPV-infected cells.

When folate-targeting nanoparticles were evaluated in preclinical settings, they showed promising results as both therapeutic and diagnostic agents [28–33]. However, there are quite a few parameters that need careful consideration to design an effective targeting system. Among them, optimisation of ligand density and ligand-conjugating polyethylene glycol (PEG) spacer length is crucial to enhance the uptake and efficiency of a drug-carrying liposomal construct. While a higher ligand density might be desirable to increase the binding avidity between the liposomes and the target antigens, it might also give rise to potential immunogenic concerns, along with the uncertainty of the effect on liposomal stability resulting from a high ligand density [34,35].

The importance of PEG spacer length in FR targeting was succinctly established by Lee and Low for KB cells using both targeted and non-targeted liposomes [36]. Liposomes with folate tethered to a long PEG spacer (250 Å) displayed a higher uptake than either non-targeted or targeted liposomes with short spacer length [36]. Additionally, in another study by Gabizon et al. (1999), placing the ligand on a longer PEG spacer length of 3350 Da, as opposed to a 2000 Da PEG layer used for steric stabilisation, increased the uptake of conjugated liposomes substantially [37]. However, it has also been speculated that, if the ligand is conjugated to a longer PEG spacer than the other PEG chains involved in the PEG brush, the extra length of the long spacer might undergo mushroom-like folding, leading to limited exposure of the ligand [35]. Therefore, it might be preferable to have the same length spacer as the surrounding PEG brush [35]. This ambiguity in the literature about the appropriate spacer PEG length requires a thorough investigation for folate conjugation in order to achieve an enhanced drug uptake in a tumour-specific manner.

As a complimentary strategy to EPR, ligand-functionalised nanoparticles offer the potential to further augment the efficacy of nanotherapeutics in treating diseases. Although there have been multiple studies reporting the enhanced uptake by FR-positive cells following folate conjugation to liposomes carrying the therapeutic payload [33,38–40], research that focuses on determining the optimal spacer length for FA conjugation on liposomal physicochemical characteristics, inhibitory potency, and in vitro performance on cervical cancer cells with differing HPV status cannot be found. The aim of this study was to assess the effect of ligand spacer length on cellular uptake and inhibitory potency against HPV-positive cervical cancer cells in vitro and to identify the most suitable liposomal formulation for delivering ATO to cervical cancer cells. The uptake and anti-tumour efficacy of the liposomal formulations were examined using two cervical cancer cell lines, HeLa (HPV-positive) and HT-3 (HPV-negative), along with two control cell lines, KB (FR-positive) and A549 (FR-negative).

2. Results

2.1. Liposome Preparation and Characterisation

Three labelled unconjugated and conjugated liposomal formulations (L1, unconjugated ATO liposomes; L2, FA-ATO liposomes with 2000 Da PEG spacer; L3, FA-ATO liposomes with 5000 Da PEG spacer) were synthesised, and the concentration of phospholipids, encapsulated arsenic, and nickel were determined by ICP-OES. All of the formulations were found to be efficiently loaded with arsenic, which was co-encapsulated with the transition metal ion Ni; loading efficiency (As/P) is depicted in Figure 1a. The loading efficiency was similar for all the liposomes, irrespective of the presence or absence of FA conjugation. The stability of the liposome suspension was analysed by storing it at 4 °C for a period of 4 weeks and calculating its loading efficiency every week (Figure 1b. Less than 5% arsenic leaked from the liposomes in this time period. All of the liposomal formulations displayed similar storage stability. The mean size of control liposomes was determined by dynamic light scattering on a Zetasizer-Nano ZS as represented in Figure 1c. The DLS size measurement of L1, L2, and L3 were 138.5 ± 1.2 nm, 134.4 ± 1.7 nm, and 142.6 ± 3.1 nm respectively. The polydispersity index for the investigated vesicles ranged from 0.1 to 0.15, indicating a homogeneous population.

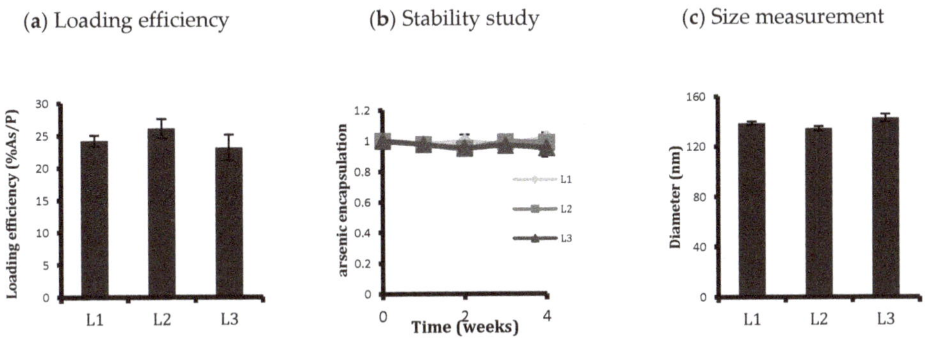

Figure 1. Characterisation of synthesised liposomes: (**a**) loading efficiency (%As/P), (**b**) ATO encapsulated in liposomes over four weeks of storage at 4 °C, and (**c**) diameter of liposomes as measured from Zetasizer with a polydispersity index of around 0.1. Conjugation or length of conjugating PEG spacer does not have any considerable effect on loading efficiency, stability over time in storage, or size of the liposomes. Data are means ± standard deviations of three replicate measurements of at least three independent experiments.

2.2. Analysing Cellular Uptake by Differing Ligand–PEG Spacer Lengths in ATO Liposomes

It has previously been reported in the literature that spacer length of the conjugated ligand impacts the efficacy of the cellular uptake of the nano-vehicle [41]. To assess this aspect, liposomes were conjugated with FA using a PEG spacer with the same length as the surrounding PEG brush (2000 Da) and a different formulation where FA was conjugated to a PEG spacer with a longer length (5000 Da) than the surrounding PEG brush (2000 Da). Previous studies of folate surface receptor expression on the two cervical cancer cell lines investigated found that HeLa (HPV+) cells positively expressed FRs while HT-3 (HPV-) cells only minimally expressed the receptor. A positive control and a negative control cell line for FR expression, KB and A549, respectively, were also included to validate the robustness of the experiments, with their receptor expression reported in the order of KB > HeLa > HT-3 > A549 cells. Subsequently, the cellular uptake of the conjugated and unconjugated liposomes was investigated with techniques including confocal microscopy, flow cytometry, plate reader analysis, and ICP-MS arsenic quantification.

Confocal laser scanning microscopy (CLSM) allowed for an evaluation of the intracellular fate of liposomes as a function of time for various cell lines. A confocal microscopic visualisation of the cellular association of DiI (1,1′-dioctadecyl-3,3,3′,3′-tetramethylindocarbocyanine-5,5′-disulfonic acid)-labelled liposomes is depicted in Figure 2. The cells were treated for 2, 6, and 24 h, fixed in 4% paraformaldehyde and stained with DAPI (4′,6-diamidino-2-phenylindole) for nuclear imaging. By keeping all the parameters for confocal microscopic visualisation the same (PMT and % laser), it was possible to draw a comparative study of the cellular uptake of liposomes, with or without ligand conjugation.

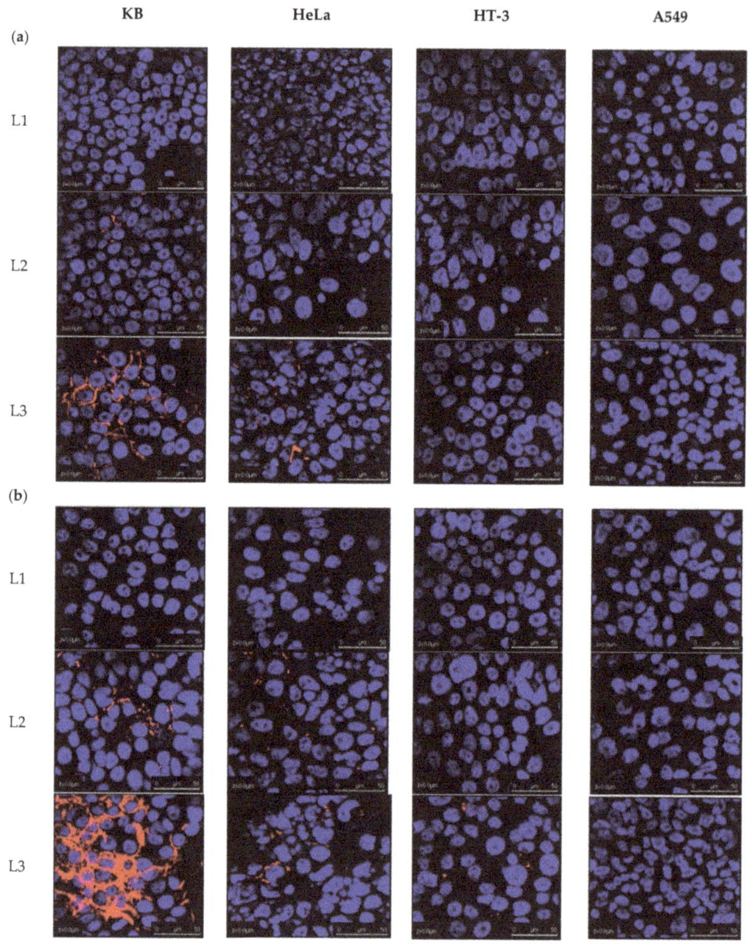

Figure 2. *Cont.*

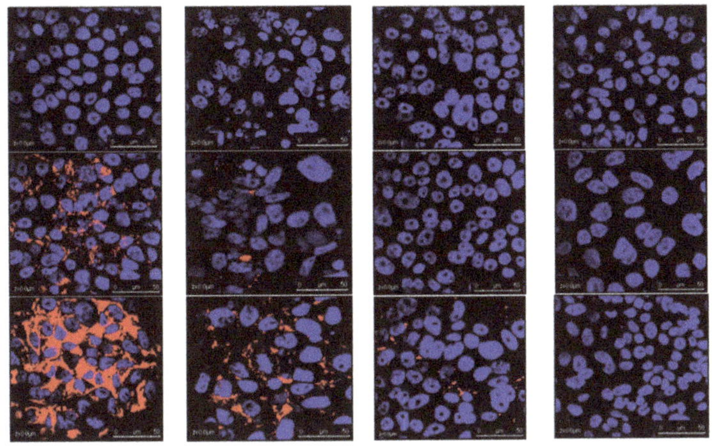

Figure 2. Confocal micrographs showing cellular uptake of DiI-labelled L1, L2, and L3 after treatment of (**a**) 2, (**b**) 6, and (**c**) 24 h. Cells were counterstained with DAPI to reveal the nuclear/DNA location. Among targeted liposomes, L3 were more efficient than L2 in being taken up by cells according to their FR expression, so the uptake in KB was highest followed by HeLa. HT-3 also took up more L3 than L2, while A549 cells had no effect on their uptake from liposomal conjugation. Scale bar: 50 μm.

Liposomes without surface functionalisation showed neither membrane accumulation nor cellular internalisation in any of the cell lines up to 24 h, as evidenced by the lack of liposomal fluorescence (Figure 2). In contrast, ligand conjugation, specifically with a 5000 Da PEG spacer length, mediated an efficient fluorescence uptake in FR-positive cell lines. After 2 h incubation, L3 was found to exhibit high cellular binding, for around 90% of KB cells, 25% of HeLa, and 5% of HT-3, as shown by the red fluorescence around the cell surface. L2 liposomes exhibited minor cellular uptake in KB cells. A549 had no effect of ligand conjugation on its liposomal uptake.

KB cells demonstrated cellular internalisation and accumulation of L3 in their cytosol at a relatively large scale from the time course study, which was shown from the red fluorescence surrounding the DAPI-stained nuclei. L2 was also seen to enter most of the KB cells at 24 h treatment, but to a much lesser extent than L3, which displayed a higher efficiency of L3 for being taken up by the FR-positive cells. A similar trend was repeated for HeLa and HT-3 cells: L3 was more efficient in being taken up by the cells than L2, with L3 entering almost all the HeLa cells by the end of 24 h. The cellular uptake of conjugated liposomes (L2 or L3) in HeLa was always higher than HT-3.

The results obtained from confocal microscopy analysis were further validated by flow cytometry (Figure 3). The cells from the four cell lines were exposed to fluorescently labelled liposomes for a period of 2, 6, and 24 h. A ratio of cells staining positive for conjugated liposomal (L2 or L3) uptake to non-conjugated (L1) liposomal uptake was calculated for each cell line to determine the efficacy of FA conjugation in enhancing the liposomal cellular internalisation. The results indicated that cellular uptake in FR+ cells increased, even with simple folate conjugation (to a PEG spacer with a length similar to that of the PEG brush)—by about 50 times for unconjugated liposomes in KB cells and to a much lesser extent in HeLa (2.1 times) and HT-3 (1.6 times) cells. When the PEG spacer length was taken into consideration and increased, i.e., from 2000 kD to 5000 kD, this further enhanced the uptake. After 24 h, L3, compared to L2, was taken up 1.2 times more in KB, twice as much in HeLa, and 1.3 times more in HT-3. A549, on the other hand, displayed a similar cellular uptake of liposomes, regardless of ligand conjugation.

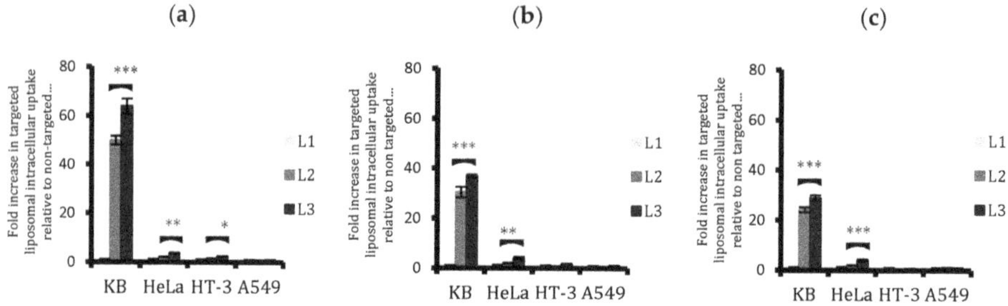

Figure 3. Comparison of the cellular uptake of non-targeted (L1) and targeted liposomes (L2 and L3) by flow cytometry after (**a**) 2, (**b**) 6, and (**c**) 24 h treatment. An unpaired t-test was used to assess any significant difference in the uptakes between the conjugated liposomes, L2 and L3. L3 uptake was significantly higher in both KB and HeLa cells. Data are means ± standard deviations of three replicate measurements of at least three independent experiments. * $p \leq 0.05$, ** $p \leq 0.01$, and *** $p \leq 0.001$.

An additional uptake assay was performed in 96 well plates by reading the fluorescence of cells incubated with fluorescent targeted and non-targeted liposomes via a microplate reader. A comparison was drawn of the differential cellular uptake by analysing the ratio of fluorescence of cells incubated with targeted liposomes to non-targeted liposomes followed by blank correction. Results corroborated the findings from confocal and flow cytometry studies as depicted in Figure 4. Conjugated liposomes (both L2 and L3) were taken up in much higher proportions than non-conjugated L1 in KB and HeLa cells, whereas A549 displayed no difference in uptake from ligand conjugation. HT-3 displayed some increase in uptake in the first six hours with L3 treatment after which the difference with L1 tapered off.

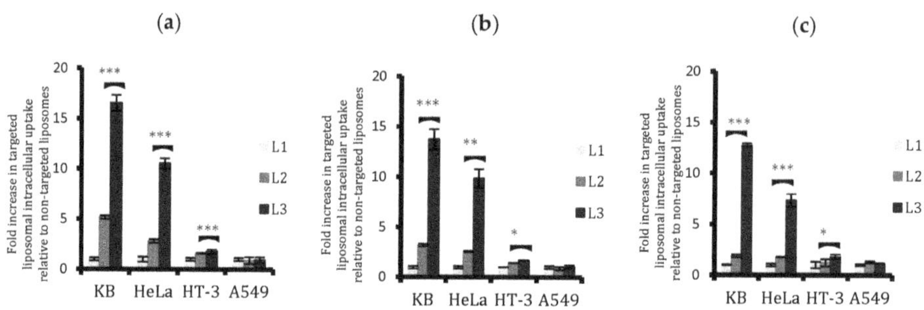

Figure 4. Comparison of cellular uptake of the three liposomal formulations L1, L2, and L3 by the four cell lines with plate reader analysis after (**a**) 2, (**b**) 6, and (**c**) 24 h treatment. L3 uptake was significantly higher than L2 in KB and HeLa. HT-3 cells also displayed a small but significantly enhanced uptake of L3 compared to L2. The difference between targeted and non-targeted liposomes decreased with time. Data are means ± standard deviations of three replicate measurements of at least three independent experiments. * $p \leq 0.05$, ** $p \leq 0.01$, and *** $p \leq 0.001$.

L3 liposomal formulation had a significantly higher uptake than L2 for FR-positive cells. They were taken up around 6.7 times more in KB cells and 4 times more in HeLa cells after 24 h. HT-3 also witnessed a 1.5 times higher uptake from L3 than L2, whereas A549 cells remained unaffected in their liposomal uptake from ligand conjugation. In fact, conjugated liposomes were taken up slightly less than the non-conjugated liposomes by a factor of 0.9 in A549 cells. Similar to the flow cytometry results, the difference between the liposomal uptakes with ligand conjugation was reduced when the treatment time was increased to 24 h. This reduction, while being true for all the cell lines investigated, is more evident from 6 to 24 h than from 2 to 6 h. It is also more obvious for KB cells than HeLa cells.

Cellular liposomal arsenic was quantified with ICP-MS after performing calibrations using arsenic ionic standards and Ga ion as an internal standard. For every experiment performed, we obtained a linear correlation for arsenic with squared correlation coefficients R^2 >0.997. With this calibration, cellular arsenic was quantified by measuring the total amount of arsenic following digestion of the cells from the four cell lines treated with media only, ATO encapsulating conjugated and unconjugated liposomes for 6, 24, and 48 h. A comparative study of the liposomal treatment was then drawn for cellular arsenic, as depicted in Figure 5.

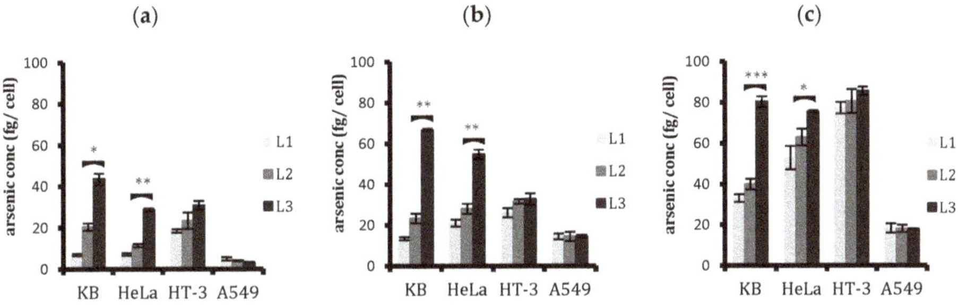

Figure 5. Arsenic concentration per cell as determined by ICP-MS in the four cell lines after (**a**) 6, (**b**) 24, and (**c**) 48 h treatment with the unconjugated (L1) and conjugated (L2 and L3) liposomes. L3 was taken up more than L2 in FR-positive KB and HeLa cells. HT-3 had a higher uptake of liposomes in general, regardless of ligand conjugation. The arsenic concentration increased with time for all cell lines. Data shown is the mean ± SD of three independent experiments. * $p \leq 0.05$, ** $p \leq 0.01$, and *** $p \leq 0.001$.

The results showed that, as the time of incubation increased, the cellular arsenic increased from all treatments for all four cell lines investigated. Quantifying arsenic clearly shows that increasing the ligand–PEG spacer length increases the liposomal uptake by FR-positive cells (Figure 5). After 6 h, L3 liposomes were taken up 2.1, 2.5, 1.3, and 0.8 times more in KB, HeLa, HT-3, and A549, respectively. It was obvious that A549 cells were not affected in their liposomal uptake by an increase in ligand-conjugating spacer length. The same trend was also observed with 24 and 48 h treatment. HT-3, despite showing a slight increase in L3 uptake compared to L2 uptake during the first six hours, reached a similar level with further prolonged treatments.

Corroborating the previous findings, the cellular uptake studies by ICP-MS are consistent with the flow cytometry studies and plate reader analysis in that the difference between the uptakes of targeted and non-targeted liposomes decreases with time in FR+ cells. As KB cells had the highest folate expression on their cell surface, this decline was more pronounced than it was in HeLa cells. On the other hand, HT-3, with minimal expression of cell surface FRs, displayed a slight difference in the first few hours; however, with an increase in time, this difference was nullified, and targeted and non-targeted liposomes were taken up by the cells equally. As for A549 cells, being FR-negative, displayed no difference in the cellular uptakes from ligand conjugation as expected. Regarding the total arsenic content in the cells from the treatment, ICP-MS studies showed that the HT-3 cell line took up more liposomal arsenic in general than all four cell lines regardless of the ligand conjugation, while A549 took up the least.

2.3. Analysing Cytotoxicity of Control Empty Liposomes with Differing Spacer Lengths of Conjugated Ligand

The results above clearly indicated the advantages of using a long PEG spacer to tether FA to the liposomes for enhancing their uptake in FR-positive cells. However, in order to employ them successfully as arsenic nanocarriers to manage cervical cancer, it becomes imperative to investigate any inherent cytotoxicity that they might possess due to the presence of longer PEG spacers. Since our focus is on cervical cancers and the presence or absence of HPV gene sequences, further experiments were carried out on HeLa, HT-3, and KB cells only. The KB cell line was included, since, despite being derived from epidermal mouth carcinoma, it had been contaminated with HeLa cells at some point. Therefore, it was positive for HPV-18 DNA sequences. Empty, unloaded liposomes of various formulations were synthesised and their cytotoxicity analysed on these cell lines via MTT assay at 48 h, as depicted in Figure 6. The dilutions employed for the empty liposomal treatment had their phospholipid concentrations corresponding to ATO liposomes with 5 μM encapsulated ATO concentration. At the dilutions used for treatment and the time period involved, none of the liposomes were toxic for any tested cell lines. With these results in consideration, L3 liposomes, out of the three, prove to be the best liposomal formulation to achieve targeting in FR-positive cell lines.

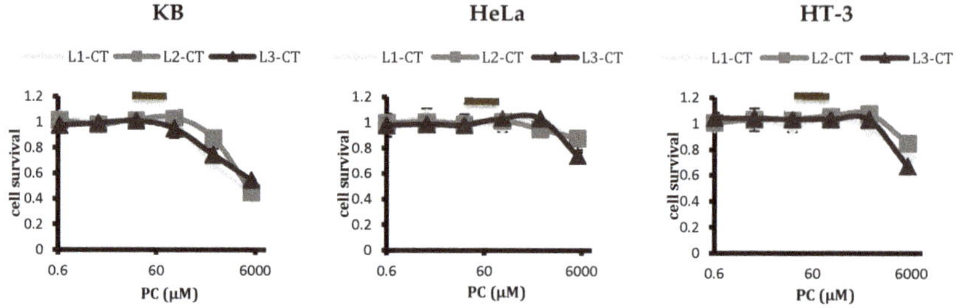

Figure 6. Toxicity studies by MTT assay of the control empty liposomes (L1, L2, and L3) on KB, HeLa, and HT-3 cell lines after 48 h treatment. Increasing the spacer length for ligand conjugation did not result in increased cytotoxicity at the phospholipid concentrations at dilutions used for the treatment. L3 liposomes were non-toxic to all cell lines tested at 48 h, similar to L1 and L2, demonstrating the suitability of L3 as a drug carrier. Points refer to the means, and bars refer to ±SD, where $n = 3$. ▬ represents the concentrations of phospholipid at dilutions used for treatment of liposomal ATO.

2.4. *Quantitative Analysis of Cellular Uptake of Arsenic with Free and Liposomal Arsenic*

After establishing the superiority of L3 liposomes in the cellular uptake by FR+ cells, they were further investigated in terms of how they compared with ATO when delivered in free form. The cells were treated with L3, L1, free ATO, and media only as control for 6, 24, and 48 h, and their arsenic intake was quantified by ICP-MS (Figure 7).

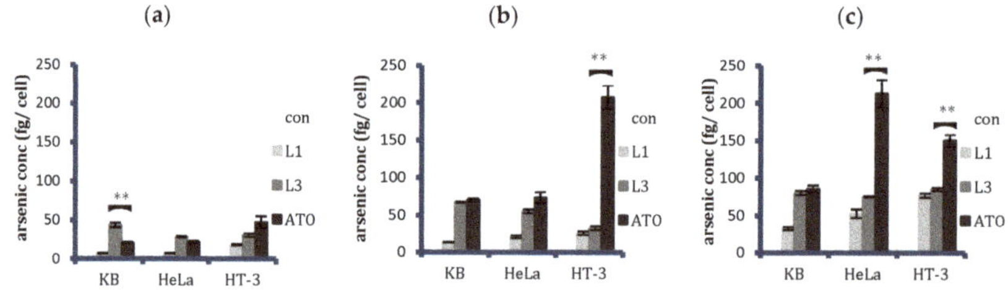

Figure 7. Arsenic concentrations per cell in KB, HeLa, and HT-3 as determined by ICP-MS after (**a**) 6, (**b**) 24, and (**c**) 48 h treatment with only media, unconjugated ATO liposomes L1, FA-conjugated ATO liposomes L3, and free ATO with a 5 μM concentration. Data displayed are mean ± SD from five independent experiments. An unpaired t-test was used to calculate any significant difference in the arsenic intake from L3 and ATO treatments; ** $p \leq 0.01$. L3 was taken up more than free ATO in KB at 6 h, after which their uptakes reached a similar level when the time of treatment was increased to 24 and 48 h. The cellular arsenic intake from free ATO was higher than L3 in HeLa at 48 h, whereas for HT-3 it became significantly higher after 24 h.

Results showed that after 6 h of treatment, targeted liposomes L3 were taken up more than free ATO in both KB and HeLa, with cellular arsenic concentration in the order of con < L1 < ATO < L3. Arsenic concentration due to L3 uptake was 2.1 and 1.3 times higher than that from ATO in KB and HeLa cells, respectively. Prolonging the incubation period to 24 h and further to 48 h resulted in reaching a similar level of cellular arsenic concentration from both the L3 and ATO treatments in KB cells. In contrast, with the prolonged incubation time, free ATO was taken up more than L3 in HeLa cells, 1.3 and 2.8 times more at 24 h and 48 h, respectively. In HT-3 cells, free ATO was always taken up more than the liposomal counterparts, both conjugated and non-conjugated, in the time periods tested. The results showed a consistent increase in the cellular arsenic concentration as the incubation time period increased for all the cell lines and the treatments, except the control non-treated sample. However, an unexpected observation was observed with the HT-3 cellular arsenic concentration following free ATO treatment, where instead of increasing from 24 to 48 h, it decreased from 207.4 fg to 150.8 fg per cell.

2.5. Selectivity of Targeted Liposomal ATO in Killing HPV-Infected Cervical Cancer Cells

With differing cellular uptakes of liposomal and free ATO, we further evaluated the cytotoxic response of cervical cancer cell lines of differing HPV statuses (HPV+ HeLa and HPV− HT-3) along with KB (HPV+) to the treatment of ATO delivered either in the free form or encapsulated within the chosen targeted liposomes with 48 h treatment (Figure 8a). The dose–response curve obtained from the MTT assay for the free drug and the liposomal drug revealed that liposomal encapsulation mitigated toxicity, except in the case of KB, where the targeted liposomes L3 induced more cell death than free ATO. At the dilutions used for treatment, i.e., at a 5 μM ATO concentration, the free drug caused more toxicity than L3 in HT-3 and HeLa cells. Cell death–uptake ratios for KB, HeLa, and HT-3 cells (Figure 8b) indicated the level of cell death induced in the cell lines per unit of arsenic taken up by the cells. Targeted liposomal ATO was more effective than the free drug in inducing cell death per unit of arsenic uptake in both folate receptor positive KB and HeLa cell lines, whereas free ATO was more effective in inducing cell death in HT-3 cells. Additionally, between KB and HeLa cells, the liposomal ATO treatment was more effective for HeLa cells than KB cells per unit arsenic.

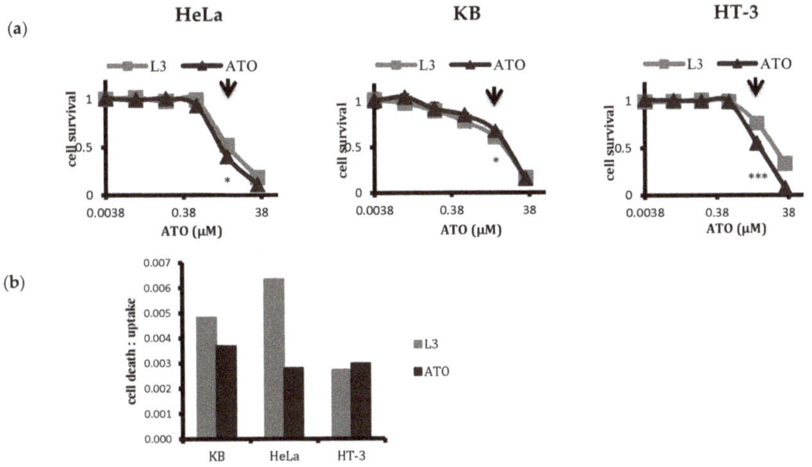

Figure 8. (**a**) MTT test analysis to compare cellular toxicity of targeted liposomal ATO with the free drug on HeLa, KB, and HT-3 cells at 48 h. Points refer to the means and bars refer to ±SD, where $n = 3$. Liposomal encapsulation attenuated drug toxicity to the cells in vitro, except for the KB cell line where L3 liposomes were more toxic than free ATO. (**b**) Ratios of cell death vs. uptake of targeted liposomal and free ATO for KB, HeLa, and HT-3 cells after 48 h treatment. Targeted liposomes induced highest toxicity per unit arsenic uptake in HeLa cells. Targeted liposomes were more effective in inducing cell death when they were corrected for intracellular ATO uptake than free ATO for both FR+ KB and HeLa cells. * $p \leq 0.05$ and *** $p \leq 0.001$. ➘ points to 5 μM ATO concentrations that were typically used for treatment.

3. Discussion

The efficiency of any actively targeting system is assessed by two important parameters: delivering capacity and targeting specificity of the drug delivery vehicle [20]. The delivering capacity depends on the structure and composition of the nanoparticles [42,43]. In order to optimize an ATO drug nanocarrier for targeting HPV-infected cervical cells, we firstly optimised ATO encapsulating liposomes, with respect to size and charge, for use as delivery vehicles for our target cells [9]. The second parameter, targeting specificity, is mainly determined from the choice of the ligand and how it interacts with the off-target molecules and cells [20]. FA was selected as the targeting agent for folate receptors, which are overexpressed on the surface of targeted HPV cervical cells and minimally expressed on non-HPV cells. In addition, FA has other advantages as an ideal targeting agent, such as being inexpensive and stable with a low molecular weight as well as having simple conjugation chemistry and relative low immunogenicity [27,44,45]. FA has a very high affinity for the FR presented on tumour's cell surface and is rapidly endocytosed [44]. Since FA is a vitamin required by the eukaryotic cells for the biosynthesis of nucleotide bases, any cargo attached to it remains within endosomes and is then released into cytoplasm rather than shuttled to the lysosome for destruction [27,44]. The folate-targeting of liposomal drug carriers has been reported to increase the therapeutic efficacy in a number of cases, but it requires careful optimisation for effective therapeutic applications for selected tumours [46,47].

The foremost necessity of synthesising an actively targeted nanoparticle is to extend its circulation times, as an increased nanoparticles' affinity for the target antigens could not compensate for the natural clearance processes in most circumstances [20]. Actively targeted nanoparticles without sufficient steric stabilisation from a PEG layer tend to lose their receptor-binding ability due to non-specific interactions with colloidal proteins [48]. Additionally, since the physical stability of the targeted liposomes without the PEG layer was also not very well explored, in the initial experiments, only the ligand-conjugating PEG spacer was included in the liposomal mix to prepare a targeted ATO delivery system, without the addition of the brush-forming PEG layer. However, this resulted in unstable liposomal formulations, which caused aggregation during the drug loading itself, regardless of the length of the ligand PEG spacer. Adding mPEG$_{2000}$-DSPE to the liposome mix stabilised the targeted formulations and prevented them from aggregating. This might be because surface-grafted PEG chains, even at low concentrations, could provide a sufficient steric barrier to prevent the fusion of liposomal membranes [49].

There was another parameter that required careful optimisation: the ligand density on the surface of the nanoparticle. Thermodynamically, the binding of a ligand to a receptor facilitates the binding of its neighbouring ligands; these multiple interactions lead to a clustering of the receptors, a wrapping of the membrane, and the eventual internalisation of the bound nanoparticle [50]. These multivalent interactions lead to an enhanced avidity of the nanoparticle towards the target cell and prevents its detachment from the cell surface. However, in vitro, a higher ligand density does not necessarily translate to an enhanced cellular uptake [51]. In some cases, it was shown that increasing the ligand density above a certain threshold resulted in saturation of the cooperative effects of the ligands, leading to unfavourable effects on cellular binding [52]. Moreover, the presence of a dense ligand covering on the nanoparticles was also known to make them more prone to being recognised by the cells of a mononuclear phagocytic system (MPS), which consequently led to a loss of their "stealth" surface characteristics [23]. Hence, in order to bypass any such negative effects and to ensure high nanoparticle avidity to the surface receptors, it is necessary to optimise ligand density in the nanoparticles for every particular cancer type.

Since this was an in vitro study, our investigation was primarily focused on the physico-chemical effects of ligand density on liposomal stability and subsequent cellular uptake. There have been reports of ligand molar ratio to range from 0.03 to 0.5% in the literature for folate-targeting liposomes [30,37,53–55]. In our experiments, it was observed that using 0.3 mol % DSPE-PEG$_{2000}$-FA and 1.7 mol % mPEG$_{2000}$-DSPE in the liposomal mixture formed stable L2 liposomes. However, the same (0.3 mol %) or higher amount of DSPE-PEG$_{5000}$-FA could not be used to prepare stable L3 liposomes, and this invariably led to the aggregation of liposomes as soon as they were synthesised, despite the presence of a stabilising PEG_{2000} brush layer. After adjusting molar ratios of components, employing 0.1 mol % DSPE-PEG_{5000}-FA and 1.9 mol % mPEG_{2000}-DSPE yielded stable L3 liposomal formulations. In addition, preliminary confocal uptake studies of L2 liposomes, synthesised with 0.3 mol % or 0.1 mol % DSPE-PEG_{2000}-FA, did not show any significant difference between cellular uptakes in HeLa cells. Henceforth, the same ratio (0.1 mol %) was used to prepare both L2 and L3 liposomes for comparison purposes. The instability of L3 liposomes with higher ligand density might be explained by studies where high folate density on the surface of nanoparticles were known to produce dimers, trimers, and quartet self-assembled folate structures, causing an aggregation of nanoparticles, with a longer ligand-conjugating PEG spacer expediting the process [56].

At a molar ratio employed for synthesising targeted liposomes, ligand conjugation had no effect on either the size of the nanoparticles, nor the arsenic loading efficiency and stability (Figure 1). All formulations (L1, L2, and L3) had similar sizes, i.e. around 140 nm, and achieved a high loading efficiency, with an arsenic-to-phospholipid molar ratio of 0.25. Our stability tests showed that all the liposomes, irrespective of ligand conjugation, were homogeneous and stable with a loss of less than 10% arsenic over four weeks of storage at 4 °C (see Figure 1).

Owing to a possibility that PEG brush might mask the targeting abilities of the ligand if it is conjugated to a PEG spacer of equal length, we synthesised L2 liposomes with an FA-PEG spacer of 2000 Da and L3 liposomes with a longer PEG-FA of 5000 Da in length with a surrounding 2000 Da PEG brush. As the previous studies employed FA-PEG3350 [57], we chose to use a longer PEG spacer length of 5000 Da for making targeted liposomes and thus to investigate any increase in cellular uptake along with any inherent toxicity that might arise from employing longer PEG spacers for ligand conjugation. We evaluated the differences in cellular uptake for the liposomal formulations using four different techniques: confocal microscopy, flow cytometry, fluorescence reading via a plate reader, and cellular arsenic quantification via ICP-MS.

Keeping all settings and parameters the same for comparative analysis, confocal microscopic images showed the superiority of using a longer PEG spacer for ligand conjugation (Figure 2). After the initial 2 h incubation, L3 liposomes clearly displayed a superior cell surface attachment in around 80% of FR+ KB cells, in contrast to 10% of KB cells with L2 liposomes. Additionally, around 20% of the HeLa cells showed membrane attachment with L3, but none was observed with either L2 or L1. Some attachment (around 5%) was also seen with L3 in HT-3 cells. A549 (an HPV-negative cell line) showed no cellular association with L3, which demonstrates the superiority of the conjugated ligand to specifically target HPV-positive cells. These results are in accordance with the hypothesis that folate-targeted arsenic loaded liposomes would be efficiently and selectively taken up by FR-positive cells and that, since FR expression in cell lines is in the order of KB > HeLa >> HT-3 > A549, the cellular uptake of targeted liposomes would also follow the same pattern. As the incubation period was increased to 4, 6, and eventually 24 h, KB cells saw a widespread internalisation and accumulation of L3 in their cytosol. L3 was also taken up in almost all of the HeLa cells and around 20% of the HT-3 cells. In contrast, L2, albeit more effective than L1, was visibly less effective than L3 in being taken up by FR+ cells. Based on the confocal images of KB and HeLa cells in particular, the cell surface association of targeted liposomes at 2 h and then the subsequent internalisation and accumulation within the cytoplasm at 24 h was significant, indicating endocytosis.

These results are consistent with results from both the flow cytometric analysis of cellular uptake of fluorescent liposomes and the fluorescence reading from the plate reader, which confirmed the superior and selective uptake of L3 liposomes by FR+ cells (Figures 3 and 4). ICP-MS studies further validated the findings and the results showed that the cellular arsenic content from L3 uptake was almost three times as much as the L2 uptake in KB cells and twice that of HeLa cells after 24 h (Figure 5). This enhanced uptake was, however, not significant in the case of HT-3 cells, owing to the weak FR expression by these cells. FR-negative A549 cells displayed a similar uptake for all the liposomes, irrespective of conjugation or non-conjugation.

This enhanced ATO cellular uptake from L3 liposomes could be explained by folate-receptor-mediated endocytosis. In order to confirm this, future work will be carried out to evaluate drug cellular uptake by blocking FRs with free folates prior to FR-Lip-ATO exposure to the cells. [36]. This would provide useful information for us to understand the uptake mechanism of folate-conjugated liposomes by the target cells. Drugs carried by delivery vehicles could be internalised in the cells via different endocytic pathways, which might prevent the drug from reaching the desired targeted cellular organelle and therefore affecting or even diminishing the therapeutic effect [58]. Of the different pathways proposed for nanomedicinal uptake, clathrin-mediated endocytosis via specific receptor–ligand interaction is the best characterised and has been reported to be the preferred pathway for particles up to 200 nm in size [59]. However, recent studies have suggested that folate-conjugated nanoparticles utilise both clathrin- and caveolae-receptor-mediated endocytosis pathways [60], with the preferred route possibly being dependent on the size of the drug carrier [58]. In one study, it was reported that the uptake of 50 nm nanoparticles followed the clathrin-mediated endocytic pathway, whilst the uptake of 250 nm particles was dominated by caveolae-mediated endocytosis [58]. Since the size of our targeted liposomes was around 140 nm in diameter, they might be predominantly employing the clathrin-mediated endocytic pathway. However, further investigations are needed to confirm this hypothesis.

It was evident from the flow cytometric and plate reader analysis that the difference between the cellular uptake of conjugated and non-conjugated liposomes decreased with prolonging the exposure time (Figures 3 and 4), which might be due to two contributing factors: an increase in non-specific uptake of unconjugated liposomes and a subsequent decrease in the uptake of conjugated liposomes due to a possible saturation of the targeted receptors on the cell surface. This effect was most pronounced for KB cells followed by HeLa. This hypothesis was corroborated after quantifying cellular arsenic via ICP-MS. Results showed that after an initial high surge of targeted liposomal arsenic uptake by FR+ cells, the rate at which they were taken up almost invariably slowed down in comparison with non-targeted liposomes. This led to a bridging of the gap between their uptakes in cells.

Results clearly demonstrated that FA conjugation to liposomes, even to a PEG spacer of similar length as the surrounding PEG brush, was able to selectively target FR-positive cells. However, their targeting efficiency increased considerably when FA was conjugated to a longer PEG spacer, making L3 the logical choice as a delivery vehicle for arsenic to FR rich HeLa and KB cells with HPV infection. However, for it to function as a successful delivery vehicle, any cytotoxicity generated needs to be evaluated. Since we intended to build on our work investigating the therapeutic potential of arsenic against HPV-infected cervical cancers and the time period chosen for our previous studies had been 48 h, all the liposomal formulations were tested for any intrinsic cytotoxicity by incubating empty liposomes for 48 h with HeLa, HT-3, and KB cells (all positive for HPV infection). The dilutions of liposomes tested for toxicity included ones that corresponded with the liposomal concentration with encapsulated 5 μM ATO and that were used for treatment. All the liposomes, including L3, were found to be non-toxic at the dilutions to be used for treatment (Figure 6). This finding was in agreement with the study by Peng et al. (2013), who observed that the conjugated docetaxel delivery systems targeting prostrate specific membrane antigen with a longer spacer length had no significant toxicity in vitro or in vivo in major organs of treated mice bearing a C4-2 tumour xenograft [41]. Kawano and Maitani [61] also reported a higher cellular association of liposomal doxorubicin with a longer ligand–PEG spacer without a significant increase in cytotoxicity.

Consequently, the inherent non-cytotoxicity of L3 liposomes along with their superior targeting abilities led us to select them as an optimised design for a targeted ATO nano-carrier. Further in vitro analysis was also carried out with L3 as the conjugated liposomes, L1 as their unconjugated counterparts, and free ATO on HPV-positive KB and HeLa cells and HPV-negative HT-3 cells.

Cellular uptake studies showed that, usually, arsenic was transported more readily within the cells when the free form was applied (Figure 7). However, owing to the very high FR expression of KB cells, these cells were shown to be most amenable towards folate targeting and consequent liposomal uptake, with L3 liposomes transporting twice as much arsenic in the first six hours of treatment. This trend was also seen with HeLa cells, albeit to a lesser extent. As treatment time increased, however, we reported an equalising of the cellular arsenic concentration between L3- and free-ATO-treated KB cells, whilst an increasing arsenic concentration with free ATO treatment in HeLa cells was surprisingly observed. In HT-3 cells, however, the free form of the drug was always taken up more so than the liposomally delivered forms, with little difference between L1 and L3 uptake.

The cells were further incubated with free and targeted liposomal ATO with a 5 μM concentration for a period of 48 h, and their cytotoxicity was assessed via MTT assay (Figure 8a). As expected, KB cells displayed maximum vulnerability towards L3 treatment, where L3 liposomes induced more cell death than the free drug. In contrast, free ATO generated a higher toxicity than the liposomal formulation for both HeLa and HT-3 cells. The cell death–uptake ratio indicates the cell death induced per unit of arsenic uptake and is an effective means of comparing the efficacy of a treatment on different cell populations. Our results indicated that, for targeted liposomal treatment, this ratio was higher for HPV-positive HeLa and KB cells than it was for HPV-negative HT-3 cells (Figure 8b). Moreover, targeted liposomes were more efficient in inducing cell death than free ATO in FR-positive cells per unit of arsenic taken up, in contrast to HT-3 cells. Interestingly, targeted liposomal ATO was found to be most effective in inducing its response towards HeLa cells, which indicates a certain susceptibility of the cell line towards liposomal treatment, an aspect that warrants further research.

4. Materials and Methods

4.1. Materials

Folate free RPMI-1640 media, Gibco McCoy's 5A (modified) medium, foetal bovine serum (FBS), penicillin–streptomycin (5000U/mL), 1,1′-dioctadecyl-3,3,3′,3′-tetramethylindocarbocyanine-5,5′-disulfonic Acid (DiI), nitric acid, Trypan Blue, isopropanol, NaH_2PO_4 (sodium dihydrogen phosphate), Na_2HPO_4 (disodium hydrogen phosphate), and Trypsin-EDTA were purchased from Fisher Scientific (Loughborough, UK). Soy phosphatidylcholine (PC), cholesterol (Chol), methoxy(polyethyleneglycol)-2000-distearoyl-phosphatidylethanolamine (mPEG-DSPE), 1,2-distearoyl-sn-glycero-3-phosphoethanolamine-N-folate(polyethylene glycol)-2000 (DSPE-PEG_{2000}-FA), and 1,2-distearoyl-sn-glycero-3-phosphoethanolamine-N-folate (polyethylene glycol)-5000 (DSPE-PEG_{5000}-FA) were purchased from Avanti Polar Lipids (Alabaster, AL, USA). VECTASHIELD Hard Set Mounting Media with DAPI were obtained from Vector Labs (Peterborough, UK). Phosphate-buffered saline (PBS), ATO, nickel acetate, dimethyl sulfoxide (DMSO), thiazolyl blue tetrazolium bromide powder (MTT), Whatman Anotop 0.1 μm syringe filters, and dialysis tubing were purchased from Sigma (Welwyn Garden City, UK). Ham's F-10 Nutrient Mix, methanol, and dichloromethane were obtained from Thermofisher (Paisley, UK).

4.2. Liposome Preparation and Characterisation

Liposomes prepared for this study were labelled with DiI and had compositions as follows: (a) untargeted liposomes (**L1**), PC/Chol/mPEG-DSPE/DiI = 52.5/45/2/0.5 mol %; (b) folate-targeted liposome with a 2000 MW PEG-FA spacer (**L2**), PC/Chol/mPEG-DSPE/DSPE-PEG_{2000}-FA/DiI = 52.5/45/1.9/0.1/0.5 mol %; folate-targeted liposomes with a 5000 MW PG-FA spacer (**L3**), PC/Chol/mPEG-DSPE/DSPE-PEG_{5000}-FA/DiI = 52.5/45/1.9/0.1/0.5 mol %. Liposomes were prepared as described elsewhere with slight modifications [62]. Briefly, the lipids were dissolved in 1:2 (v/v) methanol/dichloromethane at room temperature. The lipid mixtures were deposited on the side wall of a rotary glass vial by removing the solvent with nitrogen. The dried lipid films were hydrated in 730 mM nickel acetate ($Ni(OAc)_2$) aqueous solutions for 1 h with gentle rotation. This process led to the spontaneous formation of multilamellar PEGylated liposomes. The liposome suspension was subsequently subjected to 10 freeze–thaw cycles (freezing in liquid nitrogen for 3 min and thawing in a 37 °C water bath for 3 min and downsized via 0.1 μm Anotop filters. Extruded liposomes were dialysed overnight against 10 mM sodium phosphate buffer at pH 7 to remove unencapsulated $Ni(OAc)_2$. The nickel acetate-encapsulated liposomes were then incubated with a 20 mM ATO solution at room temperature with gentle rolling for 5 h. The unencapsulated ATO was further removed by dialysis overnight. The concentrations of phospholipids (P), encapsulated arsenic (As), and nickel (Ni) within the liposomes were determined by an inductively coupled plasma optical emission spectrometer (ICP-OES; Thermo-Scientific iCap 6500 ICP, Stafford, UK), and loading efficiency (As/P ratio) was calculated. Their mean liposomal size was calculated via dynamic light scattering on a Zetasizer-Nano ZS (Malvern Instruments, Malvern, UK). Liposomal stability over a period of a month was assessed by analysing their loading efficiency every week, while liposomes were stored at 4 °C in buffers of pH 7.4.

4.3. Cell Culture

HPV-positive and -negative cervical cancer cell lines, HeLa, and HT-3, along with FR-positive and -negative control cell lines, human nasopharyngeal epidermal carcinoma KB, and lung carcinoma A549 cells, respectively (ATCC, Middlesex, UK), were employed in this study. HeLa and KB cells were grown in folate-free RPMI-1640 media, HT-3 cells in McCoy's 5A (Modified) media, and A549 cells in Ham's F-10 nutrient mix, supplemented with 10% FBS and 1% penicillin–streptomycin for a minimum of 2 months before each experiment in 75 cm^2 flasks. The cells were grown in a humidified incubator containing 5% CO_2 and 95% air at 37 °C until they reached 90% confluence. The following experiments were set up to investigate cellular uptake following liposomal ATO exposure for different time intervals: confocal microscopic visualisation, flow cytometry, spectrophotometer analysis, and quantification of cellular arsenic uptake by ICP-MS studies. Cellular toxicity was investigated via MTT assay.

4.4. Qualitative Cellular Uptake Analysis by Confocal Microscopic Visualisation of Liposomal Arsenic

Cells were plated 24–48 h before each experiment, on sterile 22 mm coverslips inside 6-well plates at a density of 5×10^4/mL. Following attachment, the cells were exposed to DiI-labelled liposomes (L1, L2, and L3) for various time intervals at 37 °C in the incubator. The employed liposomal dilution had an arsenic concentration of 18 μmol/L. After the intended time period, the spent media were removed, and cells were washed three times with PBS and fixed with PBS-buffered 4% paraformaldehyde at 25 °C for 8 min. The coverslips were again washed with PBS and mounted on a slide using DAPI containing anti-fade ProLong Gold reagent. The fluorescence emitted from each slide was observed via a fluorescent confocal microscope at 570 nm and 460 nm for DiI and DAPI, respectively (Leica Microsystems, Wetzlar, Germany).

4.5. Flow Cytometric Analysis of Liposomal Arsenic Uptake

Cells were seeded at 5×10^5/mL in six-well culture plates and grown overnight before they were exposed to the DiI-labelled liposomal formulations with an arsenic concentration of 18 μmol/L for various time intervals at 37 °C. After the intended time interval, the cells were washed thrice with PBS, trypsinized, washed twice by PBS, and then collected into FACS tubes with 500 μL of PBS. All samples were analysed using FACSCalibur (BD, Oxford, UK). For DiI, the maximum excitation was obtained with an He–Ne laser at 555 nm, and fluorescence emission intensities were observed at 570 nm using an FL-2 filter. For each sample, a minimum of 10,000 cells were collected. Data were analysed using CellQuest Pro software from BD Biosciences. The ratio of cells staining positive with labelled FA-conjugated liposomes (L2 and L3) to cells staining positive for labelled unconjugated liposomes (L1) was calculated to yield an estimate of the differences in the uptake of liposomes of different formulations. All measurements were performed in duplicates from at least three different experiments.

4.6. Plate Reader Analysis of Liposomal Uptake by Cells

Cells were seeded at 1×10^5/mL in 96-well plates and allowed to attach overnight. Cells were further exposed to labelled liposomes at arsenic concentration of 18 μmol/L for 2, 6, and 24 h at 37 °C. Following treatment, the cells were washed twice with PBS and the fluorescence measured via BMG

LabTech FLUOstar Omega Plate Reader (Bucks, UK). As with flow cytometric analysis, the ratio of cells staining positive with FA-labelled conjugated liposomes (L2 and L3) to cells staining positive for labelled unconjugated liposomes (L1) was calculated and analysed.

4.7. Quantitative Analysis of Liposomal Drug Uptake via ICP-MS

Cellular uptake was analysed for different treatments for 6, 24, and 48 h by ICP-MS. Cells from the four different cell lines were first seeded at a density of 1×10^6 cells/flask into sixteen 75 cm^2 flasks. After a 24 h cell attachment, they were treated with liposomal formulations as mentioned above and a free drug with an arsenic concentration of 10 μmol/L. Following the incubation interval, the cells were washed with PBS, trypsinised, and counted prior to collection in Falcon tubes for further analysis. The cells were lysed by adding 2 mL of nitric acid while vortexing and heating at a temperature of 60 °C for 5 min and were topped up with 8 mL of deionised water. Arsenic concentration was analysed using ICP-MS (Thermo Fisher XSeries2, Paisley, UK) and corrected to the cell number and total volume accordingly.

4.8. In Vitro Cellular Cytotoxicity Assay

Cytotoxicity of various liposomal formulations of ATO was determined by the MTT assay as described previously [28]. Briefly, the experiment was set up in a 96-well plate, where the toxicity of control empty liposomes was investigated by taking an initial starting amount containing 0.5 mM of phospholipid concentration and diluting it in 1:10 ratio to another 6 wells. Simultaneously, the toxicity of ATO encapsulating liposomes was calculated by taking the starting amount of 30 μM of ATO concentration, diluting to a 1:6 ratio at a final working arsenic concentration of 5 μM for treating the cells. The wells were seeded with HeLa, HT-3, KB, and A549 at a density of 0.6 million cells per mL and incubated at 37 °C in the humidified incubator for 48 h. After the intended incubation, the spent media was carefully removed and 50 μL of MTT solution was added to each well. Cells were further incubated for 30 min at 37 °C. The MTT solution was then removed and 100 μL of propanol was added per well for at least 30 min incubation at 37 °C to dissolve crystals. The absorbance of this coloured solution was quantified by measuring at a wavelength of 570 nm from BMG LabTech FLUOstar Omega Plate Reader (Bucks, UK).

4.9. Statistical Analysis

Statistical analysis described in experimental sections was done using Minitab17. Statistical significance was determined by a two-sample *t*-test. $p < 0.05$ was considered significant. For flow cytometry, statistical analysis was carried out automatically though BD Calibur software provided. For confocal microscopy results, the average number of positively stained cells in a total of six fields for each sample was calculated, and average percentages were recorded.

5. Conclusions

The success of a targeted drug delivery system, tailored to the specific cancer under investigation, depends on the thorough optimisation of the nano-carrier's various parameters. In this study, we synthesised folate-targeted, ATO-encapsulated liposomes and examined the effect of increasing the ligand spacer length on cellular uptake and inhibitory potency against FR-positive cancer cells in vitro, with particular emphasis on HPV-infected FR-positive cervical cancer cells. Folate targeting, particularly when the ligand was attached to a longer PEG spacer, led to an appreciable enhancement in cellular uptake by FR-positive cancer cells. This formulation, along with being efficient and highly specific, had no inherent toxicity of its own at the dilutions used for the treatment, making it suitable to be used as the delivery vehicle of ATO to HPV-positive cervical cancer cells. Future preclinical studies are warranted to elucidate the role of EPR-mediated passive targeting and folate-active targeting in vivo and to investigate nanoparticle clearance/potential toxicity utilising tumour-bearing mice.

Our results suggest that further investigation of the molecular mechanisms behind the ATO action delivered via liposomes for targeted liposomal ATO cancer treatment is warranted.

Author Contributions: X.W. conceived the experiments; A.A. designed and executed the experiments; X.W., L.G., C.B., and D.L. contributed reagents/materials/analysis tools and provided technical assistance where required; S.X.W. provided technical expertise in liposome synthesis experiments and experimental design; A.A. analysed the data and, aided by X.W., who drafted the manuscript, wrote the paper; C.B. and L.G. participated in the initial design for the study and final draft editing. All authors read and approved the final manuscript.

Funding: This research received funding from Middlesex University, U.K.

Conflicts of Interest: The authors declare no conflict of interest.

Abbreviations

%	percentage
μg	microgram
μl	microliter
μM	micromolar
A549	human lung epithelial cancer cell line
As	arsenic
ATO	arsenic trioxide
C	centigrade
Chol	cholesterol
CLSM	confocal laser scanning microscopy
Da	dalton
DAPI	4′,6-diamidino-2-phenylindole
DiI	1,1′-dioctadecyl-3,3,3′,3′-tetramethylindocarbocyanine-5,5′-disulfonic acid
DLS	dynamic light scattering
DMSO	dimethyl sulfoxide
DNA	deoxyribonucleic acid
DSPE-PEG$_{2000}$	methoxypolyethyleneglycol-distearoyl-phosphatidylethanolamine with mPEG MW 2000 Da
DSPE-PEG$_{2000}$-Folate	1,2-distearoyl-sn-glycero-3-phosphoethanolamine-N-folate(polyethylene glycol)-2000 with mPEG MW2000 Da
DSPE-PEG$_{5000}$-Folate	1,2-distearoyl-sn-glycero-3-phosphoethanolamine-N-folate(polyethylene glycol)-5000 with mPEG MW5000 Da
EPR	enhanced permeability and retention effect
FA	folic acid
FACS	fluorescence activated cell sorting
FBS	fetal bovine serum
FITC	fluorescein isothiocyanate
fg	femtogram
FR	folate receptor
Ga	gallium
h	hour(s)
HeLa	human cervical epithelial cells of adenocarcinoma
HPV	human papilloma virus
HT-3	human cervical cancer cell line (HPV-negative)
ICP-MS	inductively coupled plasma-mass spectrometer
ICP-OES	inductively coupled plasma-optical emission spectrometer
KB	human nasopharyngeal cell line contaminated with HeLa cells
kDa	kilodalton
Lipo	liposome
mAb	monoclonal antibody
min	minute(s)
mM	millimolar

mol	molarity
MPS	mononuclear phagocytic system
MTT	3-(4,5-dimethylthiazolyl-2)-2,5-diphenyltetrazolium bromide
NaH_2PO_4	sodium dihydrogen phosphate
Na_2HPO_4	disodium hydrogen phosphate
Ni	nickle
$Ni(OAc)_2$	nickle acetate
nm	nanometre
P	phospholipids
PBS	phosphate buffered saline
PC	phosphatidyl choline
PEG	polyethylene glycol
PMT	photomultiplier tube
SD	standard deviation
UK	United Kingdom
USA	United States of America
v/v	volume by volume

References

1. Parkin, D.M.; Bray, F. The burden of HPV-related cancers. *Vaccine* **2006**, *24*, 11–25. [CrossRef] [PubMed]
2. Schiffman, M.; Castle, P.E.; Jeronimo, J.; Rodriguez, A.C.; Wacholder, S. Human papillomavirus and cervical cancer. *Lancet* **2007**, *370*, 890–907. [CrossRef]
3. Bosch, F.X.; Lorincz, A.; Munoz, N.; Meijer, C.; Shah, K.V. The causal relation between human papillomavirus and cervical cancer. *J. Clin. Pathol.* **2002**, *55*, 244–265. [CrossRef] [PubMed]
4. Walboomers, J.M.M.; Jacobs, M.V.; Manos, M.M.; Bosch, F.X.; Kummer, J.A.; Shah, K.V.; Snijders, P.J.F.; Peto, J.; Meijer, C.J.L.M.; Munoz, N. Human papillomavirus is a necessary cause of invasive cervical cancer worldwide. *J. Pathol.* **1999**, *189*, 12–19. [CrossRef]
5. Nair, S.; Pillai, M.R. Human papillomavirus and disease mechanisms: Relevance to oral and cervical cancers. *Oral Dis.* **2005**, *11*, 350–359. [CrossRef] [PubMed]
6. NeufCoeur, P.E.; Arafa, M.; Delvenne, P.; Saussez, S. Involvement of human papillomavirus in upper aero-digestive tracts cancers. *Bull. Cancer* **2009**, *96*, 941–950. [PubMed]
7. Shukla, S.; Bharti, A.C.; Mahata, S.; Hussain, S.; Kumar, R.; Hedau, S.; Das, B.C. Infection of human papillomaviruses in cancers of different human organ sites. *Indian J. Med. Res.* **2009**, *130*, 222–233.
8. Steenbergen, R.D.; de Wilde, J.; Wilting, S.M.; Brink, A.A.; Snijders, P.J.; Meijer, C.J. HPV-mediated transformation of the anogenital tract. *J. Clin. Virol.* **2005**, *32*, 25–33. [CrossRef]
9. Akhtar, A.; Wang, S.X.; Ghali, L.; Bell, C.; Wen, X. Effective Delivery of Arsenic Trioxide to HPV-Positive Cervical Cancer Cells Using Optimised Liposomes: A Size and Charge Study. *Int. J. Mol. Sci.* **2018**, *19*, 1081. [CrossRef]
10. Um, S.J.; Lee, S.Y.; Kim, E.J.; Myoung, J.; Namkoong, S.E.; Park, J.S. Down-regulation of human papillomavirus E6/E7 oncogene by arsenic trioxide in cervical carcinoma cells. *Cancer Lett.* **2002**, *181*, 11–22. [CrossRef]
11. Wen, X.; Li, D.; Zhang, Y.; Liu, S.; Ghali, L.; Iles, R.K. Arsenic trioxide induces cervical cancer apoptosis, but specifically targets human papillomavirus-infected cell populations. *Anticancer Drugs* **2012**, *23*, 280–287. [CrossRef] [PubMed]
12. Wang, Z.Y.; Chen, Z. Acute promyelocytic leukemia: From highly fatal to highly curable. *Blood* **2008**, *111*, 2505–2515. [CrossRef] [PubMed]
13. Zhu, J.; Chen, Z.; Lallemand-Breitenbach, V.; de Thé, H. How acute promyelocytic leukaemia revived arsenic. *Nat. Rev. Cancer* **2002**, *2*, 705–7113. [CrossRef]
14. Emadi, A.; Gore, S.D. Arsenic trioxide—An old drug rediscovered. *Blood Rev.* **2010**, *24*, 191–199. [CrossRef] [PubMed]
15. Larochette, N.; Decaudin, D.; Jacotot, E.; Brenner, C.; Marzo, I.; Susin, S.A.; Zamzami, N.; Xie, Z.; Reed, J.; Kroemer, G. Arsenite induces apoptosis via a direct effect on the mitochondrial permeability transition pore. *Exp. Cell Res.* **1999**, *249*, 413–421. [CrossRef] [PubMed]

16. Kroemer, G.; de Thé, H. Arsenic trioxide, a novel mitochondriotoxic anticancer agent? *J. Natl. Cancer Inst.* **1999**, *9*, 743–745. [CrossRef]
17. Platanias, L.C. Biological responses to arsenic compounds. *J. Biol. Chem.* **2009**, *284*, 18583–18587. [CrossRef] [PubMed]
18. Dilda, P.J.; Hogg, P.J. Arsenical-based cancer drugs. *Cancer Treat. Rev.* **2007**, *33*, 542–564. [CrossRef] [PubMed]
19. Liu, B.; Pan, S.; Dong, X.; Qiao, H.; Jiang, H.; Krissansen, G.W.; Sun, X. Opposing effects of arsenic trioxide on hepatocellular carcinomas in mice. *Cancer Sci.* **2006**, *97*, 675–681. [CrossRef]
20. Bertrand, N.; Wu, J.; Xu, X.; Kamaly, N.; Farokhzad, O.C. Cancer nanotechnology: The impact of passive and active targeting in the era of modern cancer biology. *Adv. Drug Deliv. Rev.* **2014**, *66*, 2–25. [CrossRef]
21. Allen, T.M.; Cullis, P.R. Drug delivery systems: Entering the mainstream. *Science* **2004**, *303*, 1818–1822. [CrossRef]
22. Cheng, Z.; Al Zaki, A.; Hui, J.Z.; Muzykantov, V.R.; Tsourkas, A. Multifunctional nanoparticles: Cost versus benefit of adding targeting and imaging capabilities. *Science* **2012**, *338*, 903–910. [CrossRef]
23. Kamaly, N.; Xiao, Z.; Valencia, P.M.; Radovic-Moreno, A.F.; Farokhzad, O.C. Targeted polymeric therapeutic nanoparticles: Design, development and clinical translation. *Chem. Soc. Rev.* **2012**, *41*, 2971–3010. [CrossRef]
24. Koshkaryev, A.; Sawant, R.; Deshpande, M.; Torchilin, V. Immunoconjugates and long circulating systems: Origins, current state of the art and future directions. *Adv. Drug Deliv. Rev.* **2013**, *65*, 24–35. [CrossRef] [PubMed]
25. Peer, D.; Karp, J.M.; Hong, S.; Farokhzad, O.C.; Margalit, R.; Langer, R. Nanocarriers as an emerging platform for cancer therapy. *Nat. Nanotechnol.* **2007**, *2*, 751–760. [CrossRef] [PubMed]
26. Shi, J.; Xiao, Z.; Kamaly, N.; Farokhzad, O.C. Self-assembled targeted nanoparticles: Evolution of technologies and bench to bedside translation. *Acc. Chem. Res.* **2011**, *44*, 1123–1134. [CrossRef]
27. Byrne, J.D.; Betancourt, T.; Brannon-Peppas, L. Active targeting schemes for nanoparticle systems in cancer therapeutics. *Adv. Drug Deliv. Rev.* **2008**, *60*, 1615–1626. [CrossRef] [PubMed]
28. Lee, R.J.; Low, P.S. Folate-mediated tumor cell targeting of liposome-entrapped doxorubicin in vitro. *Biochimica et Biophysica Acta (BBA) Biomembranes* **1995**, *1233*, 134–144. [CrossRef]
29. Ni, S.; Stephenson, S.M.; Lee, R.J. Folate receptor targeted delivery of liposomal daunorubicin into tumor cells. *Anticancer Res.* **2002**, *22*, 2131–2135. [PubMed]
30. Pan, X.Q.; Wang, H.; Shukla, S.; Sekido, M.; Adams, D.M.; Tjarks, W.; Barth, R.F.; Lee, R.J. Boron-containing folate receptor-targeted liposomes as potential delivery agents for neutron capture therapy. *Bioconju. Chem.* **2002**, *13*, 435–442. [CrossRef]
31. Sudimack, J.; Lee, R.J. Targeted drug delivery via the folate receptor. *Adv. Drug Deliv. Rev.* **2000**, *41*, 147–162. [CrossRef]
32. Werner, M.E.; Karve, S.; Sukumar, R.; Cummings, N.D.; Copp, J.A.; Chen, R.C.; Zhang, T.; Wang, A.Z. Folate-targeted nanoparticle delivery of chemo-and radiotherapeutics for the treatment of ovarian cancer peritoneal metastasis. *Biomaterials* **2011**, *32*, 8548–8554. [CrossRef]
33. Zhao, X.B.; Lee, R.J. Tumor-selective targeted delivery of genes and antisense oligodeoxyribonucleotides via the folate receptor. *Adv. Drug Deliv. Rev.* **2004**, *56*, 1193–1204. [CrossRef] [PubMed]
34. Sapra, P.; Allen, T.M. Ligand-targeted liposomal anticancer drugs. *Prog. Lipid Res.* **2003**, *42*, 439–462. [CrossRef]
35. Wang, M.; Thanou, M. Targeting nanoparticles to cancer. *Pharmacol. Res.* **2010**, *62*, 90–99. [CrossRef]
36. Lee, R.J.; Low, P.S. Delivery of liposomes into cultured KB cells via folate receptor-mediated endocytosis. *J. Biol. Chem.* **1994**, *269*, 3198–3204.
37. Gabizon, A.; Horowitz, A.T.; Goren, D.; Tzemach, D.; Mandelbaum-Shavit, F.; Qazen, M.M.; Zalipsky, S. Targeting folate receptor with folate linked to extremities of poly (ethylene glycol)-grafted liposomes: In vitro studies. *Bioconjug. Chem.* **1999**, *10*, 289–298. [CrossRef]
38. Canal, F.; Vicent, M.J.; Pasut, G.; Schiavon, O. Relevance of folic acid/polymer ratio in targeted PEG-"epirubicin conjugates". *J. Control. Release* **2010**, *146*, 388–399. [CrossRef]
39. Zhang, C.; Zhao, L.; Dong, Y.; Zhang, X.; Lin, J.; Chen, Z. Folate-mediated poly (3-hydroxybutyrate-co-3-hydroxyoctanoate) nanoparticles for targeting drug delivery. *Eur. J. Pharm. Biopharm.* **2010**, *76*, 10–16. [CrossRef] [PubMed]

40. Zhang, Z.; Jia, J.; Lai, Y.; Ma, Y.; Weng, J.; Sun, L. Conjugating folic acid to gold nanoparticles through glutathione for targeting and detecting cancer cells. *Bioorg. Med. Chem.* **2010**, *18*, 5528–5534. [CrossRef]
41. Peng, Z.H.; Sima, M.; Salama, M.E.; Kopečková, P.; Kopeček, J. Spacer length impacts the efficacy of targeted docetaxel conjugates in prostate-specific membrane antigen expressing prostate cancer. *J. Drug Target.* **2013**, *21*, 968–980. [CrossRef]
42. Gu, F.; Zhang, L.; Teply, B.A.; Mann, N.; Wang, A.; Radovic-Moreno, A.F.; Langer, R.; Farokhzad, O.C. Precise engineering of targeted nanoparticles by using self-assembled biointegrated block copolymers. *Proc. Natl. Acad. Sci. USA* **2008**, *105*, 2586–2591. [CrossRef] [PubMed]
43. Jiang, W.; Kim, B.Y.S.; Rutka, J.T.; Chan, W.C.W. Nanoparticle-mediated cellular response is size-dependent. *Nat. Nanotechnol.* **2008**, *3*, 145–150. [CrossRef]
44. Talekar, M.; Kendall, J.; Denny, W.; Garg, S. Targeting of nanoparticles in cancer: Drug delivery and diagnostics. *Anticancer Drugs* **2011**, *22*, 949–962. [CrossRef]
45. Yu, B.O.; Tai, H.C.; Xue, W.; Lee, L.J.; Lee, R.J. Receptor-targeted nanocarriers for therapeutic delivery to cancer. *Mol. Membr. Biol.* **2010**, *27*, 286–298. [CrossRef] [PubMed]
46. Stephenson, S.M.; Low, P.S.; Lee, R.J. Folate receptor-mediated targeting of liposomal drugs to cancer cells. *Methods Enzymol.* **2004**, *387*, 33–50. [PubMed]
47. Reddy, J.A.; Allagadda, V.M.; Leamon, C.P. Targeting therapeutic and imaging agents to folate receptor positive tumors. *Curr. Pharm. Biotechnol.* **2005**, *6*, 131–150. [CrossRef]
48. Salvati, A.; Pitek, A.S.; Monopoli, M.P.; Prapainop, K.; Bombelli, F.B.; Hristov, D.R.; Kelly, P.M.; Ãberg, C.; Mahon, E.; Dawson, K.A. Transferrin-functionalized nanoparticles lose their targeting capabilities when a biomolecule corona adsorbs on the surface. *Nat. Nanotechnol.* **2013**, *8*, 137–143. [CrossRef]
49. Holland, J.W.; Hui, C.; Cullis, P.R.; Madden, T.D. Poly(ethylene glycol)–lipid conjugates regulate the calcium-induced fusion of liposomes composed of phosphatidylethanolamine and phosphatidylserine. *Biochemistry* **1996**, *35*, 2618–2624. [CrossRef]
50. Mammen, M.; Choi, S.K.; Whitesides, G.M. Polyvalent interactions in biological systems: Implications for design and use of multivalent ligands and inhibitors. *Angew. Chem. Int. Ed.* **1998**, *37*, 2754–2794. [CrossRef]
51. Weissleder, R.; Kelly, K.; Sun, E.Y.; Shtatland, T.; Josephson, L. Cell-specific targeting of nanoparticles by multivalent attachment of small molecules. *Nat. Biotechnol.* **2005**, *23*, 1418–1423. [CrossRef]
52. Elias, D.R.; Poloukhtine, A.; Popik, V.; Tsourkas, A. Effect of ligand density, receptor density, and nanoparticle size on cell targeting. *Nanomedicine* **2013**, *9*, 194–201. [CrossRef] [PubMed]
53. Reddy, J.A.; Abburi, C.; Hofland, H.; Howard, S.J.; Vlahov, I.; Wils, P.; Leamon, C.P. Folate-targeted, cationic liposome-mediated gene transfer into disseminated peritoneal tumors. *Gene Ther.* **2002**, *9*, 1542–1550. [CrossRef] [PubMed]
54. Saul, J.M.; Annapragada, A.; Natarajan, J.V.; Bellamkonda, R.V. Controlled targeting of liposomal doxorubicin via the folate receptor in vitro. *J. Control. Release* **2003**, *92*, 49–67. [CrossRef]
55. Pan, X.Q.; Wang, H.; Lee, R.J. Boron delivery to a murine lung carcinoma using folate receptor-targeted liposomes. *Anticancer Res.* **2002**, *22*, 1629–1633.
56. Ohguchi, Y.; Kawano, K.; Hattori, Y.; Maitani, Y. Selective delivery of folate-PEG-linked, nanoemulsion-loaded aclacinomycin A to KB nasopharyngeal cells and xenograft: Effect of chain length and amount of folate-PEG linker. *J. Drug Target.* **2008**, *16*, 660–667. [CrossRef]
57. Chen, H.; Pazicni, S.; Krett, N.L.; Ahn, R.W.; Penner-Hahn, J.E.; Rosen, S.T.; O'Halloran, T.V. Coencapsulation of Arsenic-and Platinum-based Drugs for Targeted Cancer Treatment. *Angew. Chem. Int. Ed.* **2009**, *48*, 9295–9299. [CrossRef]
58. Suen, W.L.L.; Chau, Y. Size-dependent internalisation of folate-decorated nanoparticles via the pathways of clathrin and caveolae-mediated endocytosis in ARPE-19 cells. *J. Pharm. Pharmacol.* **2014**, *66*, 564–573. [CrossRef]
59. Rejman, J.; Oberle, V.; Zuhorn, I.S.; Hoekstra, D. Size-dependent internalization of particles via the pathways of clathrin-and caveolae-mediated endocytosis. *Biochem. J.* **2004**, *377*, 159–169. [CrossRef]
60. Turek, J.J.; Leamon, C.P.; Low, P.S. Endocytosis of folate-protein conjugates: Ultrastructural localization in KB cells. *J. Cell Sci.* **1993**, *106*, 423–430.

61. Kawano, K.; Maitani, Y. Effects of polyethylene glycol spacer length and ligand density on folate receptor targeting of liposomal Doxorubicin in vitro. *J. Drug Deliv.* **2011**, *2011*. [CrossRef] [PubMed]
62. Chen, H.; MacDonald, R.C.; Li, S.; Krett, N.L.; Rosen, S.T.; O'Halloran, T.V. Lipid encapsulation of arsenic trioxide attenuates cytotoxicity and allows for controlled anticancer drug release. *J. Am. Chem. Soc.* **2006**, *128*, 13348–13349. [CrossRef] [PubMed]

International Journal of
Molecular Sciences

MDPI

Article

Modified Carboxyl-Terminated PAMAM Dendrimers as Great Cytocompatible Nano-Based Drug Delivery System

Minh Thanh Vu [1,†], Long Giang Bach [2,3,†], Duy Chinh Nguyen [3], Minh Nhat Ho [4,5], Ngoc Hoi Nguyen [4,5], Ngoc Quyen Tran [4,5], Dai Hai Nguyen [4,5], Cuu Khoa Nguyen [4,5] and Thai Thanh Hoang Thi [6,*]

1 Institute of Chemistry and Materials, 17 Hoang Sam, Cau Giay, Hanoi 100000, Vietnam; vmthanh222@yahoo.com
2 NTT Hi-Tech Institute, Nguyen Tat Thanh University, Ho Chi Minh City 700000, Vietnam; blgiang@ntt.edu.vn
3 Center of Excellence for Functional Polymers and NanoEngineering, Nguyen Tat Thanh University, Ho Chi Minh City 700000, Vietnam; ndchinh@ntt.edu.vn
4 Institute of Applied Materials Science, Vietnam Academy of Science and Technology, Ho Chi Minh City 700000, Vietnam; nhatho.dost@gmail.com (M.N.H.); hoi83bmt@gmail.com (N.H.N.); tnquyen979@gmail.com (N.Q.T.); nguyendaihai0511@gmail.com (D.H.N.); nckhoavnn@yahoo.com (C.K.N.)
5 Graduate University of Science and Technology, Vietnam Academy of Science and Technology, Hanoi 100000, Vietnam
6 Biomaterials and Nanotechnology Research Group, Faculty of Applied Sciences, Ton Duc Thang University, Ho Chi Minh City 700000, Vietnam
* Correspondence: hoangthithaithanh@tdtu.edu.vn
† This author contributed equally to this work.

Received: 20 March 2019; Accepted: 11 April 2019; Published: 24 April 2019

Abstract: Polyamidoamine (PAMAM) dendrimers are extensively researched as potential drug delivery system thanks to their desirable features such as controlled and stable structures, and ease of functionalization onto their surface active groups. However, there have been concerns about the toxicity of full generation dendrimers and risks of premature clearance from circulation, along with other physical drawbacks presented in previous formulations, including large particle sizes and low drug loading efficiency. In our study, carboxyl-terminated PAMAM dendrimer G3.5 was grafted with poly (ethylene glycol) methyl ether (mPEG) to be employed as a nano-based drug delivery system with great cytocompatibility for the delivery of carboplatin (CPT), a widely prescribed anticancer drug with strong side effects so that the drug will be effectively entrapped and not exhibit uncontrolled outflow from the open structure of unmodified PAMAM G3.5. The particles formed were spherical in shape and had the optimal size range (around 36 nm) that accommodates high drug entrapment efficiency. Surface charge was also determined to be almost neutral and the system was cytocompatible. In vitro release patterns over 24 h showed a prolonged CPT release compared to free drug, which correlated to the cytotoxicity assay on malignant cell lines showing the lack of anticancer effect of CPT/mPEG-G3.5 compared with CPT.

Keywords: half generation polyamidoamine (PAMAM) dendrimer; carboplatin; PEGylation; drug delivery system; cancer treatment

1. Introduction

Among the second generation of platinum-containing drugs, carboplatin (CPT) is known as the most important one that is widely used in clinics for the treatment of cancer [1]. It is clinically proven to combat numerous types of malignancies such as lung, ovarian, head and neck, endometrial, esophageal cancer, etc. [2]. After crossing the cell membrane and entering cells, CPT molecules are intracellularly activated by undergoing the hydrolysis of 1,1-cyclobutanedicarboxylate, becoming positively charged [3]. This allows CPT to form reactive platinum complexes that create inter-strand and intra-strand crosslinks with DNA and protein, thereby impeding DNA replication, transcription, and translation and suppressing proliferation [4]. In spite of having a similar molecular mechanism of action in cancer cells, CPT exhibits lower reactivity and toxicity compared to the first generation platinum-containing cisplatin, with no nephrotoxicity, ototoxicity, and neurotoxicity [5]. Moreover, it showed better effectiveness to several types of cancers that are resistant to cisplatin. This is thanks to the presence of CPT's bidentate dicarboxylate as a replacement for labile chlorides of cisplatin [6,7]. However, myelosuppression, which causes the dramatic decrease in red blood cells, white blood cells, and platelets, is the main drawback of CPT [6]. Sometimes, the production levels of blood cells and platelets could be as low as 10% [8]. It also shown that a mean of 90% of administered CPT is excreted in urine within 24 h, and that CPT has an initial plasma half-life of 1.1 to 2 h [9,10]. To overcome these limitations, encapsulation of CPT in nanocarriers is a promising approach [1]. While most of the studies involved cisplatin, there are only a few investigations for CPT [11–13].

Polyamidoamine (PAMAM) dendrimer is a highly branched, globular nanostructure that has been greatly investigated for the delivery of drug molecules [14,15]. It possesses exceptional structural features including predetermined molecular weight, well-defined and stable structure, monodispersity, and high density of surface active groups that gives them the ease of functionalization [16–18]. Drugs can be encapsulated within PAMAM's large internal cavity, entrapped on the surface, or interspersed throughout the dendritic structure, thereby protecting them from the physiological degradation [17,19–21]. Additionally, PAMAM's nano-sized range is suitable for the passive targeting and accumulation of drug within the tumor site through enhanced permeation and retention (EPR) effect, thus reducing the side effects of loaded drugs [16,22]. Regardless of the many advantages, charge-associated toxicity limits the use of full generation amine-terminated PAMAM dendrimers. The electrostatic interaction between positively charged amino groups on the surface of PAMAM and negatively charged biological membranes leads to disruption of the lipid bilayer, causing cell lysis [23,24]. Moreover, positively charged amino groups also lead to the rapid clearance of amine-terminated PAMAM from blood circulation [14]. On the other hand, half generation PAMAM dendrimer with carboxylate groups on the surface does not interfere with cell membranes and is preferable for drug delivery. Some carboxyl-terminated PAMAM dendrimers were developed to deliver platinum-containing anticancer drugs [16,25–29]. Regarding CPT, Kang SJ and co-workers prepared PAMAM G3.5 loading CPT for murine retinoblastoma treatment. The formulation showed positive results with no associated toxicity. Nonetheless, the size range (>200 nm) and low drug loading capacity (47.54%) were undesirable [26]. In our previous works, carboxyl-terminated PAMAM dendrimers (G3.5 and G2.5) were prepared and utilized for cisplatin delivery. The size was effectively reduced and well-controlled, but the capacity for drug loading was dramatically decreased as well [16,25].

Polyethylene glycol (PEG) is a biocompatible, hydrophilic, and FDA-approved polymer that has gained great attention for the surface modification of PAMAM dendrimer [30–32]. Hydroxyl groups of PEG can be activated for the coupling reaction with PAMAM's surface functional groups [14]. The conjugation of PEG onto PAMAM not only increases its cavity space for drug loading but also reduces the uncontrolled outflow of drug while traveling through the circulatory system and sustains the release of drug at target sites [14,33,34]. Non-specific interaction between serum proteins and PAMAM was also prevented by PEG, thus averting the uptake of PAMAM dendrimer by the reticuloendothelial system [35], reducing renal clearance, and improving the circulating half-life [36,37]. These favorable characteristics of PEG-conjugated PAMAM dendrimer were demonstrated in several studies [14,33,38–40].

In this study, carboxyl-terminated PAMAM generation 3.5 (G3.5) dendrimer was conjugated with poly (ethylene glycol) methyl ether (mPEG) and employed as a nanocarrier with great cytocompatibility for CPT delivery so that the drug will be effectively entrapped and not have uncontrolled outflow. CPT/mPEG-G3.5 is expected to have a suitable size that accommodates high drug loading efficiency. The chemical structure and morphology of the formulation were investigated. Further, live/dead staining and the resazurin cell viability assay were used to determine the ability of CPT/mPEG-G3.5 to minimize the toxicity of CPT.

2. Results and Discussion

Therapeutic agents in chemotherapy are known to have side effects, ranging from mild to severe, in normal and non-targeted tissues without a proper delivery mechanism. Herein, PAMAM dendrimer, especially half-generation dendrimer, was chosen for its aforementioned advantages of pre-determined and controllable structure, presence of cavities and compartments for CPT encapsulation, and lack of positive charge-induced toxicity as in full generation dendrimer. Moreover, the carboxyl ends at the outermost of the half generation dendrimer are readily reactive to the amine-terminated mPEG, which is a widely used accessory molecule in drug delivery systems thanks to its comprehensive benefits (Figure 1). Theoretically, conjugation of mPEG can improve our system's capacity of drug loaded and control pre-mature drug leakage from the particles while traveling to the target site, which are demonstrated through the following results.

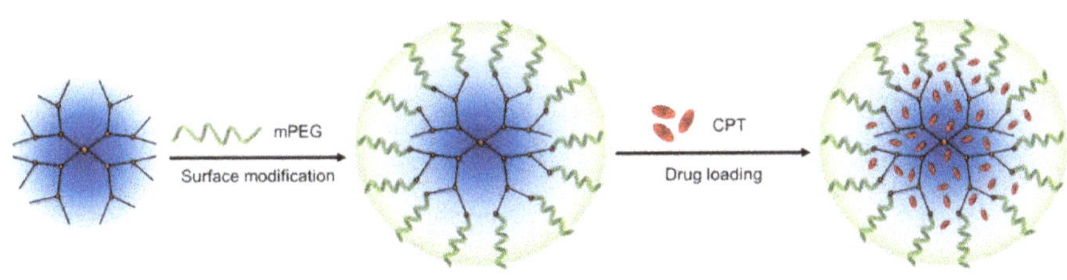

Figure 1. Illustration of carboplatin (CPT) encapsulated within mPEG-G3.5 dendrimer.

^{1}H-NMR was used to analyze the chemical structure of the synthesized complex of PAMAM G3.5 and mPEG (mPEG-G3.5) (Figure 2). As shown in Figure 2c, signals at 2.78–2.84 ppm (a), 2.39-2.41 (b), 3.22–3.27 (c), 2.57–2.63 (d), and 3.59–3.60 (f) respectively assigned to N-CH_2-CH_2-, =N-CH_2-CH_2-CO, -CO-NH-CH_2, CH_2-N, and –CO-O-CH_3 of PAMAM G3.5 were observed, which are similar to previous published studies [31]. The characteristic methyl and methylene protons at 3.44 ppm (1), 3.76 ppm (2), and 4.2 ppm (4) and a signal of N-H at 4.03-4.35 ppm (5) of mPEG-NH2 were clearly identified in the spectrum of mPEG-G3.5. The successful conjugation was shown by the absence of the resonance signals (f) at around 3.59–3.60 ppm in the spectrum of mPEG-G3.5 as compared to the spectrum of mPEG-NH_2. The number of mPEG chains was experimentally determined by ^{1}H-NMR based on the ratios of integral values for peaks assigned to mPEG (3.44 ppm) and dendrimer (3.22–3.27 ppm). Approximately 15 mPEG moieties were found to be attached to each PAMAM G3.5 dendrimer. The conjugation of PEG on the surface of G3.5 was further confirmed by Fourier transformed infrared spectroscopy (FT-IR) (Figure 3). Modified G3.5 possesses the same characteristic bands as G3.5, including the absorption at 3415–3265 cm^{-1} for -NH-, and at 1650 cm^{-1} for amide C=O, together with other absorptions at 1104 cm^{-1} of the C-O stretch and the C-H stretching at 2883 cm^{-1}, which indicates the presence of mPEG in the synthesized complex.

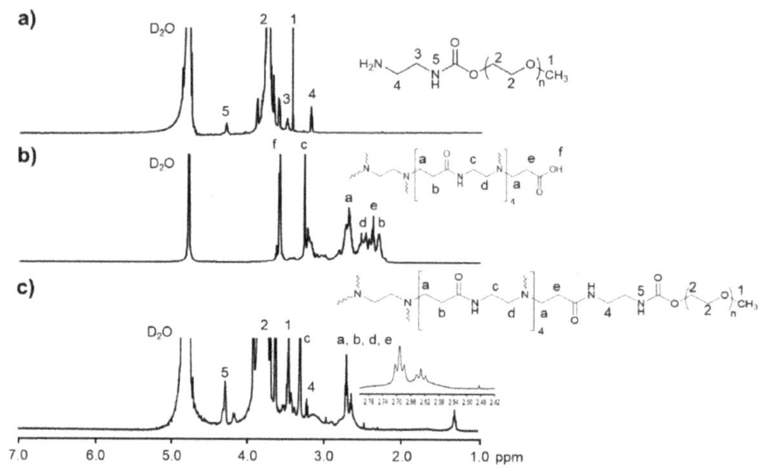

Figure 2. ^{1}H-NMR spectrum (D_2O, δ in ppm) of (**a**) mPEG-NH_2, (**b**) polyamidoamine (PAMAM) G3.5, and (**c**) mPEG-G3.5 dendrimer.

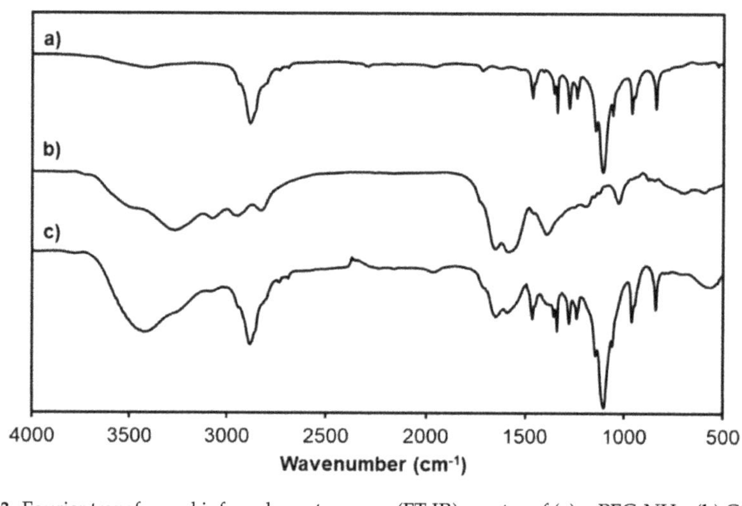

Figure 3. Fourier transformed infrared spectroscopy (FT-IR) spectra of (**a**) mPEG-NH_2, (**b**) G3.5, and (**c**) mPEG-G3.5.

GC analysis of blank run, references of ethylenediamine (EDA), MA, and methanol (MeOH), and mPEG-G3.5 were carried out. In the mPEG-G3.5 sample, diethyl ether as solvent forms the major peaks. Reference GC analysis detected the presence of MeOH, MA, and EDA at approximately the 3.0–3.1, 3.8, and 4.15-min marks on the retention timeline, which cannot be seen in the GC run of mPEG-G3.5. This result proved the purity of synthesized mPEG-G3.5, with nearly complete removal of excess EDA, MA, and MeOH solvent in the final product.

The size and morphology of the unconjugated PAMAM, as well as the empty and loaded synthesized particles, were depicted in the TEM images (Figure 4). The fate of particles after being introduced into the body, including cellular uptake and targeting, has been proven to be size-dependent. The pharmacokinetics of the nanoparticles can be either enhanced or hindered with particles in different size ranges. Specifically, the enhanced permeability and retention (EPR) effect takes place within the nanoscale since the tumor capillary system recruited as tumor cells grow for sufficient supply of nutrients and oxygen usually lacks a complete endothelial lining, resulting in significantly higher permeability and hydraulic conductivity. However, the leakage rate from the vessels is quite slow, which means the particles must possess a long enough circulation time to sufficiently accumulate [41–43]. What is more, particles of size higher than 200 nm are often taken up selectively in the liver and spleen, and are likely to be cleared out of the circulation by the mononuclear phagocyte system (PMS) [31]. However, too small particles will be eliminated rapidly by either the kidneys or renal filtration, which both have an effective size cutoff at about 10 nm (i.e., particles less than 10 nm in size are subjected to elimination by these mechanisms) [44]. Post-modification G3.5 particles showed a clear increase in size, from 5.68 ± 0.06 nm to 22.7 ± 0.4 nm, which visibly confirms the successful conjugation of mPEG onto the surface of PAMAM. After loading of CPT, the particle size saw another increase to 36.0 ± 0.2 nm, which falls into the optimal size range for nanocarriers. This indicates an improvement in controlling the size of dendrimer particles, as previous study also on PEG-modified G3.5 resulted in size that was bigger than desired [26], while another one resulted in size that was too small [25]. The conjugation of mPEG, therefore, can be utilized to control the size of the dendrimer particles. Moreover, particles at this size can also be expected to have prolonged circulation time thanks to bypassing both PMS and kidney or renal clearance, ensuring a therapeutic level of drug to build up at the tumor site by EPR effect.

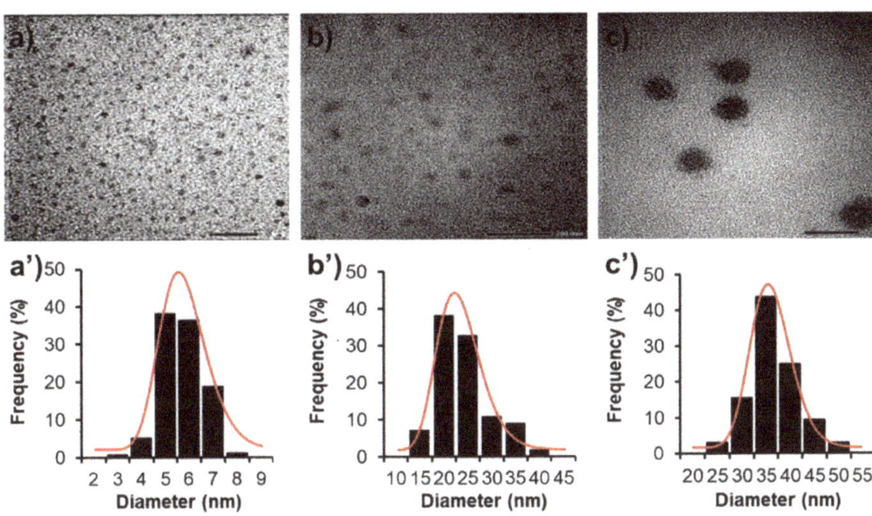

Figure 4. (**a–c**) Transmission electron microscopy micrographs and (**a'–c'**) size distribution of G3.5, mPEG-G3.5, and CPT/mPEG-G3.5 (respectively), fitted by log-normal distribution function (red line). Scale bar for a–c are 20, 200, and 50 nm, respectively.

Apart from particle size, surface charge is another parameter frequently used in characterizing of nanocarriers. Cationic charged particles have been found triggering non-specific interactions with cell membranes or opsonizing proteins owing to electrostatic bindings, which can result in unexpected cytotoxicity [31]. Moreover, various in vivo studies have associated more negative zeta potential with higher clearance rate from the blood, possibly also due to undesirable bindings with opsonizing proteins [45–47]. Thus, less negatively charged or almost neutral particles may be considered more rational [48]. Figure 5 illustrates the zeta potentials of the proposed particles, uncoated, PEG-coated, and CPT-loaded. As can be seen, naked PAMAM G3.5 had a deeply negative charge of −50.07 mV, which was to be expected since half generated dendrimers are coated with negatively charged carboxylate groups. However, after being coated with PEG, the surface charge increased to nearly neutral (1.49 mV). The reason for this increase could be due to the capping of PEG over the carboxylate groups, which in a sense further proves the conjugation of PEG. The zeta potential of the final product, CPT/mPEG-G3.5, did not show any significant difference from the previous value (1.32 mV). The near to neutral surface charge not only lessens the chance of toxicity caused by high cationic charge but also prevents premature clearance of the carriers.

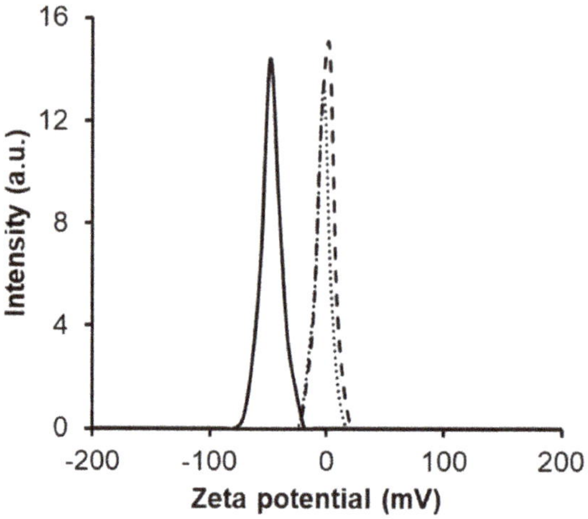

Figure 5. Zeta potential of PAMAM G3.5 (solid line), mPEG-G3.5 (dashed line), and CPT/mPEG-G3.5 (dotted line).

Drug loading efficacy is considered to be essential in the design of any drug delivery system, since it relates more or less directly to the therapeutic activity of the system. In our study, the DLE of the proposed system reached 85.37% ± 7.87%, which was exceptionally high compared to previous studies on a similar system of G3.5 but without mPEG conjugation [26]. Therefore, it is conclusive that the mPEG on the surface of the particles can play a significant role in providing more space for drug loading, and also capping the cavities to keep in the drug after loading.

As shown in the drug release profiles of free CPT and loaded CPT (Figure 6), a burst release up to more than 60% of the free drug was observed in the first hour, whereas the loaded system showed a much more prolonged behavior. This could be seen as an improvement in comparison to our prior formulation, which experienced more rapid release [22]. However, this formulation still has room to improve, since the accumulated drug released after a longer period of time should display a gradual rise to be controlled. Yet, a sustainable release of CPT is essentially desirable, as CPT has a very short initial plasma half-life (only up to 2 h) [6], and is rapidly eliminated through urination. Moreover, after administration, the particles together with their cargos need time to reach their target sites, not to mention that in the case of tumor passive targeting, the EPR effect also takes time to accumulate, due to the low leakage rate as mentioned above. During this time, preventing unwanted outflow of drug out of the delivery particles is necessarily important, not only to minimalize side effects to untargeted tissue but also to ensure the amount of drug reaching the target site. Therefore, the design of an antitumor agent delivery system needs to demonstrate an initial non-leakage time.

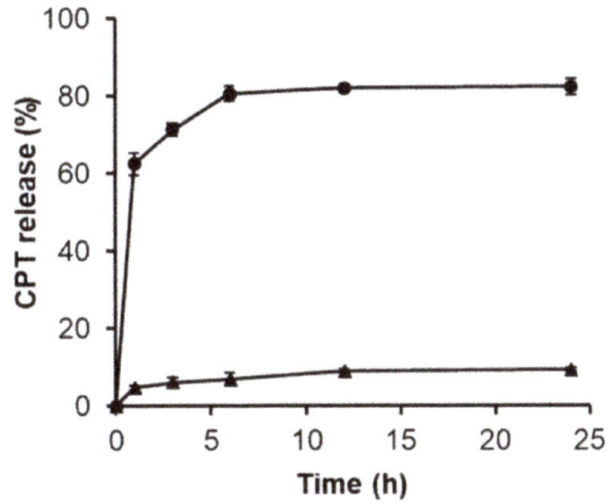

Figure 6. In vitro release profiles over the course of 24 h of free CPT (round) and CPT from mPEG-G3.5 (triangle).

The cytotoxicity of our synthesized system, empty and loaded (mPEG-G3.5 and CPT/mPEG-G3.5) towards normal cells was demonstrated by resazurin assay with normal fibroblasts (L292 line) (Figure 7C-a). As can be seen, up to the concentration of 500 µg/mL for 48 h, the system showed no toxicity to normal cells, which proves the safety of mPEG-G3.5 systems at the tested concentration. Cell viability assay results of free CPT and CPT/mPEG-G3.5 particles on two malignant cell lines, HeLa and A549, were illustrated by live/dead assay (Figures 6B and 7A) and resazurin assay (Figure 7C-b,c). Within the course of 24 h, free CPT established restricted toxicity towards HeLa and A549 cell lines, with a clear concentration-dependent tendency, while CPT/mPEG-G3.5 has not shown significant toxicity. This can be accounted for by two possible explanations: the mechanism of CPT itself and the release profile of CPT/mPEG-G3.5 [18]. CPT exerts toxicity on cells through binding with nuclear DNA forming DNA adducts, including mono- and di-adducts, and inducing apoptosis. However, it is generally accepted that the 1,2-intrastrand di-adduct is responsible for this antitumor effect, due to the fact that it is specifically recognized by High Mobility Group proteins, and less effectively repaired through nucleotide excision [49]. A previous study showed that mono-adducts accumulate in nuclear DNA over 2 days at least, due to a slow diffusion of CPT to nuclear DNA in cells [50], which may interfere with the cytotoxicity rate of both free CPT and encapsulated CPT in our assay, as lower concentrations of drug may establish a lower diffusive gradient, and thus require more time for sufficient accumulation in nuclear DNA. Moreover, as the above release profile implies, the amount of accumulated CPT after 24 h of release also corresponds to the low death rate in the first 24 h. Further in vitro and in vivo research should be carried out in future studies to gain more detail about the therapeutic efficacy of developed formulations.

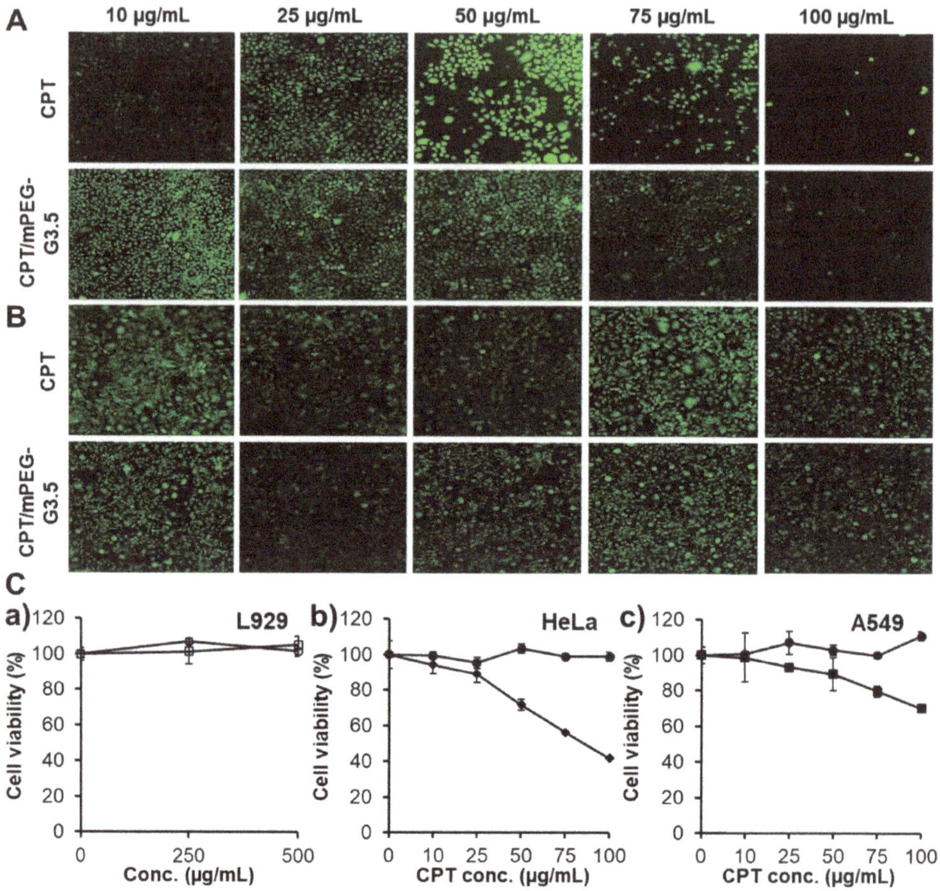

Figure 7. Cell viability after 24 h in percentage and fluorescent signals of HeLa (**A**,**C-b**) and A549 (**B**,**C-c**) with CPT (square) and CPT/mPEG-G3.5-CPT (round); Cell viability in percentage after 24 h of L929 (**C-a**) with mPEG-G3.5 (white dot) and CPT/mPEG-G3.5 (black dot)

3. Materials and Methods

3.1. Materials

Carboplatin (CPT, Mw 371.26 Da) was purchased from TCI (Tokyo, Japan). Ethylenediamine (EDA) and toluene were purchased from Merck (Darmstadt, Germany). Methyl acrylate (MA, Mw 86.09 Da), 4-Nitrophenyl chloroformate (NPC, Mw 201.56 Da), poly(ethylene glycol) methyl ether (mPEG, Mw: 5 kDa), tetrahydrofuran (THF), 1-ethyl-3-(3-dimethylaminopropyl)carbodiimide (EDC), and diethyl ether were supplied by Sigma-Aldrich (St. Louis, MO, USA). Methanol was received from Fisher Scientific. Spectra/Por®® Dialysis Membrane (MWCO 3.5 kDa, 6–8 kDa, and 12–14 kDa) was purchased from Spectrum Laboratories, Inc. (Rancho Dominguez, CA, USA). All reagents and solvents were used as received without further purification.

3.2. Synthesis of Carboxyl-Terminated PAMAM G3.5 Dendrimer

Half-generation carboxyl-terminated PAMAM G3.5 was synthesized from the EDA core by a divergent approach that employs two consecutive chain-forming reactions, including Michael addition reaction and amidation reaction, as reported previously by Donald Tomalia with minor modification (Scheme 1) [31]. The Michael addition reaction between primary amine groups of EDA and excess acrylate groups of MA gives the half generation carboxyl-terminated PAMAM denoted by Gn.5, followed by the amidation reaction between methyl propionate groups of carboxyl-terminated PAMAM with excess EDA to generate full generation amine-terminated PAMAM, designated Gn. Shortly, 20 mL of EDA was added to 150 mL of MA dissolved in methanol and kept under constant stirring (3 h at 0 °C and then 2 days at room temperature). Solvent and unreacted starting materials were discarded by rotary vacuum evaporator (Strike 300, Lancashire, PR6 0RA, UK) to collect the core precursor G-0.5. After that, a slow addition of G-0.5 (20 g in 10 mL methanol) to 130 mL EDA solution was performed to obtain PAMAM G0.0. The mixture was stirred at room temperature for 4 days, rotated under vacuum using mixed solvent toluene: methanol (9:1 *v*/*v*), and dialyzed by dialysis membrane (MWCO 3.5 kDa) against methanol to eliminate excess EDA and toluene. Finally, methanol was removed from the obtained product by drying under vacuum. These two reactions reiterated continuously to give the next higher generation of PAMAM dendrimer.

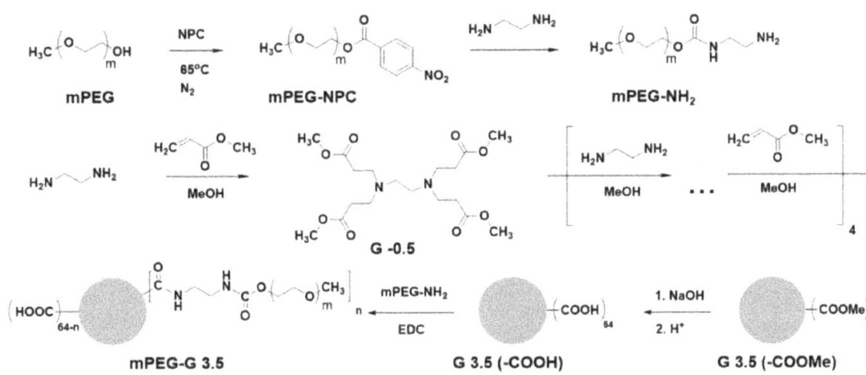

Scheme 1. Synthetic scheme of amine-terminated poly (ethylene glycol) methyl ether (mPEG) and mPEG-G3.5 dendrimers. The terminal hydroxyl group of mPEG was converted to a primary amine by using 4-Nitrophenyl chloroformate (NPC), followed by the reaction with ethylenediamine (EDA). Amine-terminated mPEG was attached onto G3.5 dendrimer, utilizing 1-ethyl-3-(3-dimethylaminopropyl)carbodiimide (EDC) as a carboxyl activating agent.

3.3. Amination of mPEG

Terminal hydroxyl group of mPEG was converted to primary amine using the following procedure. Firstly, NPC (0.4837 g, 2.4 mmol) was added to molten mPEG (10 g, 2 mmol) at 65 °C and stirred for 5 h under nitrogen atmosphere. The mixture was cooled down to 40 °C, followed by the addition of 15 mL of THF and then maintaining the constant stirring for 12 h at room temperature. Next, the obtained solution was added dropwise into diethyl ether for the precipitation, filtered, and dried under vacuum to achieve the powdery form of activated mPEG-NPC. Thereafter, mPEG-NPC (7 g, 1.355 mmol) dissolved in distilled water was added dropwise to EDA solution (1.8 mL). The mixture was stirred at room temperature for 24 h. Finally, the resulting solution was dialyzed against distilled water (dialysis membrane MWCO 3.5 kDa) for 3 days and lyophilized to give amine-terminated mPEG.

3.4. Conjugation of mPEG to PAMAM G3.5 (mPEG-G3.5)

For the conjugation of mPEG to PAMAM G3.5, synthesized PAMAM G3.5 was mixed with EDC 5% and then added into mPEG-NH_2 (6.0 g) prepared in distilled water. After 24 h of stirring at room temperature, the resulting solution was put into a dialysis bag (MWCO 12–14 kDa) and dialyzed for

3 days against distilled water to eliminate the impurities. Finally, purified product was lyophilized to obtain powder form for future use.

3.5. Characterizations

FT-IR/NIR Spectroscopy Frontier (Perkin Elmer, Waltham, MA, USA) and Bruker AC 500 MHz (Bruker Co., Billerica, MA, USA) were used to analyze the chemical structure of synthesized products. To obtain the FT-IR spectra, samples (1–2 mg) were prepared by mixing them with dried KBr (100–200 mg) and then pressing the mixture into a pellet. The measurement acquired in range of 4000–500 cm^{-1} with a resolution of 4 cm^{-1}. Gas chromatography (GC) (PerkinElmer Clarus 680) analysis of final product was carried out. Synthesized mPEG-G3.5 was dissolved in diethyl ether as solvent for GC run. TEM (JEM-1400, Tokyo, Japan) with an accelerating voltage of 100 kV was used to depict the size and morphology of the products. Sample at a concentration of 1 mg/mL was placed on a carbon-copper grid (300-mesh, Ted Pella Inc., Redding, CA, USA) and air-dried for 10 min before the measurement. Zeta potential and hydrodynamic diameter of the products were determined by Nano Particle Size SZ-100 (Horiba, Kyoto, Japan). Samples were prepared with deionized water at a concentration 1 mg/mL, filtered (pore size 0.45 μm), and sonicated for 5 min. Each sample was measured 3 times.

3.6. Preparation of CPT/mPEG-G3.5 and Drug Loading Capacity

Briefly, 10 mg of CPT was dissolved in methanol, kept under constant stirring for 1 h to completely dissolve the drug, and added slowly to the 10 mL of mPEG-G3.5. The mixture was carried out at room temperature for 24 h, followed by the removal of solvent under vacuum. The resulting mixture was then put into the dialysis bag (MWCO 3.5 kDa) for the dialysis against deionized water in 1 h. The dialyzed solution was freeze-dried to yield the product in powder form while the total volume of solution outside the dialysis bag that contains unloaded CPT was collected to determine the drug entrapment efficiency. The entrapment efficiency (%) of CPT/mPEG-G3.5 was determined by the high performance liquid chromatography (HPLC) method regarding the initial fed CPT and loaded CPT, which was calculated indirectly from the unloaded CPT.

3.7. In Vitro Release Study

For the in vitro CPT release study, a 1 mL suspension of CPT/mPEG-G3.5 (CPT content, 0.2 mg/mL) in phosphate buffer medium was loaded into a dialysis bag (MWCO 3.5 kDa). The dialysis bag was immersed into a vial containing 14 mL of PBS (0.01 M, pH 7.4), followed by placing the vials in an orbital shaker bath for temperature maintaining (37 °C) and continuously shaking (100 rpm). At predetermined time intervals, the release medium (14 mL) was collected, filtered (pore size = 0.22 μm), and replaced with an equivalent volume of fresh media. After lyophilizing the collected media, the amount of released CPT was measured by HPLC (Prominenece LC-20A, Shimadzu, Kyoto, Japan).

3.8. Cell Viability Tests

Resazurin assay was performed to investigate the cell viability of synthesized products against HeLa cells, MCF-7 cells, and normal mouse fibroblast cells L929. Cells (1.5 × 105 cells/well) were seeded in a 96-well plate containing 100 μL of Dulbecco's Modified Eagle's medium (DMEM, Gibco, Invitrogen) supplemented with 10% fetal bovine serum and 1% penicillin/streptomycin and cultured for 1 day under humidified atmosphere (98%) containing 5% CO_2, at 37 °C. Then, the media were discarded and replaced with fresh media containing free CPT (10–100 μg/mL) or CPT/mPEG-G3.5 dendrimer containing equivalent CPT concentration for the next 24 h of incubation. In control wells, cells were treated with medium only and assigned to 100% survival. After removing the supernatant, cells were washed twice with PBS, treated with resazurin (10 μL, 0.2 mg/mL), and incubated for 4 h. The fluorescent signal was detected in a micro-plate reader (Varioskan™ LUX, Thermo Scientific,

Waltham, MA, USA) at Ex/Em 560/590 nm. The relative cell viability was calculated by normalizing the fluorescence intensity of samples to that of control group as following equation:

$$\text{Cell viability } (\%) = \frac{\left([\text{Abs}]_{\text{sample}} - [\text{Abs}]_{\text{blank}}\right)}{\left([\text{Abs}]_{\text{control}} - [\text{Abs}]_{\text{blank}}\right)} \times 100 \quad (1)$$

In addition to the resazurin assay, cell viability assessment of the products against HeLa and MCF-7 cells was visualized by live/dead staining. In short, cells in each well were mixed with 25 μL of fluorescein diacetate (FDA) (10 mM) and ethidium bromide (EB) (7.5 mM). After 3 min of incubation, cells were rinsed several times with PBS. The results were observed using microscope (Eclipse Ti-E Inverted Microscope System, Nikon, Tokyo, Japan) at excitation wavelength of 485 nm for FDA and 530 nm for EB.

4. Conclusions

Spherical mPEG-G3.5 was successfully prepared with the size within the optimal range for nanocarriers and nearly neutral surface charge. The drug loading efficiency was highly improved, especially the entrapment efficiency, reaching up to more than 80%. Cell viability assay results with three cell lines correspond with the sustained release profile of the prepared particles. Even though mPEG-G3.5 has great cytocompatibility, the lack of anticancer effect of CPT/mPEG-G3.5 compared with CPT on tested malignant cell lines implied that CPT/mPEG-G3.5 is not an appropriate solution as anticancer therapy.

Author Contributions: Investigation, M.N.H., N.H.N., N.Q.T., C.K.N. and D.H.N.; writing—original draft preparation, M.T.V. and L.G.B.; writing—review and editing, D.C.N.; supervision, T.T.H.T.

Funding: This research received no external funding.

Conflicts of Interest: The authors declare no conflict of interest.

References

1. Zhang, W.; Li, C.; Shen, C.; Liu, Y.; Zhao, X.; Liu, Y.; Zou, D.; Gao, Z.; Yue, C. Prodrug-based nano-drug delivery system for co-encapsulate paclitaxel and carboplatin for lung cancer treatment. *Drug Deliv.* **2016**, *23*, 2575–2580. [CrossRef] [PubMed]
2. Dasari, S.; Tchounwou, P.B. Cisplatin in cancer therapy: Molecular mechanisms of action. *Eur. J. Pharmacol.* **2014**, *740*, 364–378. [CrossRef] [PubMed]
3. De Sousa, G.F.; Wlodarczyk, S.R.; Monteiro, G. Carboplatin: Molecular mechanisms of action associated with chemoresistance. *Braz. J. Pharm. Sci.* **2014**, *50*, 693–701. [CrossRef]
4. Baldwin, C.T.; Zwahlen, C.H.; Kirschner, S.; Nakamura, R.K. Evaluation of carboplatin sustained-release delivery system in dogs with cancer. *Collect. Vet. Med. Sci.* **2016**, *2*, 147–153. [CrossRef]
5. Adams, M.; Kerby, I.J.; Rocker, I.; Evans, A.; Johansen, K.; Franks, C.R. A comparison of the toxicity and efficacy of cisplatin and carboplatin in advanced ovarian cancer. *Acta Oncol.* **1989**, *28*, 57–60. [CrossRef]
6. Marques, M.P. Platinum and palladium polyamine complexes as anticancer agents: The structural factor. *Int. Sch. Res. Not. Spectrosc.* **2013**, *2013*, 29. [CrossRef]
7. Dogliotti, L.; Carteni, G.; Siena, S.; Bertetto, O.; Martoni, A.; Bono, A.; Amadori, D.; Onat, H.; Marini, L. Gemcitabine plus cisplatin versus gemcitabine plus carboplatin as first-line chemotherapy in advanced transitional cell carcinoma of the urothelium: Results of a randomized phase 2 trial. *Eur. Urol.* **2007**, *52*, 134–141. [CrossRef]
8. Zalba, S.; Garrido, M.J. Liposomes, a promising strategy for clinical application of platinum derivatives. *Expert Opin. Drug Deliv.* **2013**, *10*, 829–844. [CrossRef] [PubMed]
9. Van der Vijgh, W.J. Clinical pharmacokinetics of carboplatin. *Clin. Pharmacokinet.* **1991**, *21*, 242–261. [CrossRef]
10. North, S.M.; Banks, T.A. *Small Animal Oncology: An Introduction*; Elsevier Health Sciences: Amsterdam, The Netherlands, 2009.

11. Ebrahinimaf, M.; Nili-Ahmadabadi, A.; Akbarzadeh, A.; Shahemabadi, H.E.; Hasanzadegan, M.; Moradi-Sardareh, H.; Madadizadeh, H.; Rezaee-diyan, J. Preparation, characterization and cytotoxic effects of pegylated nanoliposomal containing carboplatin on ovarian cancer cell lines. *Idian J. Clin. Biochem.* **2016**, *32*, 230–234.
12. Alex, A.T.; Joseph, A.; Shavi, G.; Rao, J.V.; Udupa, N. Development and evaluation of carboplatin-loaded pcl nanoparticles for intranasal delivery. *Drug Deliv.* **2016**, *23*, 2144–2153. [CrossRef] [PubMed]
13. Arlt, M.; Haase, D.; Hampel, S.; Oswald, S.; Bachmatiuk, A.; Klingeler, R.; Schulze, R.; Ritschel, M.; Leonhardt, A.; Fuessel, S.; et al. Delivery of carboplatin by carbon-based nanocontainers mediates increased cancer cell death. *Nanotechnology* **2010**, *21*, 335101. [CrossRef] [PubMed]
14. Luong, D.; Kesharwani, P.; Deshmukh, R.; Amin, M.C.I.M.; Gupta, U.; Greish, K.; Iyer, A.K. Pegylated pamam dendrimers: Enhancing efficacy and mitigating toxicity for effective anticancer drug and gene delivery. *Acta Biomater.* **2016**, *43*, 14–29. [CrossRef]
15. Nguyen, T.T.C.; Nguyen, C.K.; Nguyen, T.H.; Tran, N.Q. Highly lipophilic pluronics-conjugated polyamidoamine dendrimer nanocarriers as potential delivery system for hydrophobic drugs. *Mater. Sci. Eng. C* **2017**, *70*, 992–999. [CrossRef] [PubMed]
16. Tran, N.Q.; Nguyen, C.K.; Nguyen, T.P. Dendrimer-based nanocarriers demonstrating a high efficiency for loading and releasing anticancer drugs against cancer cells in vitro and in vivo. *Adv. Nat. Sci. Nanosci. Nanotechnol.* **2013**, *4*, 7. [CrossRef]
17. Jiang, Y.-Y.; Tang, G.-T.; Zhang, L.-H.; Kong, S.-Y.; Zhu, S.-J.; Pei, Y.-Y. Pegylated pamam dendrimers as a potential drug delivery carrier: In vitro and in vivo comparative evaluation of covalently conjugated drug and noncovalent drug inclusion complex. *J. Drug Target.* **2010**, *18*, 389–403. [CrossRef] [PubMed]
18. Singh, S.K.; Lohiya, G.K.; Limburkar, P.P.; Dharbale, N.B.; Mourya, V.K. Dendrimer a versatile polymer in drug delivery. *Asian J. Parm.* **2009**, *3*, 178–186. [CrossRef]
19. Sommerfeld, N.S.; Hejl, M.; Klose, M.H.M.; Schreiber-Brynzak, E.; Bileck, A.; Meier, S.M.; Gerner, C.; Jakupec, M.A.; Galanski, M.; Keppler, B.K. Low-generation polyamidoamine dendrimers as drug carriers for platinum(IV) complexes. *Eur. J. Inorg. Chem.* **2016**, *2017*, 1713–1720. [CrossRef]
20. Tripathy, S.; Das, M.K. Dendrimers and their applications as novel drug delivery carriers. *J. Appl. Pharm. Sci.* **2013**, *3*, 142–149.
21. Liao, H.; Liu, H.; Li, Y.; Zhang, M.; Tomas, H.; She, M.; Shi, X. Antitumor efficacy of doxorubicin encapsulated within pegylated poly(amidoamine) dendrimers. *J. Appl. Polym. Sci.* **2014**, *131*, 40358. [CrossRef]
22. Nguyen, C.K.; Tran, N.Q.; Nguyen, T.P.; Nguyen, D.H. Biocompatible nanomaterials based on dendrimers, hydrogels and hydrogel nanocomposites for use in biomedicine. *Adv. Nat. Sci. Nanosci. Nanotechnol.* **2017**, *8*, 015001. [CrossRef]
23. Zeng, Y.; Kurokawa, Y.; Win-Shwe, T.-T.; Zeng, Q.; Hirano, S.; Zhang, Z.; Sone, H. Effects of pamam dendrimers with various surface functional groups and multiple generations on cytotoxicy and neuronal differentiation using human neural progenitor cells. *J. Toxicol. Sci.* **2016**, *41*, 351–370. [CrossRef] [PubMed]
24. Hu, W.; Qiu, L.; Cheng, L.; Hu, Q.; Liu, Y.; Hu, Z.; Cheng, L. Redox and ph dual responsive poly (amidoamine) dendrimer-poly (ethylene glycol) conjugates for intracellular delivery of doxorubicin. *Acta Biomater.* **2016**, *36*, 241–253. [CrossRef] [PubMed]
25. Nguyen, H.; Nguyen, N.H.; Tran, N.Q.; Nguyen, C.K. Improved method for preparing cisplatin-dendrimer nanocomplex and its behavior against nci-h460 lung cancer cell. *J. Nanosci. Nanotechnol.* **2015**, *15*, 4106–4110. [CrossRef]
26. Kang, S.J.; Durairaj, C.; Kompella, U.B.; O'Brien, J.M.; Grossniklaus, H.E. Subconjunctival nanoparticle carboplatin in the treatment of murine retinoblastoma. *Arch. Ophthalmol.* **2009**, *127*, 1043–1047. [CrossRef]
27. Kulhari, H.; Pooja, D.; Singh, M.K.; Chauhan, A.S. Optimization of carboxylate-terminated poly(amidoamine) dendrimer-mediated cisplatin formulation. *Drug Dev. Ind. Pharm.* **2015**, *41*, 232–238. [CrossRef]
28. Malik, N.; Evagorou, E.G.; Duncan, R. Dendrimer-platinate: A novel approach to cancer chemotherapy. *Anticancer Drugs* **1999**, *10*, 767–776. [CrossRef]
29. Kirkpatrick, G.J.; Plumb, J.A.; Sutcliffe, O.B.; Flint, D.J.; Wheate, N.J. Evaluation of anionic half generation 3.5–6.5 poly(amidoamine) dendrimers as delivery vehicles for the active component of the anticancer drug cisplatin. *J. Inorg. Biochem.* **2011**, *105*, 1115–1122. [CrossRef] [PubMed]

30. Zhu, S.; Hong, M.; Zhang, L.; Tang, G.; Jiang, Y.; Pei, Y. Pegylated pamam dendrimer-doxorubicin conjugates: In vitro evaluation and in vivo tumor accumulation. *Pharm. Res.* **2010**, *27*, 161–174. [CrossRef] [PubMed]
31. Thanh, V.M.; Nguyen, T.H.; Tran, T.V.; Ngoc, U.P.; Ho, M.N.; Nguyen, T.T.; Chau, Y.N.T.; Le, V.T.; Tran, N.Q.; Nguyen, C.K.; et al. Low systemic toxicity nanocarriers fabricated from heparin-mpeg and pamam dendrimers for controlled drug release. *Mater. Sci. Eng. C Mater. Biol. Appl.* **2018**, *82*, 291–298. [CrossRef]
32. Bao, B.Q.; Le, N.H.; Nguyen, D.H.T.; Tran, T.V.; Pham, L.P.T.; Bach, L.G.; Nguyen, T.H.; Nguyen, D.H. Evolution and present scenario of multifunctionalized mesoporous nanosilica platform: A mini review. *Mater. Sci. Eng. C* **2018**, *91*, 912–928. [CrossRef] [PubMed]
33. Ly, T.U.; Tran, N.Q.; Hoang, T.K.D.; Phan, K.N.; Truong, H.N.; Nguyen, C.K. Pegylated dendrimer and its effect in fluorouracil loading and release for enhancing antitumor activity. *J. Biomed. Nanotechnol.* **2013**, *9*, 213–220. [CrossRef]
34. Karthikeyan, R.; Koushik, O.S.; Kumar, V.P. Surface modification of cationic dendrimers eases drug delivery of anticancer drugs. *Nano Sci. Nano Technol. Indian J.* **2016**, *10*, 109.
35. Babu, A.; Periasamy, J.; Gunasekaran, A.; Kumaresan, G.; Naicker, S.; Gunasekaran, P.; Murugesan, R. Polyethylene glycol-modified gelatin/polylactic acid nanoparticles for enhanced photodynamic efficacy of a hypocrellin derivative in vitro. *J. Biomed. Nanotechnol.* **2013**, *9*, 177–192. [CrossRef] [PubMed]
36. Kim, Y.; Klutz, A.M.; Jacobson, K.A. Systematic investigation of polyamidoamine dendrimers surface-modified with poly(ethylene glycol) for drug delivery applications: Synthesis, characterization, and evaluation of cytotoxicity. *Bioconjug. Chem.* **2008**, *19*, 1660–1672. [CrossRef]
37. Sarkar, K.; Yang, H. Encapsulation and extended release of anti-cancer anastrozole by stealth nanoparticles. *Drug Deliv.* **2008**, *15*, 343–346. [CrossRef]
38. Kumar, P.D.; Kumar, P.V.; Selvam, T.P.; Rao, K.R.S.S. Prolonged drug delivery system of pegylated pamam dendrimers with a anti-HIV drug. *Res. Pharm.* **2013**, *3*, 8–17.
39. Li, Y.; He, H.; Lu, W.; Jia, X. A poly(amidoamine) dendrimer-based drug carrier for delivering dox to gliomas cells. *RSC Adv.* **2017**, *7*, 15475–15481. [CrossRef]
40. Yang, H.; Lopina, S.T.; DiPersio, L.P.; Schmidt, S.P. Stealth dendrimers for drug delivery: Correlation between pegylation, cytocompatibility, and drug payload. *J. Mater. Sci. Mater. Med.* **2008**, *19*, 1991–1997. [CrossRef]
41. Nakamura, Y.; Mochida, A.; Choyke, P.L.; Kobayashi, H. Nanodrug delivery: Is the enhanced permeability and retention effect sufficient for curing cancer? *Bioconjug. Chem.* **2016**, *27*, 2225–2238. [CrossRef]
42. Nguyen, D.H.; Choi, J.H.; Joung, Y.K.; Park, K.D. Disulfide-crosslinked heparin-pluronic nanogels as a redox-sensitive nanocarrier for intracellular protein delivery. *J. Bioact. Compat. Polym.* **2011**, *26*, 287–300. [CrossRef]
43. Nguyen, D.H.; Bae, J.W.; Choi, J.H.; Lee, J.S.; Park, K.D. Bioreducible cross-linked pluronic micelles: Ph-triggered release of doxorubicin and folate-mediated cellular uptake. *J. Bioact. Compat. Polym.* **2013**, *28*, 341–354. [CrossRef]
44. Dreaden, E.C.; Austin, L.A.; Mackey, M.A.; El-Sayed, M.A. Size matters: Gold nanoparticles in targeted cancer drug delivery. *Ther. Deliv.* **2012**, *3*, 457–478. [CrossRef]
45. He, C.; Hua, Y.; Yin, L.; Tang, C.; Yin, C. Effects of particle size and surface charge on cellular uptake and biodistribution of polymeric nanoparticles. *Biomaterials* **2010**, *31*, 3657–3666. [CrossRef] [PubMed]
46. Levchenko, T.S.; Rammohan, R.; Lukyanov, A.N.; Whiteman, K.R.; Torchilin, V.P. Liposome clearance in mice: The effect of a separate and combined presence of surface charge and polymer coating. *Int. J. Pharm.* **2002**, *240*, 95–102. [CrossRef]
47. Alexis, F.; Pridgen, E.; Molnar, L.K.; Farokhzad, O.C. Factors affecting the clearance and biodistribution of polymeric nanoparticles. *Mol. Pharm.* **2008**, *5*, 505–515. [CrossRef]
48. Sadat, S.M.A.; Jahan, S.T.; Haddadi, A. Effects of size and surface charge of polymeric nanoparticles on in vitro and in vivo applications. *J. Biomater. Nanobiotechnol.* **2016**, *7*, 91–108. [CrossRef]

49. Fuertesa, M.A.; Castillab, J.; Alonsoa, C.; Pérezc, J.M. Cisplatin biochemical mechanism of action: From cytotoxicity to induction of cell death through interconnections between apoptotic and necrotic pathways. *Curr. Med. Chem.* **2003**, *10*, 256–266. [CrossRef]
50. Hah, S.S.; Stivers, K.M.; de Vere White, R.W.; Henderson, P.T. Kinetics of carboplatin-DNA binding in genomic DNA and bladder cancer cells as determined by accelerator mass spectrometry. *Chem. Res. Toxicol.* **2006**, *19*, 622–626. [CrossRef]

International Journal of
Molecular Sciences

MDPI

Article

Development of Pectin-Type B Gelatin Polyelectrolyte Complex for Curcumin Delivery in Anticancer Therapy

Fu-Ying Shih [1,†], Ih-Jen Su [2,3,4,†], Li-Lun Chu [5], Xiaojie Lin [6], Sheng-Chu Kuo [7,8], Yu-Chi Hou [5] and Yi-Ting Chiang [5,*]

1 Ph. D. Program for Biotech Pharmaceutical Industry, School of Pharmacy, China Medical University, Taichung 404, Taiwan; u106308001@cmu.edu.tw
2 Department of Biotechnology, Southern Taiwan University of Science and Technology, Tainan 710, Taiwan; suihjen0704@stust.edu.tw
3 Department of Pathology, National Cheng Kung University Hospital, Tainan 704, Taiwan
4 National Institute of Infectious Diseases and Vaccinology, National Health Research Institutes, Tainan 704, Taiwan
5 School of Pharmacy, China Medical University, Taichung 404, Taiwan; u103003505@cmu.edu.tw (L.-L.C.); houyc@mail.cmu.edu.tw (Y.-C.H.)
6 Department of Chemical Engineering, University of Washington, Seattle, WA 98195, USA; xjlin@uw.edu
7 Chinese Medicine Research Center, China Medical University, Taichung 404, Taiwan; sckuo@mail.cmu.edu.tw
8 Research Center for Chinese Herbal Medicine, China Medical University, Taichung 404, Taiwan
* Correspondence: ytchiang@mail.cmu.edu.tw; Tel.: +886-4-2205-3366 (ext. 5611)
† These authors contributed equally to this manuscript.

Received: 11 November 2018; Accepted: 14 November 2018; Published: 17 November 2018

Abstract: Curcumin has been proven to be a potent agent in colon cancer treatment. However, its hydrophobicity and low oral bioavailability hampered its clinical application. These limitations could be improved through appropriate formulations such as using polyelectrolyte complexes (PECs). PECs were self-assembled with polycations and polyanions in polar solvents. In this study, a novel pectin-type B gelatin PEC was developed for use in curcumin formulation. At pH 4.0, natural polyanions pectin and polycations type B gelatin spontaneously formed PECs in ethanol/water solution, whereas under mimetic gastrointestinal tract (GI tract) conditions, at pH 2.0 and 8.0, pectin and type B gelatin were electrically neutralized, and the PECs swelled to allow payload release. After being transferred to pH 7.0 condition, as in the colon environment, PECs were internalized into colon carcinomas. Thus, pectin-type B gelatin PECs were successfully prepared, and their constituent ratio and drug-loading process were also optimized. The optimum particle size of the PECs was 264.0 ± 3.1 nm and they could swell as the zeta potential was altered at either pH 2.0 or 8.0. The optimum drug content and loading efficiency were 40% and 53%, respectively. At pH 2.0, curcumin was rapidly released from curcumin-loaded PECs, whereas at pH 8.0, curcumin-loaded PECs showed a sustained-release of curcumin. The bare PECs showed very low toxicity toward human normal cells, whereas curcumin-loaded PECs, after incubation at pH 2.0 for 2 h and at pH 8.0 for 4 h, induced cell cycle arrest and exhibited cytotoxic effect to HCT116 human colon cancer cells, even though these loaded PECs were pretreated with mimetic GI tract conditions. Our pectin-type B gelatin PECs were shown to be a promising oral formulation for curcumin delivery in anticancer therapy.

Keywords: polyelectrolyte complex; curcumin; anticancer therapy

1. Introduction

Polyelectrolyte complexes (PECs) are made from two or more macromolecules carrying opposite electrical charges, which can be ionized and electrically complexed with each other in a polar solution [1]. The polymeric acids (polyanions) or bases (polycations) respond to environmental pH value for their ionization in the pKa interval [2]. Therefore, in the field of pharmaceutics, PECs have been developed for decades to encapsulate drugs in oral-controlled or sustained-release system, in response to the environmental pH of the gastrointestinal tract (GI tract) [3]. Besides, PECs are usually formed in polar solvents, such as water or ethanol, to improve the solubility of hydrophobic drugs in pharmaceutics [4]. Among all developed polymeric materials, natural polymers attract much attention for application in drug delivery system because of their excellent biocompatibility and biodegradability [5]. The commonly used polyanions are polysaccharides, whose side chains are appended with carboxylic acids, such as pectin, alginates, and acacia gum [6]. Proteins, such as zein and gelatin, and modified polysaccharides, such as chitosan, are widely utilized as natural polycationic materials [7–9]. In addition, natural polymers as PECs have been applied in drug delivery systems to improve the water solubility or for controlled release designs in many studies [1]. For example, D. Wurster et al. have prepared natural polymers containing cationic type A gelatin and anionic acacia gum to generate PECs for encapsulating hydrophobic agent vitamin A palmitate. The mixing ratios and fabricating methods have been studied and optimized. The PECs could efficiently improve the water solubility of vitamin A palmitate [10]. B. Stokke et al. have built and optimized PECs with natural polymers containing cationic chitosan (pKa 6.5–6.6) and anionic alginates (pKa 3.4–3.7) at the pKa intervals, and they measured the average diameters upon environmental pH changes from pH 3.0 to 8.0. Moreover, they have found that particle sizes would significantly change at above pH 7.0 for the zeta potential alteration (above the pKa of chitosan), that would be potential in controlled releasing pharmaceutical designs [8].

Curcumin (diferuloylmethane) is a phenolic component extracted from the rhizome of turmeric [11]. As a bioactive compound, it has proven its medical efficacy in several tumors [12–14], including gastric [15] and colorectal cancers [16,17]. However, its low water solubility and bioavailability limit its clinical uses via oral administration [18,19]. Besides, curcumin in a buffer decomposes in a pH-dependent manner, and this degradation accelerates at neutral-basic conditions in preservation, strongly diminishing its pharmacological activity [20]. Studies have suggested that curcumin could be well formulated to improve its solubility and bioavailability via a reliable controlled- or sustained-release system [21]. PEC is one of several promising approaches and several attempts to generate PECs for curcumin delivery have been conducted [22,23]. K. Hu et al. have fabricated nano-sized zein-pectin PEC for curcumin delivery [7]. P. Sarika et al. have prepared a cationized gelatin and sodium alginate PEC to improve the water solubility of curcumin, allowing sustained-release of curcumin and enhancing its cytotoxicity in vitro [24]. A. Anitha et al. have prepared curcumin-loaded dextran sulfate-chitosan PECs to increase the water solubility of curcumin, and the curcumin content also exhibited cytotoxicity toward cancer cells [25]. Although the principle of PEC formation and studies have proved that PECs would be an applicable formulation for curcumin, it is still difficult to optimize their manufacturing process, including determination of constituent ratios and pH for aggregation [26]. These manufacturing factors strongly affect the particle size, morphology, loading efficiency, and drug-releasing behaviors of PEC systems [27]. Thus, it is essential to develop a PEC system with an optimized manufacturing process.

In this study, we proposed a novel PEC composed of natural polymers, and its fabrication was optimized based on its performance in mimetic GI environment. We assembled the natural polycation type B gelatin and the polyanion pectin into a PEC system at an appropriate pH in a polar solvent system, as shown in Scheme 1. Several studies have previously reported pectin-gelatin PECs, yet type A gelatin was mostly chosen for use in their systems [26,28]. Type B gelatin, to our best knowledge, was first introduced in PEC for curcumin delivery. Type B gelatin is derived from collagen after lime hydrolyzation, and its isoelectric point (pI) was 4.8–5.5. The cationized type B gelatin would yield below pH 4.8, such as at pH 4.0. Apart from the electrical properties of type B gelatin, the pI of type A gelatin is approximately 9.0; thus, it is positively charged under this broadline [29]. Importantly, the pKa of polysaccharides pectin is 3.5 [30]. This pectin has been utilized to fabricate PEC because it could be electrically neutralized to force payload release in a stomach environment, which has a pH of 2.0 [31]. Besides, recent studies have pointed out the potential of pectin in colon cancer treatment because its backbone could only be cleaved by colon microorganisms [32]. Herein, PECs composed of pectin and type B gelatin were prepared at pH 4.0, which could reduce the decomposition of curcumin during manufacture or preservation. The curcumin-releasing profiles of the PECs were further investigated in the mimetic GI environmental pH value and transit periods (pH 2.0 for 2 h to simulate conditions in the stomach, and pH 8.0 for 4 h to mimic pH conditions in the small intestine) [33]. The remaining curcumin-loaded PECs after pretreatment with mimetic GI tract conditions would be internalized into the colon carcinoma. The biosafety and cytotoxic effect of curcumin-loaded PECs to human normal cells and colon cancer were examined in this study to evaluate the potentials of PECs as a curcumin formulation for future clinical anticancer treatment.

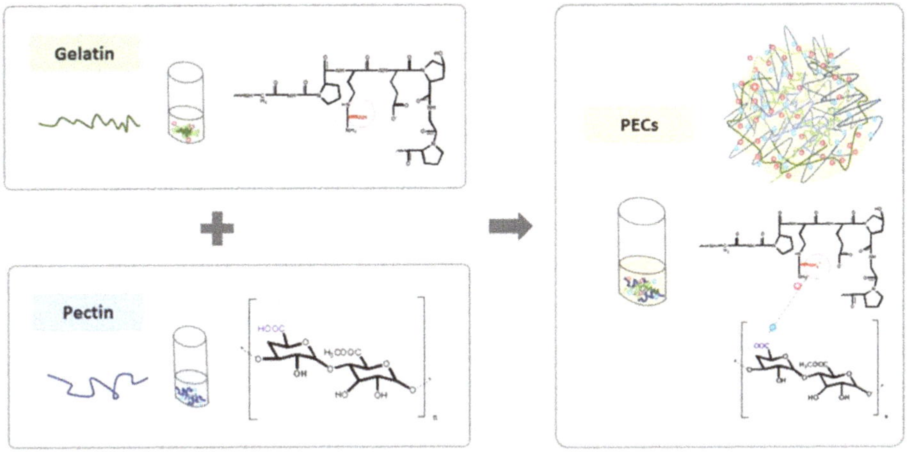

Scheme 1. Preparation of polyelectrolyte complexes (PECs). Pectin and type B gelatin were first prepared into diluted polymer solutions in ethanol/water (1:10, v/v). At pH 4.0, pectin and type B gelatin were mixed together, spontaneously constituting PECs by their charged side chains.

2. Results

2.1. Preparation and Characterization of Pectin-Type B Gelatin PECs

PECs were spontaneously assembled in the vicinity of the pKa or pI intervals of the two constituting polymers. The pI of type B gelatin was 4.8–5.5, indicating that the amine groups appended in type B gelatin would be electrically cationic below pH 4.8, whereas pectin, whose pKa is 3.5, would be anionized at above pH 3.5 owing to the ionization of the carboxylic groups appended on its side chains [30]. To fabricate PECs in this study, we first prepared type B gelatin and pectin into 0.1% and 0.02% w/w diluted aqueous solutions, respectively, to prevent rapid aggregation and formation of large particles. Next, we adjusted the pH value into 4.0 to allow simultaneous cationization of type B gelatin and anionization of pectin. Opposing electrical charges would spontaneously drive the type B gelatin and pectin into PECs in the polar solvent. Considering the subsequent curcumin-loading process, in this study we used ethanol (10% v/v) as a polar solvent in our PEC system. Lastly, the cationic type B gelatin, anionic pectin, and ethanol would be adjusted to optimize our PEC system.

To optimize the PEC system, the components (mainly pectin and type B gelatin) were assembled various ratios into the PECs, and particle size and distribution were measured by dynamic laser scattering (DLS). The results are presented in Table 1. Table 1 indicated that particle size significantly decreased along with reduction in type B gelatin content until only 60% of type B gelatin remained the PECs. For example, the particle size of the P1G9 sample (10% mole of pectin and 90% of type B gelatin) was 2745.0 ± 12.8 nm, while that of the P4G6 sample (40% mole of pectin and 60% of type B gelatin) decreased to 237.0 ± 3.3 nm. Because the type B gelatin content decreased by less than 60%, the particle size would not be dramatically altered. However, the polydispersity indexes (PDIs), which represents size distribution, would gradually increase with the continual decrease of type B gelatin content. To further investigate the components' impact on electrical properties and environmental pH, we chose to use P4G6 (40% mole of pectin and 60% mole of type B gelatin) and P6G4 (60% mole of pectin and 40% mole of type B gelatin), owing to their similar particle sizes and PDIs.

Table 1. The particle size and distribution of the PECs containing various ratios of constituents (to a final concentration of 1 mL).

Code	Components		Particle Size (nm)	PDI
	Pectin (mol %)	Type B Gelatin (mol %)		
P1G9	10	90	2745.0 ± 12.8	0.15 ± 0.08
P2G8	20	80	1502.8 ± 243.0	0.40 ± 0.10
P3G7	30	70	364.7 ± 6.4	0.13 ± 0.09
P4G6	40	60	264.0 ± 3.1	0.23± 0.02
P5G5	50	50	311.0 ± 57.8	0.18 ± 0.02
P6G4	60	40	243.1 ± 18.8	0.23 ± 0.10
P7G3	70	30	259.3 ± 21.6	0.17 ± 0.12
P8G2	80	20	276.5 ± 28.0	0.21 ± 0.07
P9G1	90	10	278.9 ± 57.1	0.25 ± 0.08

2.2. pH Responsiveness of the PECs

The particle sizes of P4G6 and P6G4 were 264.7 ± 3.1 nm and 243.1 ± 18.8 nm, respective, whereas their PDIs were both approximately 0.23. Their morphologies were observed using transmission electron microscopy (TEM) and scanning electron microscopy (SEM). The TEM images are shown in Figure 1 and they show that the observed particle sizes of P4G6 and P6G4 corresponded to their hydrodynamic diameter, as measured by DLS. However, their morphologies were different. The TEM images indicated that P4G6 particles exhibited spherical structure, whereas P6G4 particles showed irregularly spherical structure, probably as a result from the differences in their components. The morphologies of P4G6 and P6G4 PECs were also observed by a scanning electron microscopy (SEM) after the PECs were freezed dried. The SEM images of P4G6 and P6G4 were respectively present in Figure S1a,b in Supplementary Materials. P4G6 PECs exhibited dense nanosperical structure, while P6G4 PECs was getting loose after being freeze-dried. That further confirmed the components of PECs might affect the PEC structure. Both TEM and SEM images suggested that P4G6 PECs exist the stronger electrical interactions to maintain their structure, even after being freeze-dried. However, both P4G6 and P6G4 PECS could also be observed deformation by the SEM after being freeze-dried, that also suggested our PEC system should be phramaceutically delivered in a solution form. Herein, the following experiments would be perfomed and evaluated as PEC is delivered in the solution form.

By analyzing the TEM images, we identified that the components affected the particles' particular shapes. Responsiveness to environmental pH would be influenced by the PEC components. In this study, we mimicked pH changes in the digesting process of drugs following oral administration. In the stomach, the pH is 2.0 on average and maintained for 2 h; in the small intestine, the pH increases to 8.0, which is retained for 4 h [33]. Herein, we simulated the pH values and periods of drug or food digestion in the GI tract following oral administration to investigate the pH responsiveness of PECs with respect to particle size, and then optimized the PEC components. The PECs were first incubated at pH 2.0 for 2 h; afterward, the PECs were incubated at pH 8.0 for 4 h. The resulting particle size and distribution are presented in Figure 2. Figure 2a,b indicated the particle size and distribution of P4G6 and P6G4 PECs, respectively, at pH 2.0 and 8.0. The trends in size and PDI alterations were marked. The P4G6 PECs showed a steady increase in particle size and PDI values at either pH 2.0 or 8.0. P6G4 PECs exhibited an abrupt increase in size and PDI at pH 2.0, although these PECs also significantly changed their hydrodynamic diameter at pH 8.0. The different performances of particle alteration could be attributed to the different ratios of pectin and type B gelatin within the PECs, resulting in different electrical properties in various pH conditions.

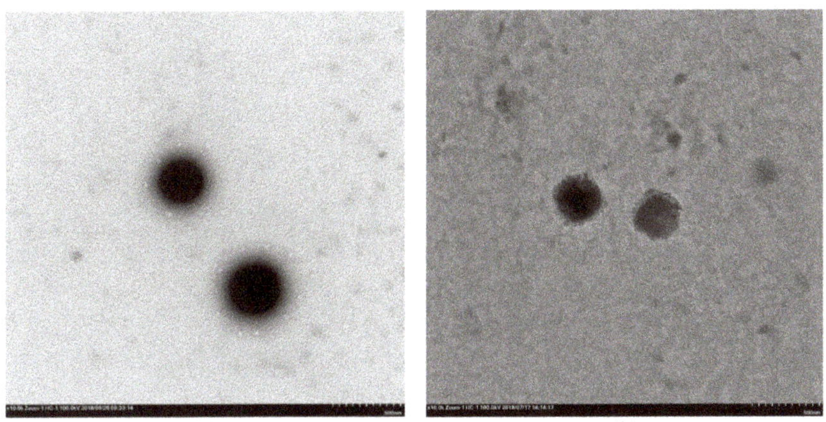

(a) (b)

Figure 1. TEM images of the P4G6 (**a**) and P6G4 (**b**) PECs. The PECs were dropped to carbon-coated cooper grids. After the PECs were attached onto the carbon cooper grids, excess PECs were removed. The attached PECs were then stained with 1% sodium phosphotungstate aqueous (PTA) solution. The 1% PTA solution was dropped to the grids and excess PTA solution was removed after staining. The TEM grids were dried in a vacuum oven and stored at room temperature until observation (scale bar: 500 nm).

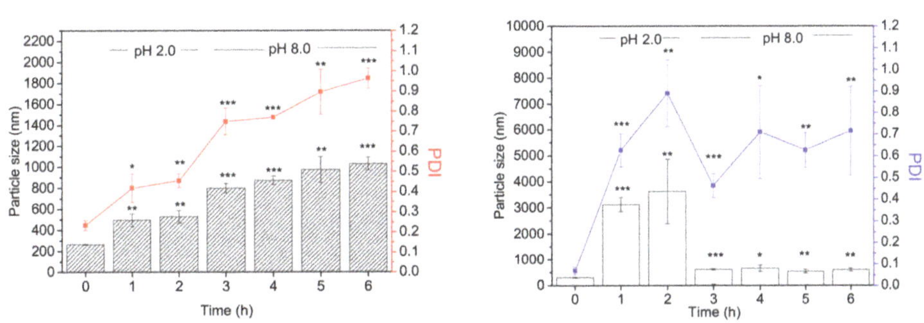

(a) (b)

Figure 2. Particle size and distribution of the P4G6 (**a**) and P6G4 (**b**) PECs. The PECs were incubated at 37 °C and pH 2.0 for 2 h, and the pH was subsequently adjusted to 8.0 for 4 h to mimic drug transits in the GI tract. At each hour, particle size and PDI were determined by DLS. The results have been statistically compared and analyzed using Student's *t*-test (Microsoft Excel 2000). Differences were recognized as statistically significant when the p values were less than 0.05, and significant differences are shown in star marks (* $p < 0.05$; ** $p < 0.01$, and *** $p < 0.001$).

Figure 3 shows the zeta potentials of P4G6 and P6G4 PECs in different pH environments, which mimicked those in the GI tract. At pH 2.0, P4G6 PECs possessed higher zeta potential than that of

P6G4 PECs, whereas P6G4 PECs were more negatively charged than P4G6 PECs at pH 8.0. That may be attributed to the different ratios in the PEC compositions. At pH 2.0, pectin was neutralized owing to the protonation of the appended carboxylic groups of pectin, whereas the type B gelatin still retained its cations [29]. P4G6 PECs, which consisted of more type B gelatin, would be more positively charged. In PECs at pH 8.0, pectin was a polyanion, whereas type B gelatin would be electrically neutralized. P4G6 PECs, which comprised of less pectin, would lead to less-negative charges. P4G6 PECs could respond to the outer pH environment via superficial electrical potentials, which would not result in destruction of PECs in either acid or alkaline condition. Herein, P4G6 would potentiate a drug delivery system with the sustained-release formulation of curcumin for colon cancer treatment.

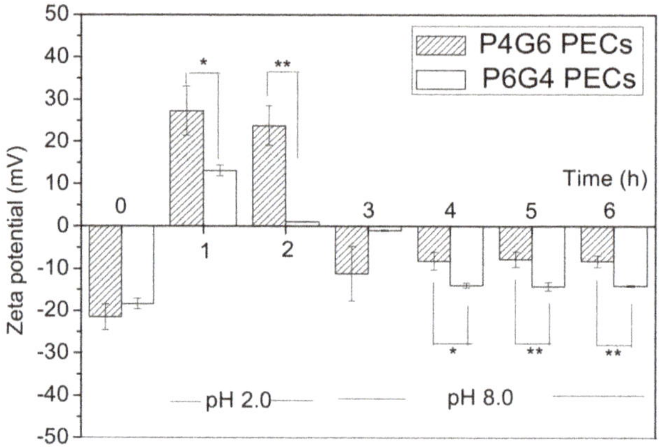

Figure 3. Zeta potentials of the PECs. P4G6 and P6G4 PECs were placed at 37 °C and pH 2.0 for 2 h, and the environmental pH was adjusted to 8.0 for 4 h to simulate drug transportation in the GI tract. At the predetermined time points (each hour), zeta potential was determined by electrophoretic light scattering (ELS). The results have been statistically analyzed using Student's *t*-test (Microsoft Excel 2000). Differences were recognized as statistically significant when the *p* values were less than 0.05, and significant differences are shown in star marks (* $p < 0.05$ and ** $p < 0.01$).

2.3. Preparation and Optimization of Curcumin-Loaded PECs

On the basis of previous results, we have chosen the P4G6 PECs as the optimum formulation. This formulation was subsequently loaded with curcumin and the loading process was also optimized. To load the hydrophobic curcumin, it was first prepared at various concentrations in ethanol, which involved our formulation as the solvent, forming 0.5, 1, 1.5, 2, and 2.5 mg/mL alcoholic solution. The alcoholic solutions were independently mixed with type B gelatin and pectin, following the abovementioned procedures. After encapsulation, the particle size and distribution of the PECs were assessed using DLS, as shown in Figure 4a. In all curcumin-loaded PECs, the particle sizes were less than 270 nm. The PECs made from 1.5 mg/mL of curcumin-loaded alcoholic solution exhibited the smallest particle size, with PDI values of 0.28, which was the highest of all PECs'. Along with the concentration of the loaded curcumin (2 and 2.5 mg/mL), particle size increased; as loading-curcumin concentration decreased (0.5 and 1 mg/mL), particle size and PDI value decreased. By ANOVA, we observed that the curcumin-loaded PECs made from 0.5 mg/mL of curcumin alcoholic solution exhibited the ideal particle size (214.67 ± 3.43 nm) and the PDI value (0.17 ± 0.01), which indicated smaller and more homogeneous particles than those of PECs without curcumin load.

Besides the particle size and PDI of the curcumin-loaded PECs, drug content and loading efficiency should also be considered to optimize curcumin encapsulation. The curcumin-loaded PECs were first freeze-dried. The entrapped curcumin was extracted using methanol from the freeze-dried powder and measured by an enzyme-linked immunosorbent assay (ELISA) reader. Drug content could be defined as the ratio of entrapped curcumin in all curcumin-loaded PECs, and loading efficiency could be determined by the ratio of loaded curcumin in comparison with the total input curcumin during the loading process. Drug content and loading efficiency are presented in Figure 4b. The highest drug content (40%) and loading efficiency (53%) were both observed in PECs made from 0.5 mg/mL of curcumin alcoholic solution. The PECs made from other loading concentrations of curcumin alcoholic solution contained less than 10% of curcumin, with loading efficiencies of lower than 35%, which resulted in a higher concentration of curcumin and interference of PEC formation [34]. The PEC made from 0.5 mg/mL of curcumin alcoholic concentration could result in not only small and homogeneous particles but also the highest drug content and loading efficiency. Herein, drug-releasing behaviors were further studied using the curcumin-loaded PECs comprising of 40% mole of pectin and 60% mole of type B gelatin, and loaded with 0.5 mg/mL of curcumin alcoholic solution.

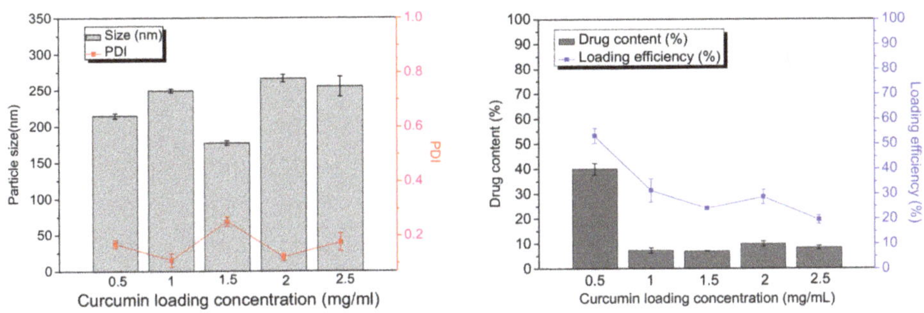

Figure 4. Drug-loading of curcumin-loaded PECs. Curcumin was prepared at various concentrations in alcohol. The curcumin alcoholic solution was mixed with type B gelatin and pectin, forming PECs at pH 4.0. The particle size and PDIs (**a**) of curcumin-loaded PECs were determined by DLS. To further determine drug contents and loading efficiency (**b**), the curcumin-loaded PECs were subsequently freeze-dried. The powdered curcumin-loaded PECs were reconstituted with methanol. Curcumin concentration in the solutions was measured by ultraviolet-visible light spectrum at 423 nm.

2.4. Drug-Releasing Profiles

The drug-releasing behaviors of curcumin-loaded PECs in mimetic GI tract conditions (pH 2.0 for 2 h and pH 8.0 for 4 h) were also studied. The drug-releasing profile is shown in Figure 5, where two releasing trends are observed based on the pH values. For the first 2 h at pH 2.0, less than 20% of curcumin was released. As the pH increased to 8.0, the PECs released curcumin sustainably and eventually a total of 60% of curcumin was released. It is worth noting that the drug-releasing curve at pH 2.0 was different from that at pH 8.0. At pH 2.0, the drug-releasing curve was close to first- order drug-releasing kinetics, whereas at pH 8.0, the drug-releasing curve was close to zero-order kinetic drug-releasing profile, releasing curcumin sustainably. Besides, almost 40% of curcumin still remained inside the PECs after 6 h of acid or alkaline treatment. The residual curcumin would be provided to treat colon cancer.

2.5. Confocal Laser Scanning Microscopic (CLSM) Observation

According to the results of drug releasing profiles (Figure 5), 60% of curcumin would be released out of the PECs in mimetic GI tract conditions. That meant 40% of the curcumin would still remain within the PECs, which has possibility to be transported to the colon lesions and uptaken by the colon cancerous cells. To confirm that the curcumin-loaded PECs could preserve curcumin and be internalized into HCT116 human colon cancer cells even after acidic or alkaline treatments, we used a CLSM for direct fluorescent observation. In order to observe the internalization of PECs to

HCT116 cells, the PECs were labeled with cyanine 5.5 monosuccinimidyl ester (Cy 5.5-NHS ester) fluorescence via amide bonds with the amine groups appended on type B gelatin. The organelle endosome/lysosome was marked with Lysotracker DND-Red and cell nuclei were stained with 4′,6-diamidino-2-phenylindole (DAPI). Curcumin was detected by its own fluorescence. Curcumin and Cy 5.5-labeled curcumin-loaded PECs were incubated in mimetic GI tract conditions and added to HCT116 cells, which were then incubated for 12 h. The fluorescence of Cy 5.5, lysotracker, curcumin, and DAPI were independently excited and detected with appropriate excitation/emission wavelengths; they are shown in grey, red, green, and blue, respectively, in Figure 6.

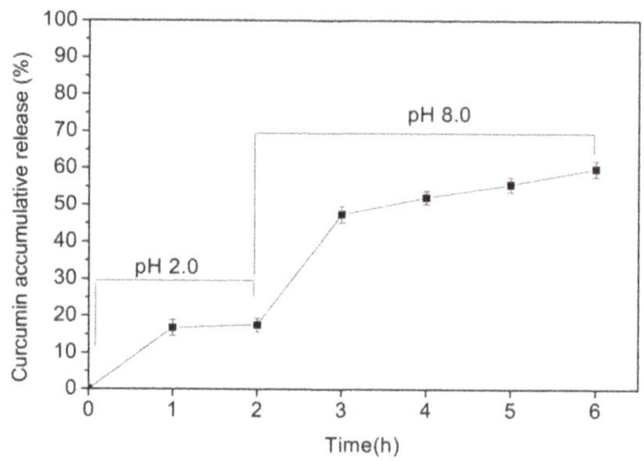

Figure 5. Drug-releasing profiles. Curcumin-loaded PECs were placed into dialysis bags (molecular weight cutoff: 6–8 k) (M.W.C.O.: 6–8 k), which were immersed to an ethanol/deionized water (1:1, v/v) solution at 37 °C. The pH of the outer medium was first adjusted to 2.0 for 2 h and then adjusted to 8.0 for 4 h to mimic drug transit in the GI tract. At the predetermined time points, curcumin release was determined by UV spectrum at a wavelength of 423 nm.

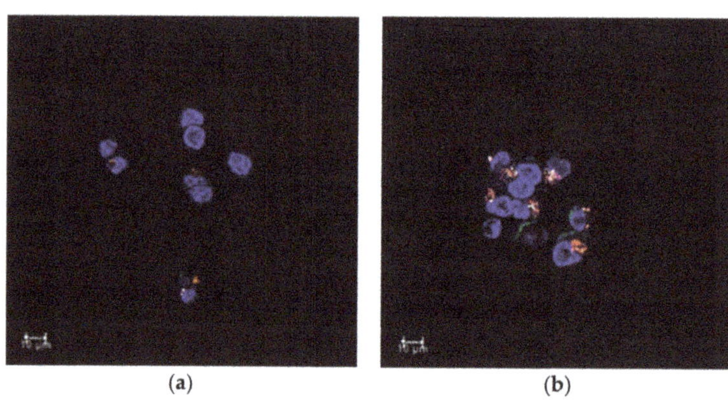

Figure 6. Confocal laser scanning microscope (CLSM) images showing internalization of curcumin (**a**) and curcumin-loaded PECs (**b**). HCT116 human colon cancer cells were incubated with curcumin or Cy5.5-labled curcumin-loaded PECs, which were pretreated with mimetic GI tract conditions. After 12 h of incubation, the cells were stained with Lysotracker DND-Red to label the endosome/lysosomes, and then fixed with paraformaldehyde. The fixed cells were then washed twice with PBS and further stained and mounted with 4′,6-diamidino-2-phenylindole (DAPI)-containing mounting medium. The fluorescence of the Cy 5.5-labeled curcumin-loaded PECs, curcumin, endosome/lysosomes, and DAPI was observed using a CLSM with appropriate excitation and emission wavelengths.

Figure 6 shows the internalization process of curcumin and curcumin-loaded PECs. Curcumin could enter the cells via diffusion and spread over the organelles within the cells. Herein, in Figure 6, the fluorescence of curcumin could be observed over the cell cytoplasm, even also be observed to overlap with that of endosome/lysosomes. Curcumin-loaded PECs were observed to be internalized to HCT116 cells, and mainly accumulated in the endosome/lysosomes. Curcumin fluorescence was also distributed within the cells, showing the intracellular drug-releasing behaviors of curcumin-loaded PECs. It is worth noting that the fluorescence of the free curcumin was weaker than that of curcumin-loaded PECs, leading probably from the chemical structure degradation of free curcumin. That would be an evidence for the protective effects of our PECs. The PECs would carry active curcumin into the colon cancer cells even through the mimetic GI tract conditions.

2.6. Cell Viability

In order to verify the therapeutic efficacy of curcumin-loaded PECs against colon cancer cells, HCT116 cells were incubated with curcumin-loaded PECs or free curcumin. The free insoluble curcumin was first prepared in 1% of dimethyl sulfoxide (DMSO), to the same concentration as curcumin concentration in the curcumin-loaded with PECs'. Both curcumin-loaded PECs and free curcumin were pretreated with mimetic GI tract conditions and transit periods, as mentioned previously. After that, curcumin and curcumin-loaded PECs were prepared at concentrations of 45 to 1.41 μM by serial dilution and respectively added to HCT116 cells. The resulting cell viability was determined by (3-(4,5-dimethylthiazol-2-yl)-2,5-diphenyltetrazolium bromide) (MTT) assay after incubation.

The viability of HCT116 cells following treatment with curcumin and curcumin-loaded PECs, which were pretreated with mimetic GI tract conditions, is shown in Figure 7. The results indicated that both curcumin and curcumin-loaded PECs exhibited concentration-dependent toxicity to HCT116 cells. The half maximal inhibitory concentrations (IC_{50}) of free curcumin and curcumin-loaded PECs were approximately 25 and 5 μM, respectively. The results directly showed that curcumin would be degraded in mimetic GI tract conditions, thereby reducing its toxicity to colon cancer cells. Curcumin-loaded PECs enabled sustained-release of their payloads in mimetic GI tract conditions, and the released payloads exhibited toxicity to cancer cells. Besides, the PECs protected curcumin from degradation due to environmental pH during treatment with mimetic GI tract conditions; hence, more bioactive ingredients were able to arrive at tumorous lesions in the colon. Herein, the curcumin-loaded PECs exhibited higher toxicity toward HCT116 human colon cancer cells than free curcumin.

Besides, the safety of the novel bare PECs was also evaluated using Detroit 551 human normal fibroblast cells, and the MTT assay was employed to determine cell viability. The concentrations of PECs were adjusted on the basis of drug contents until they were equal to those used for curcumin encapsulation. Cell viability data are presented in Figure S2 in the Supplementary Materials. Bare PECs exhibited very low toxicity toward human normal fibroblasts after they were directly cultured with the cells. Even though the bare PECs were incubated after treatment with mimetic GI tract conditions, as described above, the bare PECs exhibited very low cytotoxicity. The results clearly illustrated that our pectin-type B gelatin PECs possessed high safety to human normal cells, and that the bare PECs were digested after acid or alkaline treatments.

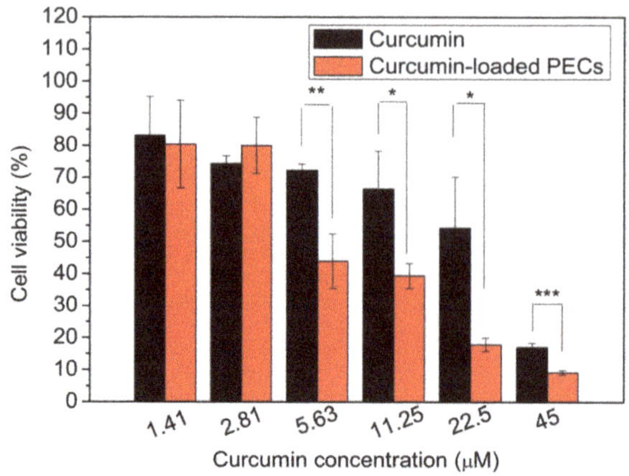

Figure 7. Cell viability. Curcumin and curcumin-loaded PECs were pretreated with pH 2.0 for 2 h and pH 8.0 for 4 h. Afterward, the pH values were adjusted back to 7.4. Next, HCT116 human colon cancer cells were incubated with the curcumin and curcumin-loaded PECs at 37 °C with 5% CO_2 supply for 12 h. The curcumin and curcumin-loaded PECs were removed and the MTT assay was utilized to determine cell viability. The results have been statistically compared and analyzed using Student's *t*-test (Microsoft Excel 2000). Differences were considered to be statistically significant when the *p* values were less than 0.05, and significant differences are shown in star marks (* $p < 0.05$; ** $p < 0.01$, and *** $p < 0.001$).

2.7. Cell Cycle Analysis

In order to further identify the relevance of the cytotoxicity and protective ability of our PECs from exteriors, HCT116 cells were treated with curcumin or curcumin-loaded PECs, which were both preincubated in mimetic GI tract conditions (pH 2.0 for 2 h and pH 8.0 for 4 h). The cell cycle of HCT116 cells was analyzed using flow cytometry after cell fixation and staining by propidium iodide (PI). The results are shown in Figure 8. Figure 8a shows that curcumin after incubation in mimetic GI tract conditions did not cause cell cycle arrest, whereas Figure 8b indicates that curcumin-loaded PECs caused G2/M phase arrest. The results were consistent with the cytotoxic results, as shown in Figure 7 Curcumin-loaded PECs could maintain curcumin bioactivity from the environment, leading to stronger cytotoxic effects of curcumin on colon cancer cells.

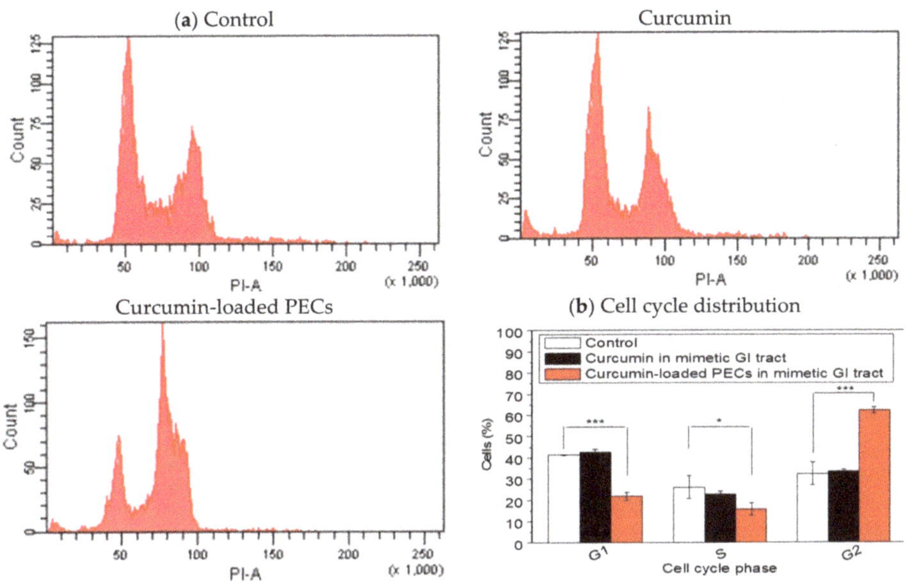

Figure 8. Cell cycle profiles of HCT116 cells treated with curcumin and curcumin-loaded PECs after incubation in mimetic GI tract conditions. (**a**) Curcumin and curcumin-loaded PECs (curcumin concentration was 6.8 μM) were pretreated with mimetic GI tract conditions, and HCT116 human colon cancer cells were then treated and the curcumin or curcumin-loaded PECs for 12 h. The cells were then fixed and stained with PI and cell cycle was analyzed using flow cytometry; (**b**) Cell cycle distribution was also analyzed using ModFit LT 3.0. The results have been statistically compared and analyzed using Student's *t*-test (Microsoft Excel 2000). Differences were recognized as statistically significant when the p values were less than 0.05, and significant differences are shown in star marks (* $p < 0.05$ and *** $p < 0.001$).

3. Discussion

Curcumin is a potent anti-cancer compound, although its clinical application is limited owing to its hydrophobicity, poor oral bioavailability, and unstable chemical structure. To overcome these challenges, we aimed to design an appropriate curcumin oral formulation for colon cancer treatment. PECs are one of the feasible approaches for achieving this formulation.

PEC is composed of two or more oppositely charged polymers in polar solvents. PEC could not only improve the water solubility of the encapsulated agents but also protect its payload from the environment. Besides, the electrical charge of PEC would change with the environments, thus potentiating oral-controlled or sustained-release drug delivery system using suitable constituents of PEC to enhance oral bioavailability of drugs. In this study, anionic pectin and cationic type B gelatin used to form PEC for curcumin formulation. Pectin-gelatin complexes have been developed [34]; however, there is a distinction between them and our pectin-type B gelatin PECs. First, our pectin-type B gelatin PECs were nano-sized (264.0 ± 3.1 nm), whereas the PECs reported by other research was micro-sized; hence, they are also called complex coacervates. This means that the aggregation that occurred during their manufacture would be different and new, requiring further exploitation. Besides, previous studies mostly reported type A gelatin as the polycation that interacts with polyanionic pectin. The pI of type A gelatin was 9.0, indicating that at pH 7.0, it is positively charged and capable to self-assemble with pectin (pKa 3.0) to form PECs. As pH increased over 9.0, type A gelatin is electrically neutralized. However, the highest pH in the small intestine is approximately 8.0, far lower than the pI of type A gelatin; thus, the drug-releasing site would be mostly only the upper GI tract, such as the stomach. However, D. Madhavi et al. have suggested sustained-release of curcumin via oral administration, which could increase 10-fold its bioavailability compared to that of unformulated curcumin [35]. The drug-releasing efficacy and relative bioavailability of pectin-type A gelatin PEC system would be considered. In this study, type B gelatin was first employed in the PEC system. The pI of type B gelatin was 4.8–5.5 until the pH decreased to 4.0, whereas pectin, whose pKa is 3.0, would be self-aggregated with type B gelatin. At pH 2.0, such as in the stomach, the pectin-type B gelatin complex would swell, as shown in Figure 2, because of the electrically neutralized pectin. The positively charged zeta potential is shown in Figure 3. As the pH increased to 8.0, type B gelatin was electrically neutralized and the zeta potential of the PECs was negatively charged, breaking the electrical balance of the PEC (Figure 3). Increased particle sizes and PDI were therefore observed (Figure 2). The accumulative drug-releasing profile of curcumin-loaded pectin-type B gelatin PECs also revealed their continuous drug-releasing behavior, as shown in Figure 5. Curcumin was not only abruptly released at pH 2.0 but also released sustainably at pH 8.0, which has not been commonly discussed in other studies. The releasing behavior of our PECs differed from that of pectin-type A gelatin PECs. I. Joseph et al. revealed that micro-sized pectin-type A gelatin PECs release their payloads following sustained release at pH 2.0 [36], whereas our PECs rapidly released curcumin at pH 2.0 within 1 h. This finding also evidenced the differences in drug-releasing profiles with different components of the PECs.

Micro-sized PECs (complex coacervates) could bear higher drug contents and loading efficacy after optimization, compared to our PECs. M. Saravanan et al. have prepared pectin-gelatin complex coacervation with particle size of 94 μm and high drug content and loading efficiency, which are 57% and 76%, respectively [26]. Pectin-gelatin complex coacervates were also fabricated by D. Silva el at., with drug content and loading efficiency of 56.7% and 89.5%, respectively [28]. In this study, the drug content and loading efficiency of pectin-type B gelatin PEC were also discussed. The optimum drug content and loading efficiency were 40% and 53%, respectively, whereas the particle size after encapsulation was 214.67 ± 3.43 nm with narrow particular distribution (PDI was 0.17), as presented in Figure 4. The low drug contents and loading efficiency were due to the small cavity of nano-sized PECs. The curcumin-loaded nano-sized PECs prepared by P. Sarika et al. also had low drug content and loading efficiency, compared to those of complex coacervates [24]. The particle size of our curcumin-loaded PECs was slightly smaller than that of bare PEC. The reasonable explanation would be the additional ethanol in the manufacture of these loaded PECs. The ethanol dissolved the curcumin, instead of dispersing it before loading [37]. Thus, the curcumin would not rapidly aggregate inside the core of PECs during manufacture, resulting in similar particle sizes of PEC after curcumin encapsulation.

The drug-releasing profiles of PECs in mimetic GI tract conditions were also investigated in our study, as shown in Figure 5. Curcumin-loaded PECs exhibited abrupt release of curcumin at pH 2.0 and the releasing curve was close to the first-order releasing profiles. At pH 8.0, the curcumin-loaded PECs exhibited similar curcumin releasing profiles with zero-order releasing mechanism. I. Joseph et al. have suggested that the constituents of PECs considerably influence their drug-releasing profile [36]. The drug-releasing profile of our PECs corresponded to the increase in particle size of the bare PECs containing the same ratios of components, as presented in Figure 2a. Particle sizes were observed to abruptly increase as the pH was adjusted to 2.0, which could be attributed to rapid drug-releasing profile at pH 2.0; the particle size became gradually larger at pH 8.0, indicating the sustained-release behavior of the curcumin-loaded PECs. Moreover, the curcumin-loaded PECs preserved almost 40% of payloads within the PECs, which could directly treat colon cancer cells in the GI tract, and the released curcumin would enter the blood via the GI tract to achieve anti-cancer effect. The remaining curcumin-loaded PECs after passing through the mimetic GI tract conditions have been identified that they could be internalized into the colon cancerous cells and further perform the intracellular drug releasing, shown in Figure 6.

The cytotoxicity of the curcumin-loaded PECs to HCT116 human cancer cells was accessed by MTT assay, as presented in Figure 7. In this study, we utilized 1% of DMSO to prepare curcumin solution, which could eliminate the cytotoxicity of DMSO itself [37]. The cell viabilities shown in Figure 7 have indicated that curcumin-loaded PECs exhibited stronger toxicity to HCT116 human colon cancer cells than free curcumin after treatment with mimetic GI tract conditions. The cytotoxicity might be attributed to the anti-cancer activity of curcumin. Y. Wang et al. have shown that active curcumin is degraded in a pH-dependent manner; thus, the alkaline (pH 8.0) treatment might reduce the anti-cancer activity of curcumin [20]. Curcumin-loaded PECs allowed sustained release of curcumin during digestion to maintain curcumin concentration in the blood. Besides, the encapsulated curcumin was protected from external conditions; therefore, the curcumin-loaded PECs could directly carry the bioactive curcumin to cancerous lesions in the colon and efficiently destroyed carcinomas, as evidenced by the results of the internalizing experiments (Figure 6). The cell viability results (Figure 7) indicated that the curcumin-loaded PECs exhibited more significant cytotoxicity to HCT116 cells than free curcumin. The results of cell cycle distribution test (Figure 8) further explained the reason. The cell cycle test results showed that curcumin-treated with mimetic GI tract conditions no longer caused any cell cycle arrest, whereas curcumin-loaded PECs triggered the G2/M phase arrest in HCT116 cells, supporting the results of the studies conducted by A. Jaiswal et al. [38] and J. Lu et al. [39]. Our results have greatly supported that our PECs efficiently preserved the therapeutic effect of curcumin as a payload until it reached carcinomas in the colon.

Curcumin-loaded PECs possessed sustained-release properties and could transport active curcumin into tumorous lesions in the colon after treatment with mimetic GI tract conditions. The results of cell viability tests also confirmed the anti-cancer efficacy of the curcumin-loaded PECs. However, the mimetic GI tract conditions in our study did not involve enzymatic effects, and those conditions were also far from simulating both fasted state gastric conditions (FaSSGF) and fasted state conditions in the small intestine (FaSSIF), as recorded in the pharmacopeia [40]. Future application of curcumin-loaded PECs as a new formulation of oral curcumin solution is still being investigated further but worth expecting.

4. Materials and Methods

4.1. Materials

Type B gelatin (molecular weight: 5 k; bloom number: 225), sodium phosphotungstate (PTA) for TEM negative staining and (3-(4,5-dimethylthiazol-2-yl)-2,5-diphenyltetrazolium bromide (MTT) powder for cell viability assay were all purchased from Sigma-Aldrich Co., Ltd. (St. Louis, MO, USA). Pectin (methylation ratio: 3%–7%) was obtained from Wako Pure Chemical Industries, Ltd. (Osaka, Japan). Curcumin was acquired from Merck Life Science (Hohenbrunn, Germany). Dialysis bags (M.W.C.O.: 6–8 k) for drug-release assessment were also purchased from Merck Life Science. Ethanol was purchased from Taiwan Sugar Corp. (Tainan, Taiwan). Methanol for drug-loading test was obtained from Alps Chem Co., Ltd. (Hsinchu, Taiwan). Reagents for pH value adjustment, hydrochloric acid and sodium hydroxide, were purchased from J. T. Baker Inc. (Center valley, PA, USA) and Uniregion Bio-tech Inc. (New Taipei, Taiwan), respectively. Biological agents, including dimethyl sulfoxide for preparing curcumin solution in cell viability tests and a Muse Cell Cycle Kit was purchased from Merck Millipore (Burlington, MA, USA). Fluorescent reagents, including Cy 5.5-NHS ester, Lysotracker DND-Red, and DAPI-containing mounting medium, were acquired from Lumiprobe (Hunt Valley, MD, USA), Thermo Fisher Scientific (Waltham, MA, USA), and Vector Laboratory (Burlingame, CA, USA), respectively.

4.2. Preparation and Characterization of Pectin-Type B Gelatin PECs

Pectin (10 mg) was weighed and dissolved in 50 mL of deionized water. Type B gelatin (50 mg) was also weighed and dissolved in 50 mL of deionized water at 25 °C. After complete dissolution, the pH of the produced type B gelatin was adjusted to 4.0 using 0.1 N HCl. Ethanol (0.1 mL) was mixed with type B gelatin solution under stirring for 30 min. Next, the pectin solution was added into the mixture, blended, and the pH of the mixture was further adjusted to 4.0. Thirty minutes later, PECs were acquired and stored at 4 °C. To optimize the preparation and component of PECs, various volumes of type B gelatin and pectin were added to form the PECs. The type B gelatin/pectin PECs composed of various amounts of pectin and type B gelatin were collected and measured by DLS (Malvern ZS 90, Malvern Instruments, Malvern, UK) for particle size, distribution, and superficial zeta potential (ξ-potential). The hydrodynamic diameter was determined by the Stokes-Einstein equation after DLS measurement. Size distribution was determined through cumulant analysis of the measured intensity autocorrelation function assumed by Gaussian distribution. Zeta potential was obtained through ELS and calculated with Henry's equation.

4.3. Observation of the Morphology of PECs

The prepared PEC sample was dropped to carbon-coated copper mesh and excess sample was removed after the PEC sample was attached to the mesh. Sodium phosphotungstate aqueous (PTA) solution 1% (*w*/*v*) was prepared and dropped to the copper grids for negative staining. Excess PTA solution was then eliminated. Next, the copper grids were placed in a vacuum oven. Twenty-four hours later, the dried copper grids were kept from light exposure at room temperature until observation. The morphology of the PECs was observed using TEM (HT7700, Hitachi, Japan) with an accelerated voltage of 120 V under high contrast mode (HC mode).

Besides, the morphology of PECs was also observed using a thermal field emission scanning electron microscope (Thermal FE-SEM) (JEOL JSM-7800F, JEOL Ltd., Tokyo, Japan). The PEC solutions were first prepared and freeze-dried using a vacuum freeze drier (Uniss MF-280, Uniss Company, New Taipei City, Taiwan). The freeze-dried powder was afterwards coated with platinum and observed by the thermal FE-SEM.

4.4. pH Responsiveness of PECs

PECs (5 mL) were placed into a sample vial and its pH was adjusted to 2.0 using HCl (0.1 N) to mimic the acidic condition in the stomach. Afterward, the PEC solution was incubated at 37 °C for 2 h in a shaker (M.BR-022UP, Taitec, Tokyo, Japan). Next, the pH of the PEC solution was adjusted to 8.0 using NaOH (0.5 N) to simulate the alkaline environment in the small intestine. The adjusted PEC solution was placed at 37 °C for 4 h in a shaker. Particle size and surface zeta potential were measured by DLS at every hour.

4.5. Preparation and Characterization of Curcumin-Loaded Pectin-Type B Gelatin PECs

Different amounts of curcumin (0.5, 1, 1.5, 2, or 2.5 mg) were precisely weighed and placed independently in a vial. Curcumin was first dissolved in ethanol (0.1 mL). Simultaneously, pectin (10 mg) and type B gelatin (50 mg) were respectively weighed and dissolved in 50 mL deionized water as stock solutions. Type B gelatin solution was treated with 0.1 N of HCl until the pH was 4.0. Curcumin alcoholic solution (0.1 mL) and type B gelatin solution (0.2 mL) were first mixed in a sample vial with stirring for 30 min. Thereafter, pectin solution (0.7 mL) was added to the mixture, which was then blended for 30 min to fabricate curcumin-loaded PECs. Particle size and distribution were determined using DLS.

4.6. Drug Contents and Loading Efficiency Determination

Curcumin-loaded PECs were prepared as abovementioned. The prepared curcumin-loaded PECs were freeze-dried using a vacuum freeze drier (Uniss MF-280). The dried sample (0.5 mg) was then weighed and dissolved in 1 mL of methanol for curcumin extraction. The extracted curcumin was quantified by an ELISA reader (BioTek Synergy HT, BioTek Instruments, Inc., Winooski, VT, USA) at 423 nm as the detected wavelength.

The drug content of the curcumin-loaded PCs was calculated according to the formula (1):

$$\text{Drug contents } (\%) = \frac{weight\ of\ encapsulated\ curcumin\ (mg)}{total\ weight\ of\ curcumin - loaded\ PECs\ (mg)} \times 100\% \quad (1)$$

Besides, drug loading efficiency was also determined and calculated according to the formula (2):

$$\text{Loading efficiency } (\%) = \frac{encapsulated\ curcumin\ weight\ (mg)}{total\ curcumin\ input\ weight\ (mg)} \times 100\% \quad (2)$$

After calculation, the optimization was analyzed with analysis of variance (ANOVA).

4.7. Determination of Drug-Releasing Profiles

Curcumin-loaded PECs (1 mL) were placed into dialysis bags, which were then immersed into 2.5 mL of methanol/deionized water solution (1:1, *v*/*v*) in a sample vial. The pH of the methanol/deionized water solutions in a sample vial was first adjusted to 2.0, and then the sample vial was incubated at 37 °C for 2 h. Afterwards, the pH value of the methanol/deionized water solution in the sample vial was adjusted to 8.0, and the sample vial was further incubated at 37 °C for 4 h. At each hour, 0.1 mL of methanol/deionized water solution was collected and absorbance wavelength at 423 nm was detected to quantify the release of curcumin.

4.8. CLSM Observation

HCT116 cells (1×10^5 cells/mL) were seeded on a slide and incubated at 37 °C with 5% CO_2 supply for 12 h. After the cells were attached, they were co-cultured with free curcumin- or Cy5.5-labeled curcumin-loaded PECs, which were pretreated with mimetic GI tract conditions for 12 h at 37 °C with 5% CO_2 supply. Excess curcumin or curcumin-loaded PECs were thereafter removed from the cells, which were then washed twice with PBS. The cells were further stained with 1 μM of Lysotracker DND-Red at 37 °C with 5% CO_2 for 1 h. Next, the lysotracker DND-Red was removed, and the cells were again washed twice with PBS and fixed with 4% paraformaldehyde for 20 min. After fixation, the cells were washed twice with PBS and the cells and their nuclei were stained and mounted with DAPI-containing mounting medium. Fluorescence was observed and detected using a confocal laser scanning microscope (Leica SP8, Leica Microsystem, Wetzlar, Germany). The fluorescence of Cy 5.5 was detected at excitation and emission wavelengths of 633 and 650 nm, respectively. The fluorescence of Lysotracker DND-Red for locating endosome/lysosomes was detected using excitation and emission wavelengths of 577 and 590 nm, respectively. The fluorescence of curcumin was observed at excitation and emission wavelengths of 488 and 520 nm, respectively. DAPI fluorescence was detected using the preset wavelength of the CLSM equipment.

4.9. Cell Viability Determination

Human colon cancer cells HCT116 (7×10^3 cells/mL) were seeded on a 96-well plate at 37 °C with 5% CO_2 supply for 12 h to allow cell attachment. Simultaneously, curcumin-loaded PECs were adjusted with the cell culture medium to the predetermined concentrations at 45 to 1.41 μM. Curcumin solutions at the same concentration series were also prepared with 1% of DMSO and cell culture medium. Afterwards, the pH of both curcumin and curcumin-loaded PECs was adjusted to 2.0 for 2 h, then the pH of the PECs was adjusted to 8.0 for 4 h. The pH of the PECs was re-adjusted to 7.4, and then HCT116 cells were treated with the PECs. The HCT116 cells were incubated at 37 °C with 5% CO_2. Twelve hours later, the PECs and cell culture medium were removed. MTT assay was utilized to evaluate cell viability.

In addition, the human normal fibroblast cells Detroit 551 (7×10^3 cells/mL) were also seeded on a 96-well plate and cultured at 37 °C with 5% CO_2 for 12 h. Bare PECs at concentrations of 112.5 to 3.52 μM were prepared with serial dilution. The bare PECs were separated into two groups: one for direct treatment with Detroit 551 cells and the other for mimetic GI tract treatment, as described above. Next, the latter bare PECs were transferred to pH 7.4 condition, and the cells were cultured with the bare PECs for 12 h. MTT assay was then applied to determine cell viability.

4.10. Cell Cycle Analysis

HCT116 cells (3×10^6 cells/mL) were seeded on a 6-well plate and cultured for 12 h at 37 °C with 5% CO_2 supply. After the cells were attached, they were exposed to curcumin or curcumin-loaded PECs pretreated in mimetic GI tract conditions (pH 2.0 for 2 h, pH 8.0 for 4 h, and back to pH 7.0), as described above, and then the cells were incubated at 37 °C with 5% CO_2 supply for 12 h. Afterward, the cells were collected, washed with PBS, and fixed with 70% of ethanol at −20 °C for 6 h. Fixed cells were re-collected using centrifugation and suspended in PBS. The cells were further stained with a flow cell cycle kit in the dark for 30 min. Cell cycles were determined using flow cytometry (BD FACSCanto, BD Biosciences, Franklin Lakes, NJ, USA) and analyzed with the software ModFit LT 3.0 (BD Biosciences, Franklin Lakes, NJ, USA).

4.11. Statistical Analysis

All experiments were repeated at least thrice and the results were averaged and presented as mean ± standard deviation (S.D.). Most results have been statistically compared and analyzed using Student's *t*-test (Microsoft Excel 2000). Differences were recognized as statistically significant when the p values were less than 0.05, and significant differences are shown in star marks (* $p < 0.05$; ** $p < 0.01$, and *** $p < 0.001$). Besides that, of all results, only the optimization of drug contents and loading efficiency were analyzed with analysis of variance (ANOVA) using OriginPro 8 (OriginLab, Northampton, MA, USA).

5. Conclusions

In this study, a novel pectin-type B gelatin PEC was fabricated for curcumin delivery and its optimum manufacturing processes, including component ratios and curcumin-loading process, were also developed. The produced PECs could increase their particle size at pH 2.0 for the neutral charges of pectin. At pH 8.0, type B gelatin would be electrically neutralized, driving the gradual swelling of the PECs. The curcumin-loaded PECs could release their payloads at both pH 2.0 and 8.0, which potentiate exploration of their oral formulation. Besides, the PECs could also preserve the bioactivity of curcumin and be internalized into colon cancer cells to allow intracellular drug-releasing behaviors. In addition, the curcumin-loaded PECs induced cell cycle arrest and excellent cytotoxicity toward the human colon cancer cells HCT116 even after treatment under mimetic GI tract conditions, whereas bare PECs by themselves showed very low toxicity to normal cells. The pectin-type B gelatin PEC that possessed high safety and outstanding anti-cancer effects is a promising oral formulation for curcumin delivery.

Supplementary Materials: Supplementary materials can be found at http://www.mdpi.com/1422-0067/19/11/3625/s1.

Author Contributions: Y.-T.C. and I.-J.S. conceived and designed the concepts and experiments. Y.-T.C. and X.L. analyzed the results and wrote the manuscript. F.-Y.S. and L.-L.C. performed the preparation and characterization of bare PECs, including particle size and pH responsiveness. F-Y.S. also optimized the drug loading and conducted drug releasing alone. F.-Y.S. conducted the cell culture and the relative experiments, such as MTT assay and cell cycle arrest analysis under the supervision by S.-C.K. and Y.-C.H.

Funding: This research was funded by Ministry of Science and Technology, Taiwan (MOST) (MOST 106-2320-B-039 -001 and 107-2314-B-039-013) and by China Medical University (CMU106-N-10 and CMU107-N-08). The research was also financially supported by the "Chinese Medicine Research Center, China Medical University" from the Featured Areas Research Center Program within the framework of the Higher Education Sprout Project by the Ministry of Education (MOE) in Taiwan (CMRC-CHM-6).

Acknowledgments: Particle size analysis was performed through the use of the Medical Research Core Facilities Center, Office of Research & Development at China medical University, Taichung, Taiwan. TEM observation was conducted by the use of the Instrument Center at National Tsing Hua University, Hsinchu, Taiwan. The authors would especially thank Ju-Chun Tai for several equipment operations, including CLSM and flow cytometry at Medical Research Core Facilities Center, Office of Research & Development at China medical University, Taichung, Taiwan.

Conflicts of Interest: The authors declare no conflict of interest.

Abbreviations

PEC	polyelectrolyte complex
GI tract	gastrointestinal tract
pI	isoelectric point
DLS	dynamic laser scattering
PDI	polydispersity index
TEM	transmission electron microscopy
ELISA	enzyme-linked immunosorbent assay
M.W.C.O.	molecular weight cutoff
MTT	(3-(4,5-dimethylthiazol-2-yl)-2,5-diphenyltetrazolium bromide)
CLSM	confocal laser scanning microscope
DAPI	4′,6-diamidino-2-phenylindole

References

1. Dakhara, S.; Anajwala, C. Polyelectrolyte Complex: A Pharmaceutical Review. *Syst. Rev. Pharm.* **2010**, *1*, 121–127. [CrossRef]
2. Lankalapalli, S.; Kolapalli, V. Polyelectrolyte Complexes: A Review of their Applicability in Drug Delivery Technology. *Indian J. Pharm. Sci.* **2009**, *71*, 481–487. [CrossRef] [PubMed]
3. George, M.; Abraham, T. Polyionic Hydrocolloids for the Intestinal Delivery of Protein Drugs: Alginate and Chitosan—A Review. *J. Control. Release* **2006**, *114*, 1–14. [CrossRef] [PubMed]
4. Lin, W.C.; Yu, D.G.; Yang, M.C. pH-Sensitive Polyelectrolyte Complex Gel Microspheres Composed of Chitosan/Sodium Tripolyphosphate/Dextran Sulfate: Swelling Kinetics and Drug Delivery Properties. *Coll. Surf. B Biointerfaces.* **2005**, *44*, 143–151. [CrossRef] [PubMed]
5. Angelova, N.; Hunkeler, D. Rationalizing the Design of Polymeric Biomaterials. *Trends Biotechnol.* **1999**, *17*, 409–421. [CrossRef]
6. Liu, Z.; Jiao, Y.; Wang, Y.; Zhou, C.; Zhang, Z. Polysaccharides-Based Nanoparticles as Drug Delivery Systems. *Adv. Drug Deliv. Rev.* **2008**, *60*, 1650–1662. [CrossRef] [PubMed]
7. Hu, K.; Huang, X.; Gao, Y.; Huang, X.; Xiao, H.; McClements, D.J. Core–Shell Biopolymer Nanoparticle Delivery Systems: Synthesis and Characterization of Curcumin Fortified Zein–Pectin Nanoparticles. *Food Chem.* **2015**, *182*, 275–281. [CrossRef] [PubMed]
8. Sæther, H.V.; Holme, H.K.; Maurstad, G.; Smidsrød, O.; Stokke, B.T. Polyelectrolyte Complex Formation Using Alginate and Chitosan. *Carbohydr. Polym.* **2008**, *74*, 813–821. [CrossRef]
9. Devi, N.; Kakati, D.K. Smart Porous Microparticles Based on Gelatin/Sodium Alginate Polyelectrolyte Complex. *J. Food Eng.* **2013**, *117*, 193–204. [CrossRef]
10. Junyaprasert, V.B.; Mitrevej, A.; Sinchaipanid, N.; Boonme, P.; Wurster, D.E. Effect of Process Variables on the Microencapsulation of Vitamin A Palmitate by Gelatin-Acacia Coacervation. *Drug Dev. Ind. Pharm.* **2001**, *27*, 561–566. [CrossRef] [PubMed]
11. Ghosh, S.; Banerjee, S.; Sil, P.C. The Beneficial Role of Curcumin on Inflammation, Diabetes and Neurodegenerative Disease: A Recent Update. *Food Chem Toxicol.* **2015**, *83*, 111–124. [CrossRef] [PubMed]
12. Darvesh, A.; Aggarwal, B.B.; Bishayee, A. Curcumin and Liver Cancer: A Review. *Curr. Pharm. Biotechnol.* **2012**, *13*, 218–228. [CrossRef] [PubMed]
13. Liu, D.; Chen, Z. The Effect of Curcumin on Breast Cancer Cells. *J. Breast Cancer* **2013**, *16*, 133–137. [CrossRef] [PubMed]
14. Dhillon, N.; Aggarwal, B.B.; Newman, R.A.; Wolff, R.A.; Kunnumakkara, A.B.; Abbruzzese, J.L.; Ng, C.S.; Badmaev, V.; Kurzrock, R. Phase II Trial of Curcumin in Patients with Advanced Pancreatic Cancer. *Clin. Cancer Res.* **2008**, *14*, 4491–4499. [CrossRef] [PubMed]

15. Cai, X.Z.; Wang, J.; Li, X.D.; Wang, G.L.; Liu, F.N.; Cheng, M.S.; Li, F. Curcumin Suppresses Proliferation and Invasion in Human Gastric Cancer Cells by Down-Regulation of PAK1 Activity and Cyclin D1 Expression. *Cancer Biol. Ther.* **2009**, *8*, 1360–1368. [CrossRef] [PubMed]
16. Goel, A.; Boland, C.R.; Chauhan, D.P. Specific Inhibition of Cyclooxygenase-2 (COX-2) Expression by Dietary Curcumin in HT-29 Human Colon Cancer Cells. *Cancer Lett.* **2001**, *172*, 111–118. [CrossRef]
17. Wang, X.; Wang, Q.; Ives, K.L.; Evers, B.M. Curcumin Inhibits Neurotensin-Mediated Interleukin-8 Production and Migration of HCT116 Human Colon Cancer Cells. *Clin. Cancer Res.* **2006**, *12*, 5346–5355. [CrossRef] [PubMed]
18. Anand, P.; Kunnumakkara, A.B.; Newman, R.A.; Aggarwal, B.B. Bioavailability of Curcumin: Problems and Promises. *Mol. Pharm.* **2007**, *4*, 807–818. [CrossRef] [PubMed]
19. Bar-Sela, G.; Epelbaum, R.; Schaffer, M. Curcumin as an Anti-Cancer Agent: Review of the Gap between Basic and Clinical Applications. *Curr. Med. Chem.* **2010**, *17*, 190–197. [CrossRef] [PubMed]
20. Wang, Y.J.; Pan, M.H.; Cheng, A.L.; Lin, L.I.; Ho, Y.S.; Hsieh, C.Y.; Lin, J.K. Stability of Curcumin in Buffer Solutions and Characterization of its Dadation Products. *J. Pharm. Biomed. Anal.* **1997**, *15*, 1867–1876. [CrossRef]
21. Naksuriya, O.; Okonogi, S.; Schiffelers, R.M.; Hennink, W.E. Curcumin Nanoformulations: A Review of Pharmaceutical Properties and Preclinical Studies and Clinical Data Related to Cancer Treatment. *Biomaterials* **2014**, *35*, 3365–3383. [CrossRef] [PubMed]
22. Siyawamwaya, M.; Choonara, Y.E.; Bijukumar, D.; Kumar, P.; Du Toit, L.C.; Pillay, V. A Review: Overview of Novel Polyelectrolyte Complexes as Prospective Drug Bioavailability Enhancers. *Int. J. Polym. Mater. Polym. Biomater.* **2015**, *64*, 955–968. [CrossRef]
23. Mahmood, K.; Zia, K.M.; Zuber, M.; Salman, M.; Anjum, M.N. Recent Developments in Curcumin and Curcumin Based Polymeric Materials for Biomedical Applications: A Review. *Int. J. Biol. Macromol.* **2015**, *81*, 877–890. [CrossRef] [PubMed]
24. Sarika, P.; James, N.R. Polyelectrolyte Complex Nanoparticles from Cationised Gelatin and Sodium Alginate for Curcumin Delivery. *Carbohydr. Polym.* **2016**, *148*, 354–361. [CrossRef] [PubMed]
25. Anitha, A.; Deepagan, V.; Rani, V.D.; Menon, D.; Nair, S.; Jayakumar, R. Preparation, Characterization, in vitro Drug Release and Biological Studies of Curcumin Loaded Dextran Sulphate–Chitosan Nanoparticles. *Carbohydr. Polym.* **2011**, *84*, 1158–1164. [CrossRef]
26. Saravanan, M.; Rao, K.P. Pectin–Gelatin and Alginate–Gelatin Complex Coacervation for Controlled Drug delivery: Influence of anionic polysaccharides and drugs being encapsulated on physicochemical properties of Microcapsules. *Carbohydr. Polym.* **2010**, *80*, 808–816. [CrossRef]
27. McMullen, J.N.; Newton, D.W.; Becker, C.H. Pectin–Gelatin Complex Coacervates II: Effect of Microencapsulated Sulfamerazine on Size, Morphology, Recovery, and Extraction of Water-Dispersible Microglobules. *J. Pharm. Sci.* **1984**, *73*, 1799–1803. [CrossRef] [PubMed]
28. Silva, D.; Favaro-Trindade, C.; Rocha, G.; Thomazini, M. Microencapsulation of Lycopene by Gelatin–Pectin Complex Coacervation. *J. Food Process. Preserv.* **2012**, *36*, 185–190. [CrossRef]
29. Du conseille, A.; Astruc, T.; Quintana, N.; Meersman, F.; Sante-Lhoutellier, V. Gelatin Structure and Composition Linked to Hard Capsule Dissolution: A Review. *Food Hydrocoll.* **2015**, *43*, 360–376. [CrossRef]
30. Kubo, W.; Konno, Y.; Miyazaki, S.; Attwood, D. In situ Gelling Pectin Formulations for Oral Sustained Delivery of Paracetamol. *Drug Dev. Ind. Pharm.* **2004**, *30*, 593–599. [CrossRef] [PubMed]
31. Ashford, M.; Fell, J.; Attwood, D.; Sharma, H.; Woodhead, P. Studies on Pectin Formulations for Colonic Drug Delivery. *J. Control. Release* **1994**, *30*, 225–232. [CrossRef]
32. Liu, L.; Fishman, M.L.; Kost, J.; Hicks, K.B. Pectin-Based Systems for Colon-Specific Drug Dlivery Via Oral Route. *Biomaterials* **2003**, *24*, 3333–3343. [CrossRef]
33. Lin, L.; Yao, W.; Rao, Y.; Lu, X.; Gao, J. pH-Responsive Carriers for Oral Drug Delivery: Challenges and Opportunities of Current Platforms. *Drug Deliv.* **2017**, *24*, 569–581. [CrossRef]
34. Plashchina, I.G.; Zhuravleva, I.L.; Antonov, Y. Phase Behavior of Gelatin in the Presence of Pectin in Water-Acid Medium. *Polym. Bull.* **2007**, *58*, 587–596. [CrossRef]
35. Madhavi, D.; Kagan, D. Bioavailability of a Sustained Release Formulation of Curcumin. *Integr. Med.* **2014**, *13*, 24–30.
36. Joseph, I.; Venkataram, S. Indomethacin Sustained Release from Alginate-Gelatin or Pectin-Gelatin Coacervates. *Int. J. Pharm.* **1995**, *126*, 161–168. [CrossRef]

37. Georges, D.; Naima, Z.; Isabelle, R.; Gérard, P.; Jean, C.C.; Philippe, A. Evaluation of the Cytotoxicity Effect of Dimethyl Sulfoxide (DMSO) on Caco2/TC7 Colon Tumor Cell Cultures. *Biol. Pharm. Bull.* **2002**, *25*, 1600–1603. [CrossRef]
38. Jaiswal, A.; Marlow, B.; Gupta, N.; Narayan, S. β-Catenin-Mediated Transactivation and Cell-Cell Adhesion Pathways are Important in Curcumin (diferuylmethan)-Induced Growth Arrest and Apoptosis in Colon Cancer Cells. *Oncogene* **2002**, *21*, 8414–8427. [CrossRef] [PubMed]
39. Lu, J.; Cai, Y.; Ding, J. Curcumin Induces DNA Damage and Caffeine-Insensitive Cell Cycle Arrest in Colorectal Carcinoma HCT116 Cells. *Mol. Cell. Biochem.* **2011**, *354*, 247–252. [CrossRef] [PubMed]
40. Rosen, R.T.; Hiserodt, R.D.; Fukuda, E.K.; Ruiz, R.J.; Zhou, Z.; Lech, J.; Rosen, S.L.; Hartman, T.G. Determination of Allicin, S-Allylcysteine and Volatile Metabolites of Garlic in Breath, Plasma or Simulated Gastric Fluids. *J. Nutr.* **2001**, *131*, 968–971. [CrossRef] [PubMed]

International Journal of
Molecular Sciences

MDPI

Article

Chemotherapeutic Efficacy of Implantable Antineoplastic-Treatment Protocols in an Optimal Mouse Model for Human Ovarian Carcinoma Cell Targeting

Jonathan M. Pantshwa [1], Khadija Rhoda [1], Sarah J. Clift [2], Priyamvada Pradeep [1], Yahya E. Choonara [1], Pradeep Kumar [1], Lisa C. du Toit [1], Clement Penny [3] and Viness Pillay [1,*]

1 Wits Advanced Drug Delivery Platform Research Unit, Department of Pharmacy and Pharmacology, Faculty of Health Sciences, School of Therapeutics Sciences, University of the Witwatersrand, Johannesburg, 7 York Road, Parktown 2193, South Africa; monwabisi.pantshwa@wits.ac.za (J.M.P.); dija.rhoda@yahoo.com (K.R.); priyamvada.pradeep@wits.ac.za (P.P.); yahya.choonara@wits.ac.za (Y.E.C.); pradeep.kumar@wits.ac.za (P.K.); lisa.dutoit@wits.ac.za (L.C.d.T.)

2 Department of Paraclinical Sciences, Faculty of Veterinary Science, University of Pretoria, Onderstepoort 0110, South Africa; sarah.clift@up.ac.za

3 Department of Medical Oncology, Division of Oncology, Faculty of Health Sciences, University of the Witwatersrand, Johannesburg, 7 York Road, Parktown 2193, South Africa; clement.penny@wits.ac.za

* Correspondence: viness.pillay@wits.ac.za; Tel.: +27-11-717-2274

Received: 23 June 2018; Accepted: 27 September 2018; Published: 4 October 2018

Abstract: The present study aimed to design and develop a nanocomposite drug delivery system employing an antineoplastic-loaded antibody functionalized nanomicelle encapsulated within a Chitosan–Poly(vinylpyrrolidone)–Poly(*N*-isopropylacrylamide) (C–P–N) hydrogel to form an in situ forming implant (ISFI), responsive to temperature and pH for cancer cell-targeting following intraperitoneal implantation. The optimum nanomicelle formulation was surface-functionalized with anti-MUC 16 (antibody) for the targeted delivery of methotrexate to human ovarian carcinoma (NIH:OVCAR-5) cells in Athymic nude mice that expressed MUC16, as a preferential form of intraperitoneal ovarian cancer (OC) chemotherapy. The cross-linked interpenetrating C–P–N hydrogel was synthesized for the preparation of an in situ-forming implant (ISFI). Subsequently, the ISFI was fabricated by encapsulating a nanocomposite comprising of anti-MUC16 (antibody) functionalized methotrexate (MTX)-loaded poly(*N*-isopropylacrylamide)-b-poly(aspartic acid) (PNIPAAm-b-PASP) nanomicelles (AF(MTX)NM's) within the cross-linked C–P–N hydrogel. This strategy enabled specificity and increased the residence time of the nanomicelles at tumor sites over a period exceeding one month, enhancing uptake of drugs and preventing recurrence and chemo-resistance. Chemotherapeutic efficacy was tested on the optimal ovarian tumor-bearing Athymic nude mouse model and the results demonstrated tumor regression including reduction in mouse weight and tumor size, as well as a significant ($p < 0.05$) reduction in mucin 16 levels in plasma and ascitic fluid, and improved survival of mice after treatment with the experimental anti-MUC16/CA125 antibody-bound nanotherapeutic implant drug delivery system (ISFI) ($p < 0.05$). The study also concluded that ISFI could potentially be considered an important immuno-chemotherapeutic agent that could be employed in human clinical trials of advanced, and/or recurring, metastatic epithelial ovarian cancer (EOC). The development of this ISFI may circumvent the treatment flaws experienced with conventional systemic therapies, effectively manage recurrent disease and ultimately prolong disease-free intervals in ovarian cancer patients.

Keywords: implant; antibody functionalized nanomicelles; epithelial ovarian cancer; chemotherapeutic drugs; nude mouse model

1. Introduction

Ovarian cancer is one of the fatal female diseases in which many patients are diagnosed at an advanced stage when the disease has already spread beyond the ovaries to the abdominal cavity [1–3]. Targeted treatment using biomarkers for OC holds great promise to improve the survival rate of patients with OC. Actively targeted chemotherapy has been a considerable advancement compared to passive-targeting based on the Enhanced Permeability and Retention (EPR) effect. To this end, actively targeted chemotherapeutic drug delivery employing nanotechnology has had a remarkable impact on cancer treatment with the following advantages: (1) Ability to deliver antineoplastic drugs to treat specific cancer metastatic sites; (2) potential reduction in the quantity of drug necessary to achieve a required therapeutic dose at the target cancer tissue; and (3) reduction in the quantity of drug delivered to healthy cells thus decreasing cell cytotoxicity [4–6]. In this context, the coating of nanomicelles with specific antibody targeting ligands can assist in conjugation of the nanomicelles to precise biomarkers and antigen receptors over-expressed on OC cells to target tumors with improved accuracy [7,8]. The flexibility of coated amphiphilic nanomicelles is amenable to modified forms of chemotherapy. Targeting cancer cells at a molecular level using nanocarriers such as nanomicelles may be the solution for more personalized chemotherapy. Antibodies are the most significant target ligands that have offered a broad-spectrum of possibilities in terms of drug targeting and accuracy of interaction. Antibody-targeted nanomicelles (or immunomicelles) can be engineered by composite conjugation of antibodies (or nanobodies) onto the trigger (exposed free terminal) of a hydrophilic segment of the nanomicelle. This ensures active targeting nanomicelle preparation without any steric-hindrance for the ligand-antibody [9,10].

Therefore, this study has innovatively focused on the use of mucins (specifically MUC16) for the advanced design of a novel intraperitoneally administered (as opposed to conventional intravenous chemotherapy) OC cell-targeting nanocomposite drug delivery system to significantly improve the chemotherapy of OC. The uniqueness of the delivery system is via the use of antibody-functionalized nanomicelles to specifically target mucin antigens known to be over-expressed in OC cells. To our knowledge, this has not been investigated before. MUC16, specifically, is an eminent cell surface antigen in OC that is shed into the serum and therefore it is also widely used clinically for the diagnosis and management of epithelial OC [11–13].

The monoclonal antibody OC125/Mab anti-MUC16 identifies this tumor-associated antigen, which is over-expressed in epithelial ovarian cancer cells, but not in the epithelium of normal fetal and mature ovaries [14]. The non-specific delivery of chemotherapeutics to healthy tissues other than the OC tumors is one of the leading cytotoxicity challenges of chemotherapy. Therefore, this study also focuses on surmounting this challenge by designing MUC16 antibody-functionalized antineoplastic drug-loaded nanomicelles to specifically target OC cells via the peritoneum. However, targeted nanomicelles have a few limitations regarding the ability to provide sustained drug release. The limitations of these nano-enabled polymeric systems include instability in the circulatory system, rapid degradation, and clearance by the immune system and a lack of controlled sustained drug release [15,16]. To overcome these limitations for clinical applications to be feasible, current research focuses on preparation of nanomicelles encapsulated with hydrogels for solid intra-tumoral release of chemotherapeutic drugs over-prolonged periods of weeks, or even months, thus diminishing the dire need for daily administration of systemic chemotherapy. Hence, prolonged circulation in the bloodstream, in-vivo stability, biodegradability, and polymer-drug compatibility with sufficient retention of the drug within the carrier system are prerequisites to successful design and preparation of an implantable drug targeting delivery system [17,18].

2. Results

2.1. Mechanism of Synthesis of the Anti-MUC16 MTX-Loaded Nanomicelles

Amphiphilic block copolymers are expected to offer improved applications in chemotherapeutics when formulated as nanomicelles functioning as nanocarriers for ionic and non-ionic drug loading [19–21]. Ionic complexes/composites involving anionic antineoplastic drugs and aspartic acid can be produced and nanomicelles further synthesized by self-assembly of the drug and amphiphilic copolymer in an aqueous medium. In this amphiphilic structure, the hydrophilic surface shields the hydrophobic interior from the interaction of the inner core with the surrounding aqueous environment, thus enabling the nanomicelles to be used as a vehicle for targeted drug delivery (Figure 1).

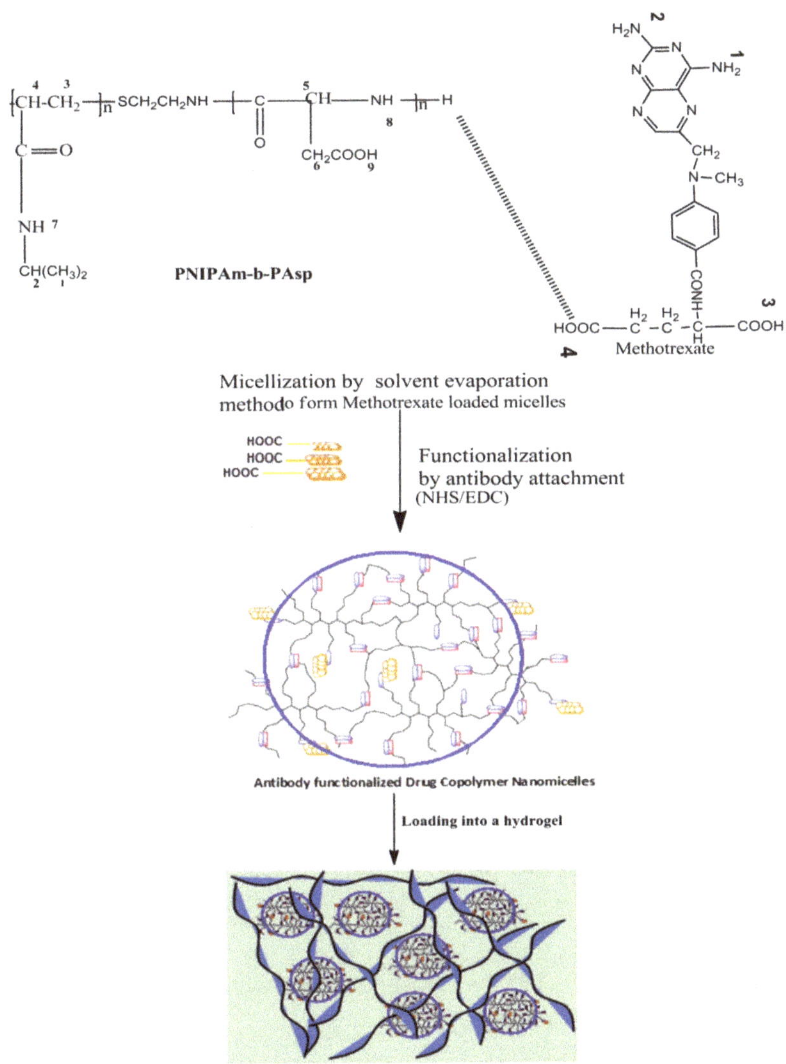

Figure 1. A schematic illustration of the synthesis of an implantable antibody functionalized methotrexate loaded nanomicelles hydrogel composite delivery system (ISFI). Blue circles refer to nanomicelle's corona.

In this study, a more efficient strategy (under moderate formulation conditions) was utilized to ensure superior targeting of the functionalized core-shell nanomicelles using a simplified synthetic approach outlined in Figure 1. The physical synthesis comprised two synthetic steps: (1) The block copolymer constituted an aspartic (ASP) backbone and poly(*N*-isopropylacrylamide) (PNIPAAm) outer surface to self-assemble the MTX-loaded core-shell nanomicelles by solvent evaporation in an aqueous medium. The polycationic backbone formed between poly(aspartic acid) (PASP) and methotrexate (MTX) generated the inner core of the nanomicelle, while the non-ionic PNIPAAm formed the outer hydrophilic shell. PASP was used to conjugate the MTX (a carboxylic drug) to its amine groups to form an amide (CO-NH) linkage (2). Furthermore, the anti-MUC16 antibody was surface-coated to the nanomicelles via carbodiimide-sulpho-NHS mediated conjugation through carboxylic (-COOH) linkage of the PNIPAAM-b-PASP copolymer. This linkage was also significant for the elucidation of the orientation of the anti-MUC16 to ensure binding affinity and stability. In this case, the MTX-loaded nanomicelles were incubated with 1-ethyl-3-(3-dimethylaminopropyl)-carbodiimide and *N*-hydroxysuccinimide for 15 min at room temperature. The resulting activated nanomicelles were then covalently linked to the anti-MUC16 antibody (1% weight compared with the polymer concentration) (Figure 1). The resultant anti-MUC16 antibody-functionalized MTX-loaded nanomicelles (the NanoComposite) were then lyophilized (Freezone 6, Model 79340, Labconco, MO, USA).

The results of this study confirmed that MTX can be loaded into the interior of the PNIPAAM-block-PASP nanocomposites and the carboxyl-amine-groups of the nanomicelles were amenable to surface coating with the anti-MUC16 antibody (Figure 1). It is well known that MUC16 is highly expressed on malignant cells and not on healthy cells and therefore the NanoComposite can impede tumor growth and even halt metastasis based on the binding to MUC16. In addition, it is important to note that OC cells also use MUC16 as a mechanism for metastasis via muco-adhesion to other organs within the peritoneum. Thus, the developed NanoComposite has far-reaching potential for the targeted treatment and management of OC employing advanced nanochemotherapeutics.

2.1.1. Analysis of the Molecular Structure Integrity of the Functionalized Nanocomposite

Fourier-transform infrared (FTIR) spectral analysis is one of the most robust analytical techniques to evaluate the molecular structure of block copolymer functional groups [22,23]. The interactions and functionalization of the MTX-loaded nanomicelles with anti-MUC16 was evaluated and described with FTIR spectral analysis and is presented in Figure 2.

As demonstrated, there were distinct chemical structure transitions for the functionalized MTX-loaded nanomicelles compared with the non-functionalized MTX-loaded nanomicelles. The FTIR spectra of the blank nanomicelles and MTX-loaded NanoComposite were congruent with those of the native polymers PNIPAAm and ASP [24,25]. This result showed that the polymer formation into nanomicelles occurred with minor chemical modification during the polymerization process. Hence, the nanomicelles possessed elemental chemical features that were representative of the native polymers.

Variations were noted in the FTIR spectrum of the blank nanomicelles and MTX-loaded nanomicelles in Figure 2. The additional peaks that were recorded in the MTX-loaded nanomicelles were due to the 1,3-replacement composite (1516.53–1451.23 cm^{-1}) and a phenyl amino compound (1647.22–1451.13 cm^{-1}). This indicated that MTX was entrapped within the hydrophobic interior of the nanomicelle structure by weak electrostatic H-bonds involving the COO- moiety of MTX and the OH-group of aspartic acid or by the charged ionic groups involving the NH_2 bonds of MTX and the COO-bonds existing in PNIPAAm-b-PASP.

MTX was dispersed within the PNIPAAm-b-PASP matrix in the amorphous form exclusive of any polymorphic transformation or alteration in the aqueous form. The anti-MUC16 amine groups (N–H) were evident by the 1660.20 cm^{-1} peak as well as the derived amine groups induced by the broad range O–H group and the C–H group observed between 1090.30–1310.11 cm^{-1} that also overlapped the 1000.12–1200.22 cm^{-1} band representative of C–O groups that were attributed to the PNIPAAm-b-PASP copolymer. These transitions indicated that there were prominent interactions between the MTX-loaded nanomicelles and the anti-MUC16 antibody that confirmed the surface coated functionalization of the MTX-loaded NanoComposite.

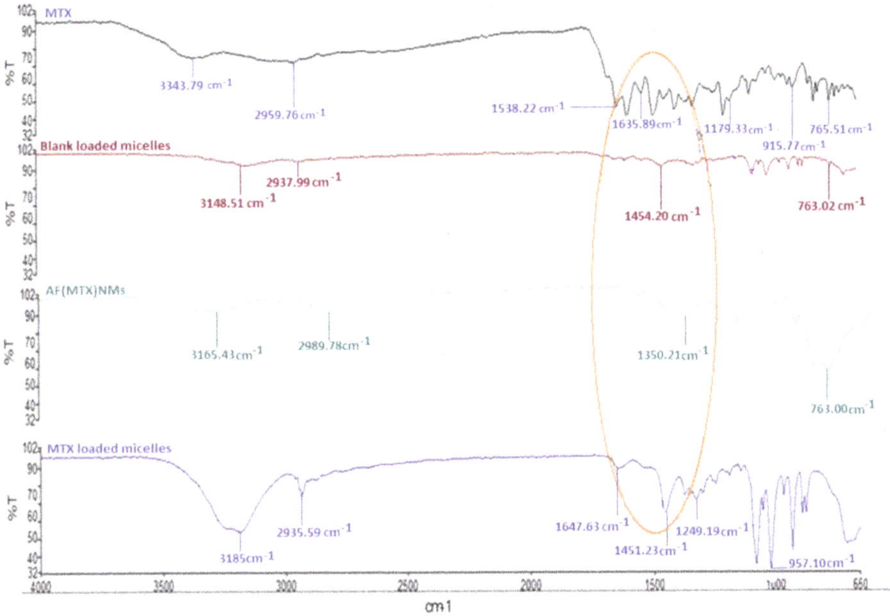

Figure 2. Fourier-transform infrared (FTIR) spectra illustrating the molecular structural transitions of MTX, blank micelles, AF(MTX)NM's and (MTX)NM's.

2.1.2. Analysis of the Thermal Features of the Drug-Free and MTX-Loaded PNIPAAm-b-PASP Nanomicelles

Differential scanning calorimetric investigations were conducted to observe the drug's physical form in the nanomicelle, since this could affect the drug release from the delivery system. DSC thermograms of MTX, blank micelles and MTX-loaded micelles are depicted in Supplementary Data Figure S1. The DSC curve of the MTX displayed a critical melting peak at 125 °C. MTX-loaded micelles displayed the melting peak for the drug at 145 °C. Incorporation of MTX within the nanomicelle thus elevated the melting point indicative of enhanced thermal stability of the drug. This would suggest that on incorporation of the partially soluble crystalline MTX within the hydrophobic core it was converted to a less crystalline form, resulting in a favorable MTX entrapment efficacy of 80.6 ± 0.3%. The glass transition temperature (T_g) of the co-polymeric nanomicelles is also evident at ~150 °C, correlating with previous findings [25].

2.2. Synthesis and Characterization of Chitosan-Poly(N-Vinylpyrrolidone)-Poly(N-Isopropylacrylamide) (C–P–N) Hydrogel

Figure 3 illustrates the constituent chemical structures and mechanism of synthesis of the C–P–N composite hydrogel structure. Physicochemical characterization via nuclear magnetic resonance (NMR) and FTIR analysis indicated the formation of chitosan-poly(*N*-vinylpyrrolidone)-poly(*N*-isopropylacrylamide) covalent bonds between the protonated amine groups of chitosan (annotated with red circles) and carboxylated groups of poly(*N*-vinylpyrrolidone) (annotated with green circles). Both covalent crosslinking and chemical structure moderated the swelling of the IPN hydrogel. The synthesized C–P–N hydrogel was also verified with chemical structural correlation and proton NMR crest peak assignments (Supplementary Materials Figure S2).

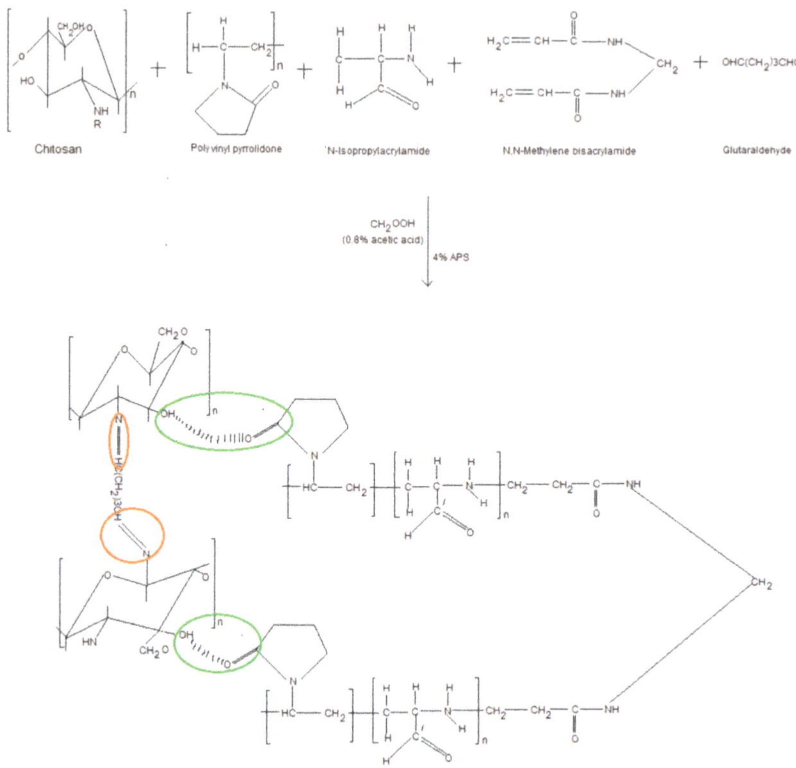

Figure 3. Illustration of constituent chemical structures and reaction mechanism forming composite hydrogel structure.

Assessment of Polymeric Structural Variations of the Cross-Linked C–P–N Hydrogel

FTIR analysis of the C–P–N hydrogel proved successful radical polymerization of chitosan (CHT), polyvinylpyrrolidone (PVP) and PNIPAAm, and interpenetration of glutaraldehyde (GA) within the cross-linked network (Supplementary Materials Figure S3). The FTIR spectrum of the C–P–N hydrogel demonstrated peaks at 3428 and 1654 cm^{-1} assigned to stretching vibrations of -NH and -OH, respectively. The peak at 1720 cm^{-1} was attributed to the stretching vibrations of carbonyl group in the PVP-PNIPAAm molecule which differentiated chitosan and the chitosan derivative (CHT-PVP). The PNIPAAm hydrogel sample exhibited significant peaks at 1654, 1551, 1385 and 1369 cm^{-1}, which were assigned to the characteristic peaks of amide I, amide II and the isopropyl group, respectively (Supplementary Materials Figure S3). For the composite hydrogel, the intensity of the absorption peak at 1654 cm^{-1} was attributed to increased amide group compared with the CHT-PVP hydrogel due the incorporated NIPAAm. The intensity of the methyl peak and isopropyl peaks decreased when compared with PNIPAAm hydrogel. This indicated successful radical polymerization of CHT-PVP-PNIPAAm and interpenetration by GA and *N,N*-methylenebisacrylamide.

2.3. Characterization of the In Situ Forming Implant (ISFI)

2.3.1. Assessment of the Gelation Temperature Using Oscillatory Rheology

The crossover of the storage and loss modulus (G′ and G″) indicates the gelation temperature (T_{gel}) of the hydrogel as illustrated in Figure 4. The storage modulus (G′) of a viscoelastic solid is associated with the solid properties or the elastic energy storage properties which indicated that the sample would return to its original state following removal of the deformational energy, while the loss modulus (G″) explains the behavior of the viscoelastic solid when it is acting as a liquid i.e., the viscous properties of the sample representing the dissipation of energy after the application of the deformational stress. Hence at the point at which the storage modulus exceeded the loss modulus, the C–P–N hydrogel was behaving more like a solid than a liquid and hence this was used to determine the thermal gelation temperature. Gelation studies were also conducted on non-crosslinked samples and used as controls (Figure 4).

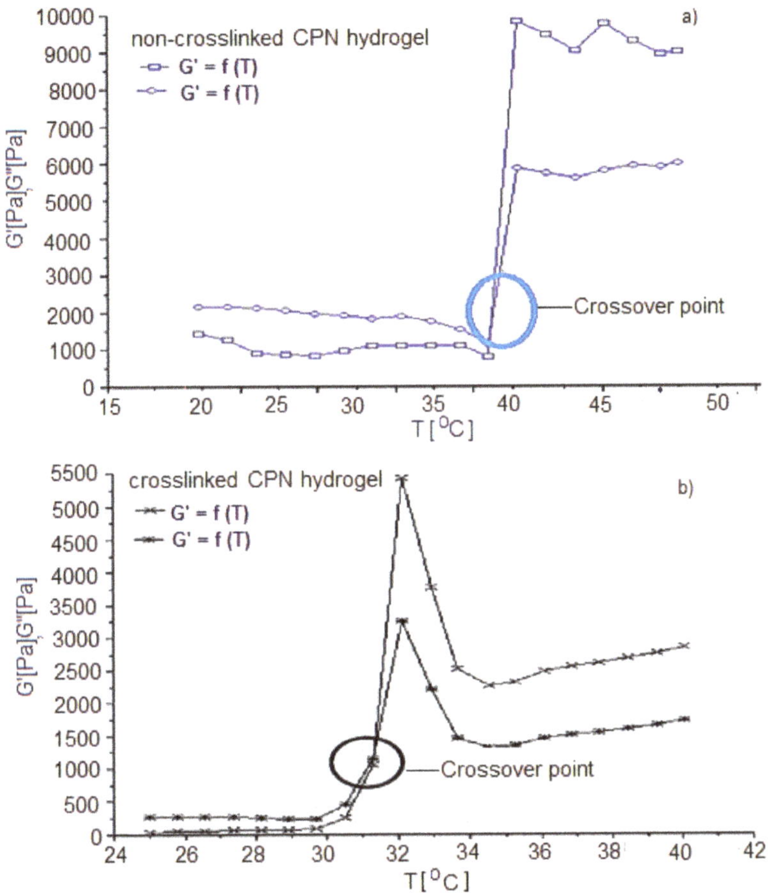

Figure 4. Typical profile obtained during oscillatory rheological testing of non-crosslinked (**a**) and crosslinked (**b**) C–P–N hydrogel samples. The black and blue circles demarcates the crossover point storage modulus (G′) and the loss modulus (G″).

2.3.2. Morphological Characterization of the ISFI

The structure and properties of the ISFI were validated by confocal laser scanning microscopy (CLSM) and scanning electron microscopy (SEM). Figure 5 markedly illustrates the morphological properties of the drug-filled coated nanocomposites after encapsulation in C–P–N gel of the ISFI. Drug-loaded coated nanocomposites have a homogenous spherical form. CLSM confirmed optical distribution of the labeled coated nanocomposites in the spatial polymeric-grounded depot platforms. Control native or unlabeled C–P–N hydrogel possessed no rhodamine activities. The overall results substantiate that drug-loaded coated nanocomposites maintain their integrity after-lyophilization for 48 h.

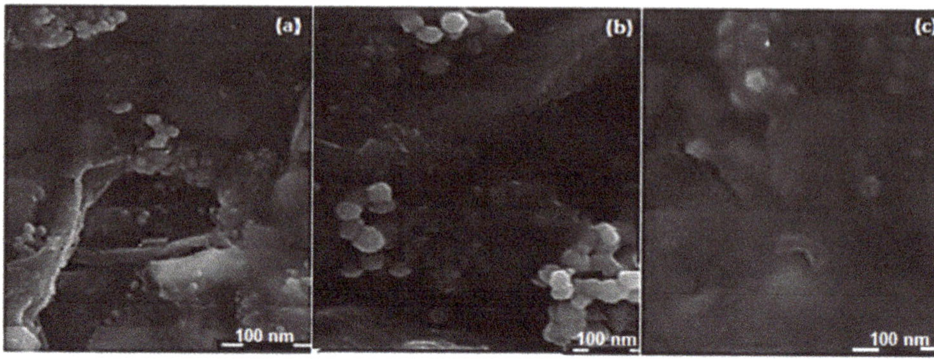

Figure 5. Typical SEM micrographs of the drug-loaded coated nanomicelles after encapsulation in the C–P–N at 50× lower magnification (**a**,**c**); and (**b**) 100× Increased magnification of the ISFI.

To fully understand the potential of utilizing the ISFI for local drug delivery to human ovarian carcinoma cell, the first factor that had to be assessed was the cytotoxicity. Investigation of in vitro cytotoxicity was conducted on the nanomicelles via MTT assay of OVCAR 5 cells in our previously published work [26]. Similar trends were observed in this study, except that the ISFI enabled a prolonged drug release of 30 days.

2.4. *In Vivo Studies in the Athymic-Nude Mice Model*

In vivo studies were conducted in three platforms: Firstly intraperitoneal (IP) induction of human ovarian carcinoma in Athymic Swiss nude mice, secondly IP implantation of various treatment protocols including ISFI on the optimal ovarian tumor-bearing Athymic nude mouse model, and thirdly in vivo evaluation of chemotherapeutic efficacy utilizing a variety of indices including whole mouse weight, tumor size (measured with calipers and sonography), quantification of mucin 16 antigen expression levels (using immunohistochemistry), and sandwich quantitative CA125 Human ELISA (on plasma and ascitic fluid), as well as the survival rate of mice post-treatment.

2.4.1. Induction of Human Ovarian Carcinoma in Athymic Swiss Nude Mice

Growth of cancer cells in the intra-peritoneal cavity and subcutaneous region of immuno-compromised mice is a common procedure for evaluating tumorigenic potential in vivo [19–21]. These procedures are also utilized to evaluate the effects of chemotherapeutic interventions on tumor cell lines. Charles River France Swiss nude mice (n = 80), were established as a simple, reproducible mouse model for the induction of human ovarian carcinoma using the IP inoculated NIH:OVCAR5 cell line, as was evidenced by the visible development in vivo of intra-abdominal tumor nodules with associated severe ascites. This occurred within 10 days of inoculation. Advanced ovarian carcinoma disease in the Athymic nude mice was consistently associated with peritoneal carcinomatosis/transcoelomic metastasis, which always preceded the formation of severe ascites. Solid tumor nodules coated all serosal surfaces, especially within the pelvis. These tumor-bearing mice were further IP implanted with various treatment protocols and chemotherapeutic efficacy was evaluated. Due to the widespread dissemination of the IP tumors, it was not feasible to analyze specific regions of the peritoneum to assess the uptake of the antibody (anti-MUC16) and drug-loaded nanomicelle delivery system in the peritoneum.

2.4.2. Chemotherapeutic Efficacy in the Treatment of Human Ovarian Carcinoma

The chemotherapeutic efficacy of the AF(D)NMs implant (the preferred chemotherapeutic model system) was evaluated against the non-functionalized (D)NMs and the comparison group (IV methotrexate/cisplatin only), as well as the control/placebo group. A variety of indices were assessed, including tumor size (measured with calipers and sonography), mouse weight, quantification of mucin 16 antigen expression levels as well as survival rate of mice post-treatment.

2.4.3. Tumor Size

The tumor size was measured with calipers and sonography (Figure 6) and two chemotherapeutic drugs utilized in this study were cisplatin (4 mg/kg) and methotrexate (15 mg/kg), as illustrated with nude mice IP-growth curves, respectively (Figure 7A,B). The average tumor size in the 3 treatment groups each with 10 mice (two experimental groups, and a comparison group (i.v group)) decreased significantly ($p < 0.05$) from day 15 after implantation of the AF(D)NMs treatment. Conversely, in the placebo group, the average tumor size increased steadily, indicating biocompatibility of the blank nanomicelle implant delivery system (placebo) in vivo. During the evaluation phase, nanomicelle implant ((D)NMs)-treated mice reached the ultimate point (of 100 mm^3 average tumor diameter) within 21 days of treatment whilst the group of AF(D)NMs-implant treated mice survived until the completion of the study. After necropsy examination of the IP inoculated mouse post treatment, the AF(D)NMs implant treatment resulted in reduced average tumor size and ascitic fluid.

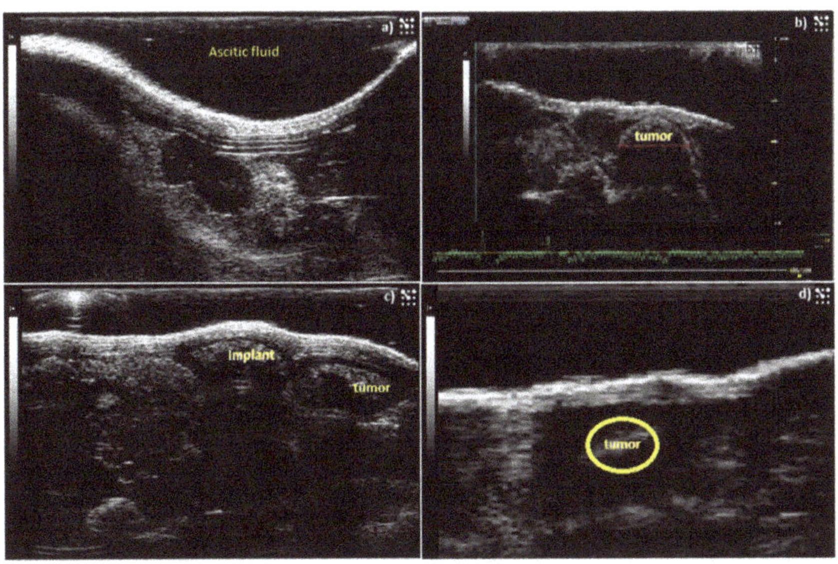

Figure 6. Sonographic representation of IP-tumor growth and response to antibody-bound drug loaded nanomicelle hydrogel composite delivery system (AF(D)NMs). (**a**) Ascitic fluid development in a nude mouse 5 days post-induction; (**b**) Tumor growth 10 days post induction with NIH:OVAR-5 cell suspension; (**c**) Chemotherapeutic implant injected IP-adjacent to tumor growth 11 days post-induction, (**d**) decrease in tumor size and only a small tumor nodule was noticeable 15 days after implementation of the IP-(AF(D)NMs) treatment.

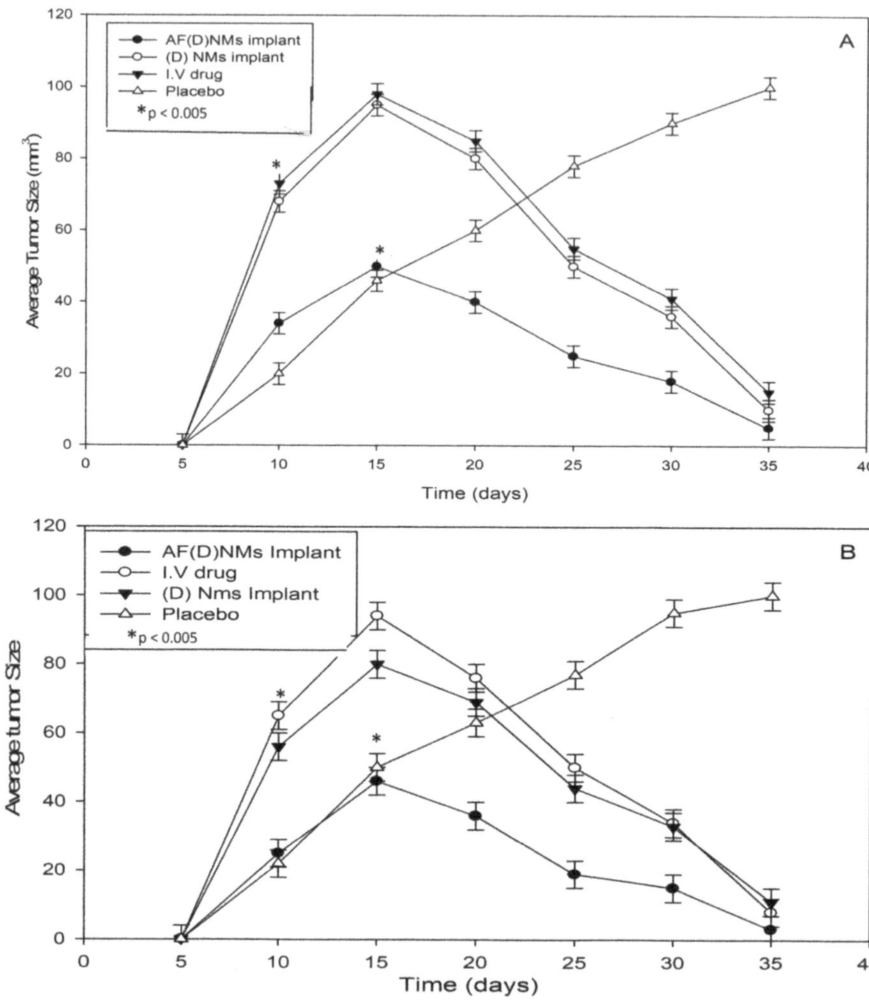

Figure 7. Nude mouse IP-tumor growth curves illustrating chemotherapeutic efficacy of the three treatment groups vs. the control/placebo group, expressed as average tumor sizes in NIH:OVCAR-5 EOC-bearing nude mice. (**A**) refers to the methotrexate and (**B**) the cisplatin-loaded nanomicelle implant delivery system in NIH:OVCAR-5 EOC-bearing nude mice. Each point depicts mean (n = 10/group); bar, ±SD. * $p < 0.005$ by t-test.

2.4.4. Whole Mouse Weight

Mouse weights in both the comparison group (IV cisplatin/methotrexate group) and the 2 experimental treatment groups decreased during this study ($p < 0.05$) (Supplementary Materials Figure S4). Conversely, in the placebo group body weights increased slightly and were normalized to baseline weight ($p < 0.05$). The final average weight of nude mice with NIH:OVCAR-5 tumor treated with AF(D)NMs formulations was 16.94 ± 0.3 g compared with 17.68 ± 0.3 g in mice treated with methotrexate/cisplatin–loaded nanomicelles (D)NMs and 26.34 ± 0.36 g in mice administered with placebo injections ($p < 0.05$). The IV drug group final average weight was 22.62 ± 0.28 g. Baseline weights were 26.00 ± 0.40 g, 24.00 ± 0.33 g and 19.1 ± 0.35 g for placebo, (D)NMs, IV drug and AF(D)NMs, respectively. The body weight of the placebo group was normalized to 26.00 g on day 30, indicative of implant biocompatibility and low levels of cytotoxicity (Supplementary Materials Figure S4).

The notable reduction in tumor size and mouse weights corresponded with decreased MUC16/CA125 antigen expression levels determined with the ELISA technique. These results indicated that there was significant difference ($p > 0.01$) in tumor burden between different chemo-treatment groups and confirmed that antibody functionalized combinational treatment significantly improved chemo-therapeutic efficacy as demonstrated by the inhibition of tumor growth ($p < 0.05$).

2.4.5. Quantification of Plasma and Ascitic Fluid MUC16/CA125 Antigen Levels

Using the Cancer Antigen CA125 Human ELISA Kit, MUC16 antigen concentrations in IV-treated mouse plasma samples were significantly lower ($p < 0.05$) than the pre-treatment group. However, MUC16 antigen concentrations in mice treated with (D)NMs and AF(D)NMs were all in a lower range, i.e., 1.8–2.4 U/mL. The MUC16 antigen concentration in the plasma samples was, typically, slightly more elevated than in the ascitic fluid, due to the AF(D)NMs site-specificity and localization in the peritoneal cavity. AF(D)NMs target MUC16 antigens expressed on ovarian carcinoma cells within tumor nodules and in ascitic fluid. This AF(D)NMs group significantly reduced MUC16/CA125 antigen concentrations in plasma, as compared to (D)NMs group and IV chemotherapeutic drug group ($p < 0.05$) (Figure 8). At the time of euthanasia, the average plasma MUC16/CA125 concentration in the AF(D)NMs group was 1.9460 U/mL, compared to 2.0180 U/mL in the (D)NMs group, 2.077 U/mL in the IV drug group and 2.368 U/mL in the placebo group. The decrease in MUC16/CA125 antigen concentration in ascitic fluid was consistent with the overall reduction in ascitic fluid production and average tumor size. These results indicate that the AF(D)NMs can specifically target MUC16/CA125 antigens on the surface of EOC cells, thereby effectively decreasing their expression.

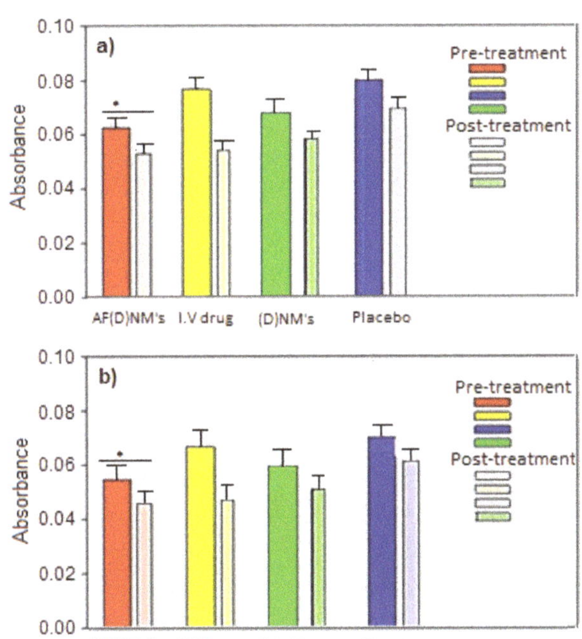

Figure 8. Point-of-euthenasia levels of MUC16/CA125 antigens in the plasma and ascitic fluid of mice in the experimental and control groups. (D) refers to methotrexate in (**a**) and cisplatin in (**b**). Thick and thin bars refer to pre-treatment and post-treatment groups, respectively. *: $p < 0.05$ by t-test.

2.4.6. IP-Tumor Histopathology

Gross necropsies were performed on all the pre-treatment animals that died or were euthanized by carbon dioxide (CO_2) inhalation after 10 days. There was no difference in histomorphological appearance of IP-inoculated tumors. Histopathology was subsequently performed in all IP nodules and in all cases, the existence of anaplastic ovarian carcinoma(s) was confirmed (Figure 9). The intra-abdominal tumors, however, were far more infiltrative (widespread transcoelomic metastases with extensive regions of necrosis and histopathology also revealed tumor emboli within lymphatic vessels). Necrosis was linked with a diffuse inflammatory response, comprising of mononuclear cells (Figure 10b). The inflammatory process on the tumor site is controlled by the neutrophil-infiltration. Macrophages are activated to release TNF-α by neutrophil chemotactic factors and reactive oxygen species resulting in the elimination of tumor cells, however an uncontrolled pattern can promote cell laceration and necrosis. After a month of chemotherapy, the mice in the post-treatment groups were euthanized in the same manner as those in the pretreatment group and full liver histopathology performed.

2.4.7. Liver Histopathology

In contrast, tissues sections of nude mice of the treated group demonstrated minor necrotic regions and a lower number of IP-tumor cells (Figure 9). The reduced necrosis was consistent with the decrease in IP tumor-growth. In this vein, histopathology performed on IP nude mice implanted with experimental IV and placebo treatments displayed multifocally coalescing neoplastic nodules throughout the peritoneal cavity, as well as multiple random foci of hepatocellular coagulative necrosis associated with bile "lakes" (so called "bile infarcts") likely due to biliary outflow obstruction by ovarian carcinomas (in several instances, carcinomatous foci were observed immediately abutting extra-hepatic biliary cysts (Figure 9a,b)). There was also evidence of neoplastic emboli in some sections in the placebo group. There was very mild bile ductile proliferation within the liver sections as well as occasional extra-hepatic biliary cysts (lined by hyperplastic epithelium), the latter also embedded within an increased fibrous connective tissue stroma. Numerous bile ductules within the liver were bile-laden and some bile ductules were severely distended in portal areas and lined by hyperplastic (in certain regions, pseudoepitheliomatous) epithelium. There was a mild portal peri-ductular infiltration of mature (small) lymphocytes and neutrophils and there was evidence of a mild to moderate multifocal portal fibrosis. There was also a venous thrombus in one section of the liver. All these observations are illustrated in Figure 9a,b.

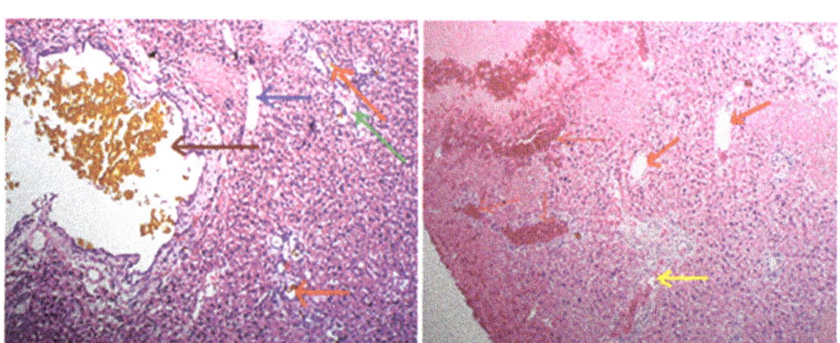

Figure 9. *Cont.*

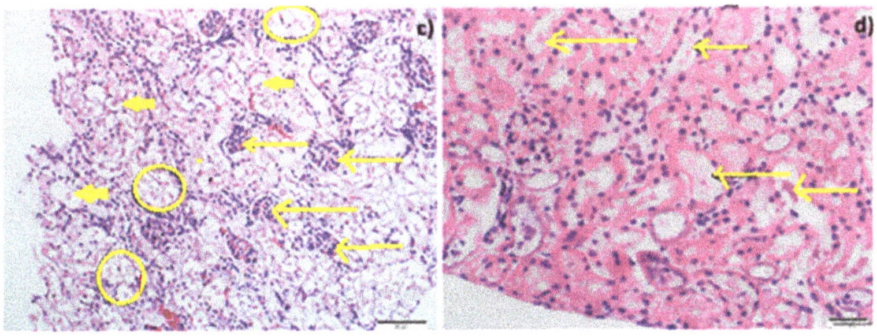

Figure 9. Microscopic view of the liver (**a**,**b**) (scale bar: 100 μm) and kidney (**c**,**d**) (scale bar: 50 μm) in IP-AF(D)NMs, IP-(D)NMs and IV-drug only treatment groups, stained with H and E. (**a**) Cystic distension of bile ductule containing bile (dark brown thick arrows), foci of hepatocellular necrosis (dark blue arrow), mild bile ductule proliferation (red arrows), and mild lymphocytic infiltration into portal areas (green arrow); (**b**) multifocal to coalescing hepatocellular coagulative necrosis with (red thin arrows) or without associated haemorrhage (red thick arrows) which infiltrated with heterophils (yellow arrow); (**c**) severe nephrosis as evidenced by ragged intraluminal clumps of cell debris (yellow rings), inflamed cells (thick short yellow arrows), thin long yellow arrows denote clusters of heterophils HE; and (**d**) basophilic (calcified), granular intra-tubular cellular detritus (yellow arrow), multifocal (segmental) karyolysis in some tubular epithelial cells, and multifocal mild distension of proximal convoluted tubular lumens (lined by slightly attenuated epithelium), HE (Hematoxylin and Eosin staining).

2.4.8. Renal Histopathology

There was widespread but only mild/moderate cortical and medullary intra-tubular cast formation evident as basophilic, granular, and variably vacuolated (derived from degenerate tubular epithelial cells) cellular detritus that filled some tubules. The outer cortex presented with multifocal (patchy) severe nephrosis-lytic necrosis of proximal convoluted tubular epithelial cells, as evidenced by clumps of ragged, mineralized cell debris (due to dystrophic calcification) and occasional karyolitic nuclei in the affected tubular epithelium. In the deeper cortex, there was widespread desquamation/shedding of proximal convoluted tubules (PCT) epithelial cells into tubular lumens, with slight condensation of nuclear chromatin in the affected epithelial cells. There was mild distension of occasional PCT lumens, which were lined by mildly attenuated epithelial cells. All these observations are illustrated in Figure 9c,d.

2.4.9. Immunohistochemistry (IHC)

MUC16/CA125 IHC revealed positive labeling of 1–10% of the area within the specified regions of interest (ROI). Generally, throughout these tumors, variably-sized clusters of neoplastic cells were labeled with the MUC16/CA125 antibody. Labeling was both cellular (cytoplasmic and cell membrane) and extracellular (around shrunken apoptotic-like cells and in tubular lumina lined by irregularly branching papillae of neoplastic epithelial cells). There was occasional distinct membranous labeling of epithelial cells lining the papillary projections in some sections. However, most labeling appeared extracellular in most tumor sections, and there was more cytoplasmic labeling of neoplastic cells compared to membranous labeling (Figure 10a–c).

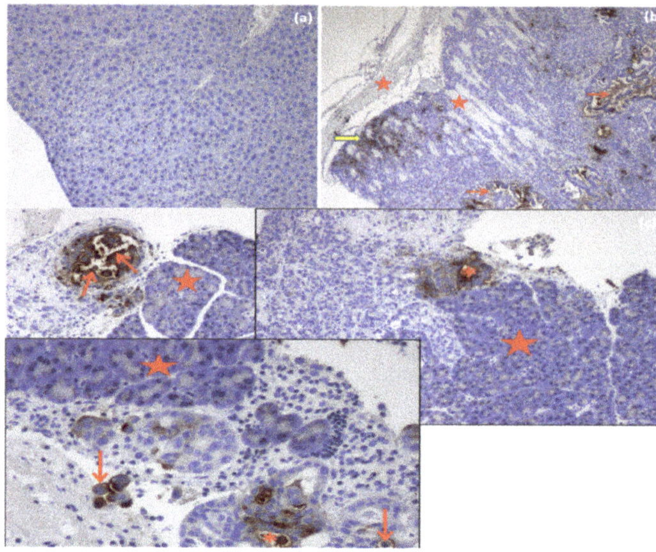

Figure 10. MUC16/CA125-positive labeling of EOC cells in mouse tissues and tumor foci in IP-inoculated nude mice (MUC 16/CA 125, scale bar: 50 µm): (**a**) Negative tissue control (normal mouse liver). MUC16/CA125 immunolabeling, DAB chromogen, hematoxylin counterstain; (**b**) IP-ovarian carcinoma infiltrating skeletal myofibrils (stars) of the body wall. EOC cell-associated (yellow arrow) and extracellular (red arrows) MUC16-positive labeling. MUC16/CA125 immunolabeling, DAB chromogen, hematoxylin counterstain; (**c**) IP-inoculated ovarian carcinoma infiltrating the pancreas (stars) with MUC16-positive labeling of cytoplasmic membranes (arrows) and extracellular (* asterisks) labeling of neoplastic cells. MUC16/CA125 immunolabeling, DAB chromogen, hematoxylin counterstain.

In addition, the MTX IP delivery by the polymeric ISFI was intracellularly taken up by OVCAR 5 tumor cells via endocytosis. The intracellular-MTX was released by a lysosomal route and distributed to the cytosol where it attached to its dihydrofolate reductase target enzyme and generated a robust S-phase portion. This modification in ovar5 cell cycle provided the anti-proliferative effect in vivo, demonstrated by the regression in tumor volume.

2.4.10. MUC16/CA125 IHC Analysis on Formalin-Fixed, Paraffin-Embedded (FFPE) Epithelial Ovarian Cancer (EOC) Tissue Sections

The percentage of MUC16-positive labeling per square centimeter of each selected EOC tissue section (one per mouse) was determined with the aid of the phase separation function for both pre and post-treatment mice. Positive labeling was identified as being brown in color and was observed in both cellular (cytoplasmic and cell membrane) and extracellular locations (in tubular lumina) throughout the neoplastic foci (Figure 11A). MUC16 expression in pre and post-treatment groups using both drugs (methotrexate and cisplatin) was up-regulated and displayed an increasing trend in the placebo (4.31–5.11%), AF(D)NMs (6.13–6.36%) and IV-drug groups (7.67–8.14%) ($p < 0.05$), whilst in the NMs (D) group, MUC16 expression was down-regulated (2.67–3.3%) (Figure 11).

AF(D)NMs reduced CA125 antigens, thereby effectively decreasing their expression. This was validated in quantification of plasma and ascitic fluid MUC16/CA125 antigen levels using the Cancer Antigen CA125 Human ELISA Kit (Figure 8). MUC16/CA125 IHC analysis on FFPE EOC tissue sections (as shown Figure 11B) proved not be a satisfactory tool in differentiation of MUC16 expression levels from pre and post-treatment groups, due to the inaccuracy, ambiguity, and superficial nature of this technique.

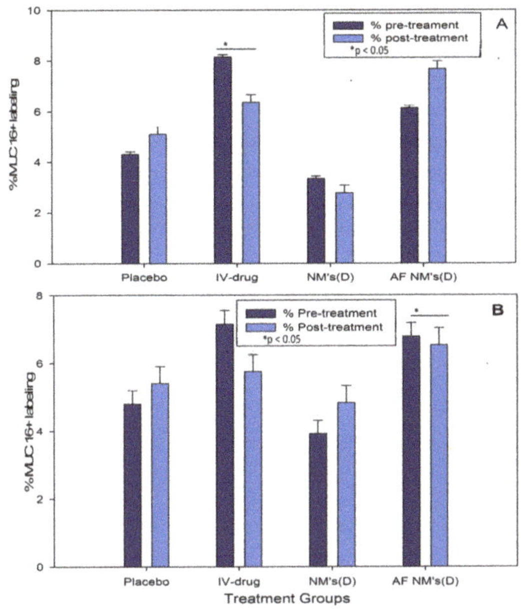

Figure 11. ROI (regions of interest) measurement of IHC images on the stained slides for each treatment group; the MUC16 density in each image was calculated as percentage of MUC16-positive labeling per square centimeter of each selected EOC tissue section (one per mouse). (D) refers to methotrexate (**A**) and cisplatin (**B**). * $p < 0.05$ by *t*-test.

The survival rate in response to various chemotherapeutic protocols was a significant index for comparing antitumor efficacy between the groups. Nude mice survival rates were significantly different ($p < 0.05$) between the experimental treatment and control/placebo and comparison (IV chemotherapy) groups, as shown by means of Kaplan-Meier analysis. Mouse survival was significantly improved (100% over 35 days) in the AF(D)NMs test group compared to the placebo group, as well as all the other groups (Figure 12). The data also indicated that the mice in the AF(D)NMs test group exhibited the greatest overall reduction in tumor size and had the longest survival times.

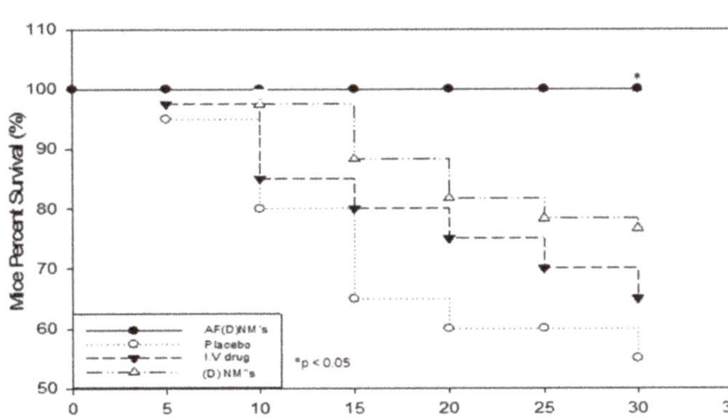

Figure 12. Kaplan-Meier mouse survival curves, showing the three chemo-treatment groups ($n = 10$) including nude mice injected with IP-AF(D)NM's, IP-(D)NM's and IV chemotherapeutic drugs and an IP-control placebo implant treatment group. The survival rate of mice in the IP-AF(D)NM's test group was significantly improved compared with the other groups. * $p < 0.05$.

2.4.11. Mechanism of Intraperitoneal ISFI Delivery for Human Ovarian Carcinoma Targeting

The ISFI was fabricated by encapsulating a nanocomposite comprising of anti-MUC16 (antibody) functionalized methotrexate (MTX)-loaded PNIPAAm-b-PASP nanomicelles (AF(MTX)NM's) within a C–P–N hydrogel. Given that the peritoneal cavity is the principal site of disease in ovarian cancer, the ISFI was injected and released nanocomposites into the peritoneal cavity. Following the release of nanocomposites from the hydrogel, the nanomicelles (formulated to circulate for prolonged periods in the peritoneal fluid) targeted specific mucin antigens significantly over-expressed on ovarian cancer cells circulating in the peritoneal fluid (when patients are usually diagnosed) and cancer cells forming nodules at distant sites in the peritoneal cavity. This targeting system reduced the tumor load responsible for adhesion at the sites of secondary metastasis (peritoneal and abdominal surfaces). The anti-MUC16 antibody conjugated nanomicelles has great potential in improvement of tumor selectivity, eliminating/reducing the tumor load, while improving the recovery and long term survival rate of most patients suffering from ovarian cancer.

3. Discussion

Ovarian cancer is an insidious female disease that is asymptomatic during its early stages and is usually diagnosed at an advanced, often untreatable stage when the disease has spread beyond the ovaries throughout the abdominal cavity and even further afield [1–3]. Thus, deaths due to ovarian cancer could be significantly lowered by developing new, ultrasensitive, yet reliable methods for early diagnosis, and by developing improved treatment protocols. Mucins are among the most promising molecular biomarkers for EOC and have proven invaluable in the diagnosis and monitoring of treatment regimes in various types of ovarian cancer. Increased mucin expression, otherwise known as "mucin switching" that occurs during the neoplastic transformation of ovarian surface epithelium to ovarian carcinoma is important in the progression of this disease [26–28]. MUC16 is an important clinical biomarker of ovarian cancer and it is a target for various immuno-chemotherapies currently under investigation [29–31].

Blood, serum, plasma, and ascitic fluid levels of MUC16 in mice and humans have been well-researched and are now commonly used as diagnostic markers when evaluating the effect of various chemo-treatments and for establishing the advent of relapse in EOC [32,33]. In this study, quantification of MUC16/CA125 concentration was conducted utilizing the highly sensitive Cancer Antigen 125 (CA125) Human ELISA Kit with a date-to-date coefficient of sensitivity >5%. Progressively increasing MUC16/CA125 values are correlated with ovarian malignancy, whilst steady MUC/CA125 values, even when raised, are correlated with some benign conditions. Significantly, in the pre- vs. post-treatment groups of mice with induced EOC, the MUC16 antigen levels in the blood were 2.368 U/mL in the placebo group, compared to 1.9460 U/mL in the (AF(D)NMs) post-treatment group (Figure 11). Our findings also showed that plasma MUC16 antigen levels were consistent with tumor growth, ascitic fluid development and mouse survival with distinct and combination chemo-treatments, indicative of MUC16 as a valuable biomarker to investigate the effect of chemo-therapeutic interventions on EOC in Athymic Swiss mice.

Histopathology was subsequently performed on all IP nodules and in all cases, the existence of anaplastic ovarian carcinoma(s) was confirmed histologically (Figure 10). Cancer linked inflammation response includes white blood cells infiltration, notably phagocytic macrophage cells, the occurrence of polypeptide representatives of inflammation (as tumor necrosis feature, interleukin-1 and chemotatic chemokines) and the presents of tissue alteration and angiogenesis [34]. This flow of episodes presents an uncontrolled pattern that can promote cell laceration and necrosis. Our findings revealed foci of coagulative necrosis but necrotic foci were most often associated with cholestasis and bile infarcts thought to be due to obstruction to bile outflow by tumor nodules. In prior studies, additional toxic side effects that were observed in association with the use of methotrexate/cisplatin included chronic interstitial, obstructive pulmonary disease, severe renal outer cortex nephrosis [35–37]. Immunohistochemical analysis of epithelial ovarian tumors in the present study confirmed an apparent over-expression MUC16 in all induced tumors. Significantly, cellular labeling of MUC16 was detected both in cytoplasmic granular and/or (to a lesser extent) cytoplasmic membrane labeling of neoplastic cells. Our results showed that MUC16 is up-regulated in most of the tumor tissue samples analyzed (Figure 11). As has been shown in previous studies, we also detected a down-regulation of MUC16 expression in ovarian tumors post-treatment as compared to the pretreatment stage. At euthanasia, MUC16 antigen expression in the ovarian carcinoma tissues significantly decreased post-treatment with the AF(D)NMs and only a few small areas of cell debris stained positive in these sections. MUC16/CA125 IHC analysis on FFPE EOC tissue sections (as shown Figure 11B) proved not be an invaluable tool in differentiation of MUC16 expression levels from pre and post-treatment groups, due to the inaccuracy, ambiguity, and superficial nature of this technique. The improved survival rate associated with the AF(D)NMs treatment was probably at least partly due to the profound decrease in ascitic fluid formation in this group ($p < 0.01$). The reduction in ascitic fluid probably occurred due to the specific and prolonged retention of AF(D)NMs hydrogels intra-peritoneally close to the tumor burden; intra-peritoneal tumor cells had close and prolonged exposure to the chemotherapeutic drugs, no doubt resulting in increased drug efficacy in situ. The novel aspect of the study was the design and implementation of an antibody-functionalized nanomicelle hydrogel composite to specifically target MUC16 antigens known to be over-expressed on ovarian cancer cells. Following the release of antibody-bound nanomicelle from the hydrogel, said nanomicelle targeted specific MUC16 antigens over-expressed on ovarian carcinoma cells within the ascitic fluid and within neoplastic nodules at distant sites within the peritoneal cavity. This MUC16 targeting system reduced the mass of neoplastic cells capable of adhering to parietal and visceral peritoneal surfaces at sites distant to the original tumor implant. Chemo-treatment with AF(D)NMs significantly inhibited growth of and transcoelomic metastasis of EOC (via reducing MUC16 antigen expression on tumor xenografts) and reduced the production of ascitic fluid, thereby possessing potential for increasing longevity in patients with ovarian carcinoma.

4. Materials and Methods

4.1. Materials

Natural and synthetic biopolymers such as Chitosan (MW = 600.000) and Poly(*N*-vinylpyrrolidone) (MW = 40.000) were purchased from Sigma-Aldrich (Steinhelm, Germany). *N*-isopropylacrylamide (MW = 113), poly(*N*-vinylpyrrolidone) (MW = 40.000), *N*,*N*-methylenebisacrylamide, glutaraldehyde (GA) (25%), 2,2′-azoisobutyronitrile, ammonium persulfate, α-amino Aspartic acid, ethanethiol hydrochloride (AET-HCl), 98% analytical grade triphosgene, methotrexate (MTX), 3.5 mg/mL penicillin, 0.1 g/L streptomycin, RPMI 1640 medium, 10% FBS and 0.25% trypsine, and 0.03% EDTA solution were procured from Sigma Aldrich (St. Louis, MO, USA). 98% DMSO, 99.8% anhydrous *N*,*N*′-dimethylformamide (DMF), 99.0 anhydrous tetrahydrofuran (THF), 98.0% diethyl ether, 75% analytical petroleum ether bp (30–60 °C), 1 M hydrochloric acid, potassium chloride (KI), disodium hydrogen phosphate (Na_2HPO_4), and potassium dihydrogen phosphate (KH_2PO_4) were from Merck Chemicals (Pty) Ltd. (Darmstadt, Germany). NIH:OVCAR-5 cancer cells were procured from director of Fox Chase Cancer Institute, Philadelphia, PA, USA (Dr. Tom C. Hamilton). RayBio® Human CA-125 (MUC16) Elisa Kit for serum, plasma, cell culture media, and urine 96-wells filter plates were procured from Biocom Biotech in (Centurion, Pretoria, South Africa). Anti-MUC16 antibody [OC125] ab693 was procured from Abcam Inc. (Cambridge, MA, USA). Purified deionized water was prepared by a Milli-Q System (Millipore Co., Billerica, MA, USA). Culture plates were purchased from Corning Inc. (Corning, NY, USA). Ovarian cancer cells were cultured in a RS Biotechnological Galaxy (Irvine, UK) incubator kept at 37 °C with 5% carbon dioxide in an entirely humidified environment. Cell experimentations were conducted in the logarithmic stage of development.

4.2. Anti-MUC16 Functionalized MTX-Loaded NanoComposite Preparation

4.2.1. Synthesis of PNIPAAm-b-PAS for NanoComposite Formulation

Firstly, a copolymer comprising PNIPAAm-b-PASP was synthesized by solvent evaporation as described previously [38,39]. Thereafter the PNIPAAm-b-PASP copolymer (0.125 g, 0.0079 mmol), anti-MUC16 antibody (0.2 mL, 0.022 mmol), NHS (0.60 mg, 0.0522 mmol), and DCC (10.8 mg, 0.0522 mmol) were dissolved in 10 mL DMF. The solution was mixed under inert conditions at room temperature in the dark for 14 h before dilution with 25 mL deionized water followed by centrifugation to extract DCU. The aqueous supernatant was further extracted by membrane dialysis against deionized water for 24 h with subsequent lyophilization.

4.2.2. Preparation of the MTX-Loaded NanoComposite

MTX-loaded nanomicelles were prepared by solvent evaporation as reported previously [31]. Briefly, 7 mg MTX and 25 mg of PNIPAAm-b-PASP were dissolved in 5 mL DMF and 20 mL distilled water was added drop wise to the solution under homogenization at 360 rpm for 5 min at room temperature. This was followed by solvent removal using a rotary vacuum evaporator to obtain a nanomicelle solution that was then filtrated through a 0.2 μm filter membrane to remove any residual MTX with subsequent lyophilization. The anti-MUC16 antibody functionalized MTX-loaded nanomicelles were prepared with modifications by introducing the anti-MUC16 functionalized block copolymer [40,41].

4.2.3. Evaluation of the Molecular Structural Integrity of the Functionalized Nanomicelles

Fourier Transform Infrared (FTIR) spectroscopy (Perkin Elmer Life spectroscopy 100, Lantrisant, Wales, UK) was utilized to differentiate the chemical structure of the block copolymer NanoComposite following nanomicelle coating with the anti-MUC16 antibody, employing a oscillating vibrational component with a diamond interior gauge. Evaluation of Pure MTX, blank nanomicelles, MTX-loaded nanomicelles and the MTX-loaded NanoComposite was conducted at an interval of 650 and 4000 cm^{-1} wavenumber with a 4 cm^{-1} resolution value and smoother 64 scans per spectrum.

4.2.4. Differential Scanning Calorimetry for Elucidation of the Thermal Events of the Methotrexate-Loaded Nanomicelles

Differential Scanning Calorimetry (DSC) studies were conducted with a Mettler Toledo advanced DSC1-STARe equipment from Schwerzenback in Switzerland. The Mettler-STARe software program (version 9, Schwerzenback, Switzerland) was utilized for DSC results acquisition and interpretation. Blank specimen pans were used as reference points and experimental scans were conducted by heating samples at intervals of −10 to 25 degrees for fifteen minutes with a continual isotherm. The lyophilized powder specimens (10 mg) were placed into DSC aluminium-pans and compress-sealed. Relative DSC scans were conducted on blank micelles, MTX-loaded micelles and pure MTX at a heating rate of 10 °C/min from −10 to 325 °C in a stable stream of nitrogen gas. Fresh sample analysis was conducted in triplicate for method development and validation purposes. DSC thermogram patterns were then analyzed for variations in thermal periods. The phase changes of the drug-free micelle and MTX were correlated with the phase changes of the MTX-loaded PNIPAAm-block-PAsp nanocomposites.

4.3. Chitosan-PVP-PNIPAAm (C–P–N) Hydrogel Synthesis

Interpenetrating polymer networks (IPN's) were fabricated by free-radical polymeri-zation mechanism of *N*-isopropylacrylamide distinctive monomers in excipient chitosan and water-soluble poly(*N*-vinylpyrrolidone) (PVP). PVP, *N*-isopropylacrylamide NIPAAm, crosslinkers-*N*,*N*′-methylenebis(acrylamide)-MBAAm, and glutaraldehyde dissolved in aqueous chitosan-CHT (0.02 L, 2%, 1.6% CH_2COOH-acetic acid), then stirred. CHT, PVP, NIPAAm employed were in 2:2:45 mass percentage. The percentages (w/w %) between glutaraldehyde-OHC(CH2)$_3$HO and CHT were 1%, 2%, 4%. 4% weight percentage existed between MBAAm with NIPAAm *N,N,N′,N′*-Tetramethylethylenediamine (TEMED)/ammonium persulfate (APS) solution was utilized as initiator in the cross-linking process of NIPAAm. Thereafter, APS-TEMED solution (24 μL, 4%) was subsequently added to this reaction mixture. A cross-linked Chitosan-PVP-PNIPAAm (C–P–N) hydrogel formed within 1 h of polymerization reaction at 25 °C. After 24 h, the IPN hydrogels were cleaned with deionized H_2O, then air/vacuum dried. The hydrogel samples prepared were called F_1 = C–P–N/1, F_2 = C–P–N/2, and F_3 = C–P–N/3 IPNs.

4.3.1. Nuclear Magnetic Resonance (NMR) Spectroscopic Analysis

I^{1}H-NMR measurements were applied to the C–P–N components for the confirmation of the copolymer structure and composition. For ^{1}H-NMR measurement, 5 mg sample vacuum dried at 50 °C for 48 h was added into a 5 mm NMR test tube, and further vacuum dried at 50 °C for 48 h, to which 500 μL D_2O solvent was introduced, and lastly the test tube was vortexed to dissolve the polymer in solution. The NMR spectrum was generated using a Bruker DRX400 spectrometer (Bruker, Germany).

4.3.2. Determination of Polymeric Structural Variations

Molecular structural changes in the polymer backbone may alter the inherent chain stability and therefore affect the physicochemical and physicomechanical properties of the selected polymer type for the intended purpose. The molecular structure of native polymers (CHT, PVP, PNIPAAm) the non-cross-linked, and cross-linked C–P–N hydrogel, blank micelles, and drug-loaded micelles were analyzed using FTIR spectroscopy to elucidate any variations in vibrational frequencies and subsequent polymeric structure resulting from drug-co-polymer interaction during nanostructure and hydrogel formation. Samples were analyzed in triplicate at high resolution with scans ranging from 4000 to 400 cm^{-1} on a PerkinElmer Spectrum 100 Series FTIR spectrometer coupled with Spectrum FTIR research grade software (Perkin Elmer Life and Analytical Sciences Inc., Hopkinton, MA, USA).

4.4. Preparation of Bio-Responsive IPN Nanomicelle/Hydrogel Composite Based Implant (ISFI)

The ISFI was fabricated by encapsulating antibody functionalized nanocomposite comprising anti-MUC16 (antibody) functionalized methotrexate (MTX)-loaded PNIPAAm-b-PASP nanomicelles within the synthesized bio-responsive C–P–N hydrogel followed by freeze drying. Anti-MUC16 functionalized nanocomposites were encapsulated into the C–P–N at a proportion of 1:5 (NM's: C–P–N hydrogel). The anti-MUC16 functionalized nanocomposites in solution were introduced drop-wise into the C–P–N gel, and syntheses were permitted to stir until a uniform mixture was reached with a mechanical vortexing system at 37 °C.

4.4.1. Determination of the Gelation Temperature of the Polymeric ISFI Utilizing Oscillatory Rheology

Dynamic rheology is one of the most extensive methods to study rheological properties of polymer hydrogels, and is also the most direct and reliable way for the determination of sol–gel transitions. Viscoelastic properties were measured with a Modular Advanced Rheometer system (ThermoHaake MARS Rheometer, Thermo Fischer Scientific, Karlsuhe, Germany) to indicate the storage modulus (G′), the loss modulus (G″), and tan δ of an aqueous solution of the sol, using cone plate geometry, where the G′ and G″ were recorded under constant deformation. The study of the flow properties was considered extremely significant to this study as it is central to the mechanisms by which the ISFI functions. At room temperature, the implant remains in the liquid state to allow delivery via an 18 G needle and at body temperature (37 °C) the implant forms a solid-like structure. To characterize and analyze the flow behavior and determine the lowest critical solution temperature (LCST) or gelation temperature of the ISFI formulations, rheology studies were conducted using a Haake Modular Advanced Rheometer System (ThermoFisher Scientific, Karlsruhe, Germany). As the polymeric material acts as a visco-elastic solid, some background on viscoelastic solids is provided. To determine the lower critical solution temperature (LCST) and hence the gelation temperature of the ISFI formulation, the temperature of samples was ramped from 20–50 °C at a rate of 0.25 °C/min while applying the predetermined stress of 13.4 Pa, at the frequency of 10 Hz. The gelation temperature was determined as the temperature at which the cross-over of G′ and G″ occurred, i.e., the point at which the formulation was no longer acting as a liquid (G″), but as a solid (G′). In all cases, a solvent trap was used to prevent sample evaporation.

4.4.2. Morphological Characterization of the ISFI

Peripheral morphology, the surface area, and cross sectional area of the ISFI post-lyophilization were observed by using scanning electron microscopy—SEM (JEOL Ltd., JSM-Japanese-Electronic-Optical-Laboratories, Tokyo, Japan). Microscopic observations of the exterior-surface of the nanocomposite configuration of the C–P–N hydrogel was conducted, by first freeze drying the hydrogel at 25 mTorr pressure (Virtis™-Gardiner, New York, NY, USA). The specimen was attached to dual-sided adhesive tape adhered to a metallic specimen stand and sputter-covered with a film of gilded gold. Each specimen was observed under different increasing magnifications at 20 keV-an accelerating-voltage.

4.5. In Vivo Studies in the Swiss Athymic-Nude Mice Model

4.5.1. Mouse Model

Four- to six-week-old female Swiss Athymic nude mice (purchased from Charles River Laboratories International Inc., Ecully, France) were housed in a specific pathogen free (SPF) facility at 25 °C with 60–70% relative humidity (RH) with a 12-h light/dark rotation and then given a week of acclimatization before the experiments commenced. Mice were fed ad libitum diet and water. All procedures were performed under sterile conditions in a laminar flow hood. The animals were monitored daily for general health status. Animal ethics for in vivo studies was approved by University of the Witwatersrand, Animal Ethics Screening Committee (AESC Number: 2012/46/05; Dated 5 December 2012) and the study protocol adhered to these guidelines and those of the South African Council on Animal Care.

4.5.2. In Vitro Cell Culture

The NIH:OVCAR-5 ovarian cancer (OC) cell line is a developed human ovarian epithelial carcinoma cell line that transmits the MUC16 autoantigen [26]. These NIH:OVCAR-5 cells were procured from the director of the Fox Chase Cancer Facility, USA-Jenkintown-CA (T. C. Hamilton). The cells were developed to 80% confluency and attached in plastic culture containers in RPMI growth medium augmented with 10% heat-inactivated FBS, 2 mM glutamine, 50 units/mL penicillin/streptomycin, and 50 units/mL insulin. OC cells were cultured in a RS Biotechnological Galaxy (Irvine, UK) incubator kept at 37 °C with 5% carbon dioxide in an entirely humidified environment. OC cells were weekly sub-cultivated, collected by dissociation of adherent cells with trypsin, twofold cleaned in 10% PBS, dead cells tainted with sterile Trypan Blue and enumerated. Cell culture experiments were performed after 1–2 sub-cultivation rotations. An aqueous 0.4% *w/v* trypan blue in sterile 10%, 0.1 M, pH 7.4 PBS was formulated for cell enumeration. A sterile hemocytometer grid was the tool employed for enumeration and a 20 µL cell mass and 60 µL trypan blue volume were collectively mixed. The ratio of live cells was computed as a determinant of the amount of cells enumerated per total number of grid quadrants counted as depicted in % live cell Equation (1). This was 1:3 proportion which generated a 4 dilution-factor.

$$\%\ Live\ cells = \frac{\text{No of cells counted}}{No\ of\ grid\ quadrants\ counted} \times DG \times 10^4\ \text{cells/mL} \quad (1)$$

where DG is the dilution-factor utilized and 10^4 is equation constant. Only specimens that demonstrated viability of more than 95% were used in following analysis.

4.5.3. Induction of Human Ovarian Carcinoma in Swiss Athymic Nude Mice—Pre-Treatment Phase

The ovarian carcinoma induction stage of this study was referred to as the pre-treatment phase. All experiments were conducted inside an Esco Frontier™ DuoFume Hood in the specific pathogen free (SPF) facility. In the present, the model was generated by injecting female Athymic 4- to 6-week-old nude mice IP with 2 × 10^8 cells/mL of NIH:OVCAR-5 cancer cell suspensions in 0.2 μL RPMI. Inoculations were performed using a 26 G gauge needle and a 1 mL syringe [26]. Human ovarian carcinomas (NIH:OVCAR-5) were induced within 10 days of inoculation. As NIH:OVCAR-5 mainly proliferated within the peritoneal cavity and over-expressed mucin 16 (MUC16/CA125) as compared to normal tissue, it was concluded that the IP growth mimicked human clinical disease. All tumor-bearing mice were further monitored until the tumors developed to their target size of 80–100 mm^3, which was regarded as the baseline-for initiation of treatment [27].

4.5.4. IP-Inoculated Mice

Mice were inspected weekly and tumor development with peritoneal carcinomatosis in the IP-inoculated mice was monitored based on overall health using the following clinical positive indicators: Presence of intra-abdominal nodular growths, distension of the abdominal cavity due to ascites, weakness, weight loss with extensive skin tenting due to dehydration, and changes in behavior due to perceived pain (Figure 13a). One of the modalities used to assess IP tumor development was a Vevo® 2100 imaging system (Visualsonics Inc., Toronto, ON, Canada), which was used to detect early tumor development (Figure 13b,c). Mice were anaesthetized with Ketamine and Isoflurane before scanning for tumors using the Vevo® 2100 Imager. Tumor-bearing mice were scanned from day 1 PI-post-inoculation (at the site of induction) until day 10 PI and thereafter mice were scanned every second day after implantation of the AF(D)NMs treatment, to evaluate treatment efficacy over the course of therapy (Figure 13b,c).

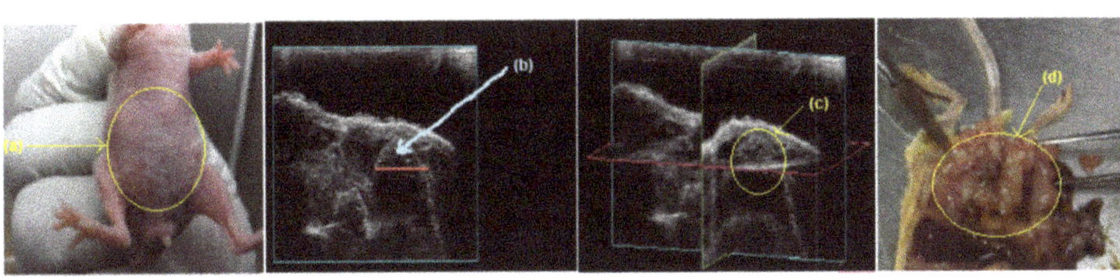

Figure 13. (**a**) An IP-inoculated nude mouse with a moderately distended abdomen due to tumor growth and IP tumor dissemination and ascites; (**b**,**c**) ultrasound images taken using the Vevo® 2100 Imager indicating a tumor diameter of 100 mm^3 and a 3D image of the circumference of the tumor, respectively; and (**d**) multiple coalescing tumor nodules disseminated throughout the peritoneal cavity (peritoneal carcinomatosis) as seen during the gross necropsy examination.

4.5.5. Experimental Design

Nude mice with growing tumors 10 days post-inoculation were then randomly allocated to five experimental groups including one pre-treatment group (ovarian carcinoma induction group), two experimental groups (a drug loaded nanomicelle ((D)NMs) implant and an anti-MUC16/CA125 functionalized drug loaded nanomicelle (AF(D)NMs) implant group), a comparison group (I.V drug only) and a placebo control group /blank nanomicelle implant group of 10 mice each (n = 10/group). Tumor-bearing mice in the pre-treatment group were euthanized by carbon dioxide (CO_2) inhalation after 10 days once the tumors have developed to their target size of 80–100 mm^3, which was regarded as the baseline-for initiation of treatment. Organs including the liver, intestines, omentum, mesentery, lungs, pancreas, uterus, oviduct and ovaries, kidney, body wall (skeletal muscle lined by parietal peritoneum), and spleen, as well as solid tumors were collected and ascitic fluid was stored in heparinized tubes (Improve® Improvacuter® Lithium Heparin collection tubes, GmbH, Hamburg, Germany) at −80 °C. The 5 mm^3 tissue and organ samples were preserved in plastic containers filled 10% neutral buffered formalin for up to 5 days, afterwards the formalin-fixed (FF) samples were submitted to the Section of Pathology (SP), Department of Paraclinical Sciences (DPS), Faculty of Veterinary Science (FVS), University of Pretoria (UP), Onderstepoort (OP) for histopathology, and MUC16/CA125 immunohistochemistry/IHC. The tumor-bearing mice in the 2 experimental, as well as the control and placebo groups, were further subjected to chemotherapeutic efficacy studies. Survival curves were calculated utilizing the Kaplan-Meier method. Survival rate was evaluated as the sum of the days conceded between the introduction of a chemo-therapeutic intervention and mice euthanasia, and percentage (%) nude mouse survival was mice still alive in test groups following introduction of chemo-treatment.

4.5.6. Chemotherapeutic Efficacy Studies in EOC-Inoculated Nude (NU/NU) Mice

Different chemotherapeutic drugs and formulations were used in each of the two experimental groups/models. The two chemotherapeutic drugs utilized in this study were cisplatin (4 mg/kg) and methotrexate (15 mg/kg). In intervention study 1, anti-mucin 16/CA125antibody-functionalized cisplatin-loaded PEG-PBLG-PF68 nanomicelle implants were utilized, whereas in intervention study 2, anti-mucin 16/CA125antibody-functionalized methotrexate-loaded PNIPAAm-b-PASP nanomicelle implants were used for the targeted treatment of ovarian carcinoma in mice. PEG-PBLG-PF68, PNIPAAm-b-PASP copolymers were utilized in the formulation of the drug-loaded nanomicelle and further encapsulated in hydrogel based implants. These implants were subsequently injected into the mice at a volume of 0.2 mL—(4 mg/kg for cisplatin and 15 mg/kg for Methotrexate) directly into a palpable tumor mass within the peritoneal cavity. Each mouse in the experimental group received a once-off implant treatment monitored for a period of 30 days whilst the comparison group (IV drug only) was administered treatment at 11-day intervals over the period of a month. Mice in all 4 groups were routinely weighed and tumor size was determined sonographically every five days. Mouse masses were consistently correlated with masses at first day of therapeutic administration, to ascertain the ratio mass variation. After a month of treatment, the mice in the post-treatment groups were euthanized in the same manner as those in the pretreatment group. There was no difference in post mortal sample collection post-treatment compared to pre-treatment.

4.5.7. Quantification of MUC16/CA125 Levels in Plasma and Ascitic Fluid

Quantification of the MUC16/CA125 antigen in plasma and ascitic fluid was performed by means of the Cancer Antigen CA125 Human ELISA Kit (Code No. ab108653, Abcam, Cambridge, MA, USA), which is based on the solid-phase assay system. The minimum measurable concentration of mucin 16/CA125 in this Elisa assay was 5 units per mL. For Elisa-analysis, ascitic fluid and whole blood (obtained from cardiac punctures) were collected in heparinized tubes for plasma (Improve® Improvacuter® Lithium Heparin collection tubes, GmbH, Hamburg, Germany). The tubes were centrifuged at 3000 rpm for 5 min to separate the cells from the plasma using a desktop centrifuge (Model TD5A-WS, Shanghai Luxiangyi Centrifuge Instrument Co., Ltd., Shanghai, China). The selected Elisa employs a mouse monoclonal anti-MUC16/CA125 antibody (Mab) against a distinctive antigenic determinant on the intact MUC16/CA125 molecule. The Mab was utilized for solid phase assay immobilization (on the microtiter-wells). Following incubation at 37 °C for 90 min, the microtiter-wells were rinsed with Wash-Buffer to clear unbound-labeled-antibodies. Tetramethylbenzidine (TMB) reagent solution was introduced and incubated for 20 min, resulting in the formation of a blue color. The color-formation was quenched with the introduction of Stop Solution transforming the color to yellow. The concentration of CA125 was directly proportional to the color intensity of the test sample. Absorbance was measured spectrophotometrically at 450 nm.

4.5.8. Histopathology and IHC

Formalin-fixed, paraffin-embedded (FFPE) organs from pre and post-treatment mice (of both cisplatin-and methotrexate-treated nude mouse groups) and IP ovarian carcinomas were sectioned at 3–4 μm and routinely stained with Haematoxylin and Eosin (H&E). Additional 3–4 μm-thick sections were submitted for IHC (specifically the immunoperoxidase labeling technique) to detect membrane-bound and extracellular/shed MUC16. Immunohistochemistry was performed, by hand, following validated protocols. The standard immunoperoxidase procedure for the detection of MUC16 included deparaffinization and hydration of slides, incubation with 3% hydrogen peroxide in methanol for 15 min to quench endogenous peroxidase activity, heat-induced epitope retrieval/HIER (in a microwave using citrate buffer, pH of 6.0 for 14 min at 96 °C), followed by non-specific immunoglobulin binding (Goat blood serum (code nmber G9023 from Sigma Chemical Corporation, St. Louis, MO, USA), PBS diluted to 1:5 factor, pH 7.6, 0.1% BSA for 20 min in 25 °C humidified chamber), incubation for 40 min with the mouse monoclonal anti-MUC16 (Anti-MUC16 antibody [X325] (code no. ab10033, Abcam, Cambridge, UK)) antibody (diluted 1:50 in buffer (0.1 N PBS, pH 7.6, 0.1% BSA)), with subsequent application of the Envision Polymer Detection System (DakoREALEnVision HRP Rabbit/Mouse (ENV) (reference no. K5007, DakoCytomation, Glostrup, Denmark)) according to the manufacturer's instructions. The reaction product was developed by incubating the tissue sections with a liquid 3,3′-diaminobenzidine (DAB) substrate/chromogen (included in the Envision Immunodetection System) for 1–2 min. Thereafter the portions were counter stainted with Lilly Mayer's hematoxylin for 20 s, rinsed with water for 10 min, routinely dehydrated through increasing alcohol concentrations and xylol, mounted using Entellan™ (Code No. 1076, Merck Millipore, Darmastadt, Germany) and cover slipped for screening using an Olympus BX43 light microscope. Positive-tissue controls included sections of mouse-inoculated human ovarian carcinoma (NIH:OVCAR-5 cell line) and negative-mouse tissue controls included normal spleen, liver, omental and mesenteric fat, uterus, oviduct, ovary, and pancreas. For negative reagent control purposes, buffer[c] was substituted for the primary antibody.

4.5.9. Immunohistochemical Quantification of MUC16/CA125 Antigens in FFPE Tissue Sections

All IHC-labeled tissue slides (one slide per mouse) were scanned utilizing the Olympus dotSlide scanner (VS120-S6-W slide loader system in the Department of Anatomical Pathology at the Medical School, University of the Witwatersrand) for the generation of virtual slide images. The single neoplastic nodule with the most MUC16-specific positive labeling (as assessed with the naked eye) per slide was selected for quantification. The dimension count and measure module from the corresponding Olympus-Cell Sens software version 1.12 (Wirsam-Scientific and Precision Equipment-PTY LTD., Johannesburg, South Africa) was utilized to delineate and measure the area of each selected region of interest (ROI) per slide, and the calculation of the percentage IHC positive labeling within each ROI (per mouse) was performed with the assistance of the phase separation function (Figure 14A,B).

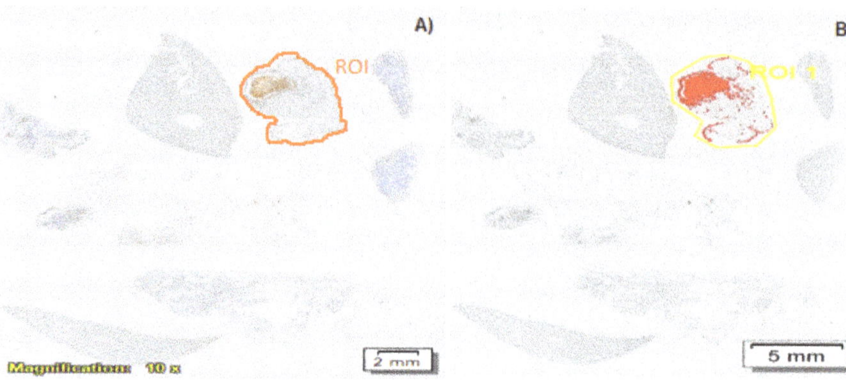

Figure 14. (**A**) Scanned IP-implanted, AF(D)NMs-treated nude mouse tissue section showing a section delineated (brown line) of ovarian carcinoma expressing MUC16/CA125 antigens (brown staining). MUC16/CA125 IHC, DAB, chromogen, hematoxylin counterstain. (**B**) Shows the region of interest/ROI delineated (yellow line) and MUC16-positive signal (red) within the ROI for quantification using the Olympus Count and Measure function (Olympus-Cell Sens software).

4.5.10. Experimental Data Statistical Analysis

Experimental data were evaluated as the mean/average of the data acquired, and the margin of error calculated via the standard-deviation (SD). Statistical analysis was performed by the repeated measures ANOVA. Significant differences between means of treated and untreated groups were analyzed for statistical significance using the two-tail student's *t*-test for paired/unpaired observations. Two-sided p-values < 0.05 were statistically significant. Kaplan-Meier mouse survival curves were prepared and were utilized to ascertain whether MUC16 expression correlated with mouse survival. Mouse Survival rate was computed as the number of days between commencement of the therapeutic regime and mice euthanasia, and mice survival percentage (%) was the total mice existing in each therapeutic group after conception of appropriate therapeutic regime. Kaplan-Meier Survival curves were evaluated, and the variance in mice survival was assessed for statistical significance. Sigma plot 11 graphing software excel (Systat Software, Inc., Richmond, CA, USA) was used for experimental data statistical exploration.

5. Conclusions

The present study has demonstrated for the first time that combining anti-MUC16 antibodies with drug-loaded nanomicellles in a hydrogel composite can inhibit intra-peritoneal tumor growth (and therefore peritoneal carcinomatosis), reducing the consequent production of ascites, resulting in increased survival rate of the Athymic Swiss nude mouse model. Our strategy of employing monoclonal antibodies for targeting ovarian cancer cells generated specificity to cancer cells that express MUC16. This functionalized nanomicelle may offer the groundwork for minimizing the cisplatin or methotrexate quantity utilized, thereby decreasing potentially harmful (toxic) side effects without significantly impacting on treatment efficacy. The actively targeted nature of this drug delivery system lends applicability to other solid tumors since targeting modalities that are specific to ovarian cancer have been employed. Hence, potential for use in other solid tumors should be investigated. As a result, this antibody-bound nanotherapeutic implantable drug delivery system may be a potent immuno-chemotherapeutic treatment that can be effectively employed in cases of advanced, and/or recurring, metastatic EOC. In conclusion, in vivo studies showed promising results with the prolonged release of drug from the implant exceeding a month and improved biocompatibility of the ISFI.

Supplementary Materials: Supplementary materials can be found at http://www.mdpi.com/1422-0067/19/10/3030/s1.

Author Contributions: P.K., Y.E.C. and V.P. conceived and designed the experiments; J.M.P. performed the experiments; J.M.P. and C.P. performed the preliminary cell culture experiments; J.M.P. and K.R. performed the in vivo experiments; S.J.C. assisted with the histological analysis; J.M.P., K.R., P.P., L.C.d.T. analyzed the data and authored the paper.

Acknowledgments: This study was funded by the National Research Foundation (NRF) of South Africa and the Cancer Association of South Africa (CANSA).

Conflicts of Interest: The authors declare no conflict of interest.

References

1. Whitehouse, C.; Solomon, E. Current status of the molecular characterization of the ovarian cancer antigen CA125 and implications for its use in clinical screening. *J. Gynecol. Oncol.* **2003**, *88*, 152–157. [CrossRef]
2. Wang, L.; Chen, H.; Pourgholami, M.H.; Beretov, J.; Hao, J. Anti-MUC1 Monoclonal Antibody (C595) and Docetaxel Markedly Reduce Tumor Burden and Ascites, and Prolong Survival in an in vivo Ovarian Cancer Model. *J. Comb. Ther. Hum. Ovarian Cancer* **2011**, *6*, e24405. [CrossRef] [PubMed]
3. Cho, H.; Lai, T.C.; Kwon, G.S. Poly (ethylene glycol)-block-poly(ε-caprolactone) micelles for combination drug delivery: Evaluation of paclitaxel, cyclopamine and gossypol in intraperitoneal xenograft models of ovarian cancer. *J. Control. Release* **2013**, *166*, 1–9. [CrossRef] [PubMed]
4. Sawant, R.R.; Jhaveri, A.M.; Torchilin, V.P.; Jhaveri, A.M.; Torchilin, V.P. Immunomicelles for advancing personalized therapy. *Adv. Drug Deliv. Rev.* **2012**, *64*, 1436–1446. [CrossRef] [PubMed]
5. Song, H.; He, R.; Wang, K.; Ruan, J.; Bao, C.; Li, N.; Ji, J.; Cui, D. Anti-HIF-1a antibody-conjugated pluronic triblock copolymers encapsulated with Paclitaxel for tumor targeting therapy. *Biomaterials* **2010**, *31*, 2302–2312. [CrossRef] [PubMed]
6. William, W.N., Jr.; Heymach, J.V.; Kim, E.S.; Lippman, S.M. Molecular targets for cancer chemoprevention. *Nat. Rev. Drug Discov.* **2009**, *8*, 213–225. [CrossRef] [PubMed]
7. Wang, Y.Z.; Li, Y.J.; Han, L.M.; Sha, X.Y.; Fang, X.L. Difunctional Pluronic copolymer micelles for paclitaxel delivery: Synergistic effect of folate-mediated targeting and Pluronic-mediated overcoming multidrug resistance in tumor cell lines. *Int. J. Pharm.* **2007**, *337*, 63–73. [CrossRef] [PubMed]
8. Batrakova, E.V.; Li, S.; Li, Y.L.; Alakhov, V.Y.; Elmquist, W.F.; Kabanov, A.V. Ditribution kinetics of a micelle-forming block copolymer Pluronic P85. *J. Control. Release* **2010**, *100*, 389–397. [CrossRef] [PubMed]
9. Xie, J.; Lee, S.; Chen, X. Nanoparticle-based theranostic agents. *Adv. Drug Deliv. Rev.* **2010**, *62*, 1064–1079. [CrossRef] [PubMed]
10. Torchilin, V.P. PEG-based micelles as carriers of contrast agents for different imaging modalities. *Adv. Drug Deliv. Rev.* **2002**, *54*, 235–252. [CrossRef]
11. Jager, K.; Wu, G.; Sel, S.; Garreis, F.; Brauer, L.; Paulsen, F.P. MUC16 in the lacrimal apparatus. *Histochem. Cell Biol.* **2007**, *127*, 433–438. [CrossRef] [PubMed]
12. Felder, M.; Arvinder Kapur, A.; Gonzalez-Bosquet, J.; Horibata, S.; Heintz, J.; Albrecht, R.; Fass, L.; Kaur, J.; Hu, K.; Shojaei, H.; et al. MUC16 (CA125): Tumor biomarker to cancer therapy, a work in progress. *J. Mol. Cancer* **2014**, *13*, 129. [CrossRef] [PubMed]
13. Teicher, B.A. Antibody-Drug Conjugate Targets. *Curr. Cancer Drug Targets* **2009**, *9*, 982–1004. [CrossRef] [PubMed]
14. Bast, R.C.; Klug, T.L.; St John, E.; Jenison, E.; Niloff, J.M. A radioimmunoassay using a monoclonal antibody to monitor the course of epithelial ovarian cancer. *N. Engl. J. Med.* **1983**, *309*, 883–887. [CrossRef] [PubMed]
15. Díaz, M.R.; Vivas-Mejia, P.E. Nanoparticles as Drug Delivery Systems in Cancer Medicine: Emphasis on RNAi-Containing Nanoliposomes. *J. Pharm.* **2013**, *6*, 1361–1380. [CrossRef] [PubMed]
16. Babu, A.; Templeton, A.K.; Munshi, A.; Ramesh, R. Nanoparticle-Based Drug Delivery for Therapy of Lung Cancer: Progress and Challenges. *J. Nanomater.* **2013**, *2013*, 14. [CrossRef]
17. Ma, D.; Zhang, H.; Tua, K.; Zhang, L. Novel supramolecular hydrogel/micelle composite for co-delivery of anticancer drug and growth factor. *Soft Matter* **2012**, *8*, 3665–3672. [CrossRef]
18. Lee, S.C.; Chang, Y.; Yoon, J.S.; Kim, C.; Kwon, I.C.; Kim, Y.H.; Jeong, S.J. Synthesis and Micellar Characterization of Amphiphilic Diblock Copolymers Based on Poly(2-ethyl-2-oxazoline) and Aliphatic Polyesters. *Macromolecules* **1999**, *32*, 1847–1852. [CrossRef]

19. Hamilton, T.C.; Robert, C.; Young Karen, G.; Louie Brent, C.; Wilma BMMcKoy, R.G.; Ozols, R.F. Characterization of a Xenograft Model of Human Ovarian Carcinoma Which Produces Ascites and Intraabdominal Carcinomatosis in Mice. *Cancer Res.* **1984**, *44*, 5286–5290. [PubMed]
20. Abolmaali, S.S.; Tamaddon, A.M.; Dinarvand, R. A review of therapeutic challenges and achievements of methotrexate delivery systems for treatment of cancer and rheumatoid arthritis. *Cancer Chemother. Pharmacol.* **2013**, *71*, 1115–1130. [CrossRef] [PubMed]
21. Bastakoti, B.P.; Wu, K.C.W.; Inoue, M.; Yusa, S.; Nakashima, K.; Yamauchi, Y. Multifunctional Core-Shell-Corona-Type Polymeric Micelles for Anticancer Drug-Delivery and Imaging. *Chem. Eur. J.* **2013**, *19*, 4812–4817. [CrossRef] [PubMed]
22. Zhang, W.; Shi, Y.; Chen, Y.Z.; Yu, S.Y.; Hao, J.G.; Luo, J.Q.; Sha, S.; Fang, X. Enhanced antitumor efficacy by Paclitaxel-loaded Pluronic P123/F127 mixed micelles against nonsmall cell lung cancer based on passive tumor targeting and modulation of drug resistance. *Eur. J. Pharm. Biopharm.* **2010**, *75*, 341–353. [CrossRef] [PubMed]
23. Weers, J.G.; Scheuing, D.R. Characterization of viscoelastic surfactant mixtures, I: Fourier transform infrared spectroscopic studies. *Colloid. Surf. B* **1991**, *55*, 41–56.
24. Rimmer, S.; Carter, S.; Rutkaite, R.; Haycock, J.W.; Swanson, L. Highly branched poly-(*N*-isopropylacrylamide)s with arginine–glycine–aspartic acid (RGD)- or COOH-chain ends that form sub-micron stimulus-responsive particles above the critical solution temperature. *J. Soft Matter* **2007**, *3*, 971–973. [CrossRef]
25. Lin, J.; Li, Y.; Li, Y.; Cui, F.; Yu, F.; Wu, H.; Xie, L.; Luo, F.; Hou, Z.; Lin, C. Self-targeted, bacillus-shaped, and controlled-release methotrexate prodrug polymeric nanoparticles for intratumoral administration with improved therapeutic efficacy in tumor-bearing mice. *J. Mater. Chem. B* **2015**, *3*, 7707–7717. [CrossRef]
26. Pisano, C.; Vesci, L.; Foderà, R.; Ferrara, F.F.; Rossi, C.; De Cesare, M.; Zuco, V.; Pratesi, G.; Supino, R.; Zunino, F. Antitumor activity of the combination of synthetic retinoid ST1926 and cisplatin in ovarian carcinoma models. *Ann. Oncol.* **2007**, *18*, 1500–1505. [CrossRef] [PubMed]
27. Daman, Z.; Ostad, S.N.; Amini, M.; Gilani, K. Preparation, optimization and in vitro characterization of stearoyl-gemcitabine polymeric micelles: A comparison with its self-assembled nanoparticles. *Int. J. Pharm.* **2014**, *468*, 142–151. [CrossRef] [PubMed]
28. Pantshwa, J.; Choonara, Y.E.; Kumar, P.; du Toit, L.C.; Penny, C.; Pillay, V. Synthesis of novel amphiphilic poly(*N*-isopropylacrylamide)-b-poly(aspartic acid) nanomicelles for potential targeted chemotherapy in ovarian cancer. *J. Drug Deliv. Sci. Technol.* **2017**, *39*, 308–323. [CrossRef]
29. Wei, Z.; Hao, J.; Yuan, S.; Li, Y.; Juan, W.; Sha, X.; Fang, X. Paclitaxel-loaded Pluronic P123/F127 mixed polymeric micelles: Formulation, optimization and in vitro characterization. *Int. J. Pharm.* **2009**, *376*, 176–185. [CrossRef] [PubMed]
30. Bae, Y.; Nishiyama, N.; Kataoka, K. In vivo antitumor activity of the folate conjugated pH-sensitive polymeric micelle selectively releasing adrinamycin in the intracellular acidic compartments. *Bioconj. Chem.* **2007**, *18*, 1131–1139. [CrossRef] [PubMed]
31. Connolly, D.C. Animal Models of Ovarian Cancer. In *Ovarian Cancer, Cancer Treatment and Research*; Stack, M.S., Fishman, D.A., Eds.; Fox Chase Cancer Center: Philadelphia, PA, USA, 2009; Volume 149.
32. Orsulic, S.; Li, Y.; Soslow, R.A.; Vitale-Cross, L.A.; Gutkind, J.S.; Varmus, H.E. Induction of ovarian cancer by defined multiple geneticchanges in a mouse model system. *Cancer Cell* **2002**, *1*, 53–62. [CrossRef]
33. Roberts, D.; Williams, S.J.; Cvetkovic, D.; Weinstein, J.K.; Godwin, A.K.; Johnson, S.W.; Hamilton, T.C. Decreased expression of retinolbindingproteins is associated with malignant transformationof the ovarian surface epithelium. *DNA Cell Biol.* **2002**, *21*, 11–19. [CrossRef] [PubMed]
34. Vendramini-Costa, D.B.; Castro, I.B.D.; Ruiz, A.L.T.G. Effect of goniothalamin on the development of Ehrlich solid tumor in mice. *Bioorg. Med. Chem.* **2008**, *18*, 6742–6747. [CrossRef] [PubMed]
35. Kievit, E.; Van Gog, F.B.; Schluper, H.M.M.; Van Dongen, G.A.M.S.; Herbert, M.; Pinedo, H.M.; Boven, E. Comparison of the biodistribution and the efficacy of monoclonal antibody labeled with either ^{131}I or ^{186}Re in human ovarian cancer xenografts. *Int. J. Radiat. Oncol. Biol. Phys.* **1997**, *38*, 813–823. [CrossRef]
36. Qiao, P.; Niu, Q.; Wang, Z.; Cao, D. Synthesis of thermosensitive micelles based on poly(*N*-isopropylacrylamide) and poly(L-alanine) for controlled release of adriamycin. *J. Chem. Eng.* **2010**, *159*, 257–263. [CrossRef]
37. Rump, A.; Morikawa, Y.; Tanaka, M.; Minami, S.; Umesak, N.; Takeuchi, M.; Miyajima, A. Binding of ovarian cancer antigen CA125/MUC16 to mesothelin mediates cell adhesion. *J. Biol. Chem.* **2004**, *279*, 9190–9198. [CrossRef] [PubMed]

38. Gubbels, J.A.; Belisle, J.; Onda, M.; Rancourt, C.; Migneault, M.; Ho, M.; Bera, T.K.; Connor, J.; Sathyanarayana, B.K.; Lee, B.; et al. Mesothelin-MUC16 binding is a high affinity, N-glycan dependent interaction that facilitates peritoneal metastasis of ovarian tumors. *Mol. Cancer* **2006**, *5*, 50. [CrossRef] [PubMed]
39. Tamada, Y.; Takeuchi, H.; Suzuki, N.; Susumu, N.; Aoki, D.; Irimura, T. Biological and therapeutic significance of MUC1 with sialoglycans in clear cell adenocarcinoma of the ovary. *Cancer Sci.* **2007**, *98*, 1586–1591. [CrossRef] [PubMed]
40. Gulley, J.L.; Arlen, P.M.; Tsang, K.Y.; Yokokawa, J.; Palena, C. Pilot study of vaccination with recombinant CEA-MUC-1-TRICOM poxviral-based vaccines in patients with metastatic carcinoma. *Clin. Cancer Res.* **2008**, *14*, 3060–3069. [CrossRef] [PubMed]
41. Oei, A.L.; Moreno, M.; Verheijen, R.H.; Sweep, F.C.; Thomas, C.M. Induction of IgG antibodies to MUC1 and survival in patients with epithelial ovarian cancer. *Int. J. Cancer* **2008**, *123*, 1848–1853. [CrossRef] [PubMed]

International Journal of *Molecular Sciences*

MDPI

Article

Targeting Glioblastoma Cells Expressing CD44 with Liposomes Encapsulating Doxorubicin and Displaying Chlorotoxin-IgG Fc Fusion Protein

Hafizah Mahmud [1,†], Tomonari Kasai [1,†], Apriliana Cahya Khayrani [1], Mami Asakura [1], Aung Ko Ko Oo [1], Juan Du [1], Arun Vaidyanath [1], Samah El-Ghlban [2], Akifumi Mizutani [1], Akimasa Seno [1], Hiroshi Murakami [1], Junko Masuda [1] and Masaharu Seno [1,*]

1. Department of Medical Bioengineering, Graduate School of Natural Science and Technology, Okayama University, Okayama 700-0080, Japan; hfzhmhmd@gmail.com (H.M.); t-kasai@okayama-u.ac.jp (T.K.); apriliana41@gmail.com (A.C.K.); asakura@okayama-u.ac.jp (M.A.); kokooo.aung@gmail.com (A.K.K.O.); djmail@yeah.net (J.D.); arunvnath@okayama-u.ac.jp (A.V.); mizut-a@cc.okayama-u.ac.jp (A.M.); aseno@okayama-u.ac.jp (A.S.); muraka-h@cc.okayama-u.ac.jp (H.M.); junkomasuda@okayama-u.ac.jp (J.M.)
2. Department of Chemistry, Biochemistry Division, Faculty of Science, El Menoufia University, Shebin El Kom, Menoufia 32511, Egypt; S_elghlban@yahoo.com

* Correspondence: mseno@okayama-u.ac.jp; Tel.: +81-86-251-8216

† These authors contributed equally to this work.

Received: 27 December 2017; Accepted: 23 February 2018; Published: 26 February 2018

Abstract: We recently have established a successful xenograft model of human glioblastoma cells by enriching hyaluronic acid-dependent spheroid-forming populations termed U251MG-P1 cells from U251MG cells. Since U251MG-P1 cells have been confirmed to express CD44 along with principal stemness marker genes, *OCT3/4*, *SOX2*, *KLF4* and *Nanog*, this CD44 expressing population appeared to majorly consist of undifferentiated cells. Evaluating the sensitivity to anti-cancer agents, we found U251MG-P1 cells were sensitive to doxorubicin with IC_{50} at 200 nM. Although doxorubicin has serious side-effects, establishment of an efficient therapy targeting undifferentiated glioblastoma cell population is necessary. We previously designed a chlorotoxin peptide fused to human IgG Fc region without hinge sequence (M-CTX-Fc), which exhibited a stronger growth inhibitory effect on the glioblastoma cell line A172 than an original chlorotoxin peptide. Combining these results together, we designed M-CTX-Fc conjugated liposomes encapsulating doxorubicin and used U251MG-P1 cells as the target model in this study. The liposome modified with M-CTX-Fc was designed with a diameter of approximately 100–150 nm and showed high encapsulation efficiency, adequate loading capacity of anticancer drug, enhanced antitumor effects demonstrating increasing uptake into the cells in vitro; M-CTX-Fc-L-Dox shows great promise in its ability to suppress tumor growth in vivo and it could serve as a template for targeted delivery of other therapeutics.

Keywords: M-CTX-Fc; liposome; glioblastoma cells; specific targeting; doxorubicin; MMP-2

---

1. Introduction

Glioblastoma is a highly invasive cancer where the cells demonstrate their infiltrative growth to diffuse into the brain tissue [1]. This is the reason why the treatment of glioblastoma requires a multidisciplinary approach. Current standard of care for glioblastoma includes maximal safe surgical resection followed by radiotherapy and chemotherapy with the alkylating agent Temozolomide. The extensive and complete surgical treatment involves difficulties in glioblastoma with high degree of invasion because simultaneous removal of the surrounding normal areas will impair the function of brain controlling speech, motor function, sense and personality [2]. On that occasion, developing of brain tumor specific targeting drug delivery systems, which increase drug accumulation in the tumor region with less toxicity to the adjacent normal brain tissue, would significantly be a great approach for brain tumor treatments. The drug delivery to brain faces has been considered difficult as the agents need to across the blood–brain barrier (BBB). However, a recent study has shown that the tight junction of the BBB loses the integrity by increasing permeability of the capillary endothelium [3].

On the other hand, recurrence of relapsing cancer is currently the central big issue to be studied. This is considered to occur due to the residual subpopulation of cancer cells after the treatment because they are believed to be resistant to chemotherapy and radiotherapy even though they are the minor population in the cancer tissues [4]. In this context, chemical agents toxically effective against cancer cells should properly be chosen to design an effective drug delivery system.

As for the efficacy of drug delivery, targeting ability is another strong issue to be designed. Some cell surface antigens including receptors specific to gliomas and/or neovasculatures will be crucial markers to be targeted [5]. Hence, several approaches to treat brain cancer employ ligands specific to the tumor cells targeting cell surface markers, which are overexpressed in cancer cells but low or not expressed in normal cells.

Matrix metalloproteinase-2 (MMP-2) is an extracellular matrix degrading enzyme, which plays an important role in tumor invasion and is highly expressed in related cancer cells [6]. As for MMP-2 in glioblastoma, the activity is increased along with the tumor grade and the expression is significantly higher than that in normal brain tissue. The increment is associated with poor prognosis and the overall short-life of survivors. MMP-2 is secreted as an inactive zymogen and prior to its activation, it will bind to tissue inhibitor of metalloproteinase-2 associated with membrane type matrix metalloproteinase-1, which localizes on the cell surface of the tumors [7,8]. Membrane type matrix metalloproteinase-1 is replenished by auto degradation or clathrin-dependent internalization. Collectively, MMP-2 has been considered as a target for cancer therapy.

Chlorotoxin (CTX) is a peptide derived from Egyptian scorpion venom, which has initially been characterized as an MMP-2 inhibitor and also as a voltage-gated chloride channel blocker [9,10]. CTX exhibited high specificity, selectivity and affinity for glioma and other tumors of neuroectodermal origin [11]. Following the discovery, CTX has been extensively developed as a ligand for active targeting to deliver cytotoxic agents, fluorescent dyes for imaging and iodine for labeling tumor cells [12–14]. Recent findings suggested that the delivery of CTX-modified liposomes was mediated by MMP-2 but not correlated with the chloride channel CIC-3 when targeting U87 glioma cells [15].

In our previous report, the CTX fused to human IgG Fc domain without hinge region in monomeric form (M-CTX-Fc) showed inhibition of mortality of the glioma cell line A172 [16]. This inhibitory effect was enhanced when compared to the original CTX peptide. The similar effect was observed in pancreatic cancer cell PANC-1 cells [17]. Indicating M-CTX-Fc could be a potential ligand for active targeting of glioblastoma cells, the target-dependent internalization of bionanocapsules displaying M-CTX-Fc on the surface into cells was described [16].

Very recently, we condensed the population overexpressing CD44 in U251MG cells exploiting the preferential affinity for hyaluronic acid as U251MG-P1 cells [18]. Since CD44 is well known as a common marker of cancer stem cells, we confirmed the expression of stemness markers as well as the tumor-initiating capacity in U251MG-P1 cells.

In this study, we designed a liposomal drug delivery system of doxorubicin to evaluate the ability of M-CTX-Fc as an effective moiety to target glioblastoma cells expressing stemness markers using U251MG-P1 cells as the target model.

2. Results and Discussion

2.1. Sensitivity of U251MG-P1 Cells to Doxorubicin

First of all, we assessed the effect of doxorubicin on U251MG-P1 cells (Figure 1). We chose doxorubicin as the first priority because the liposomal formulation is well established by pH gradient and ammonium sulfate gradient method and has clinically been tested in glioblastoma cancer therapy [19–22]. As a result, the MTT assay showed an IC_{50} at around 200 nM, which was a feasible concentration available as an agent for cancer chemotherapy.

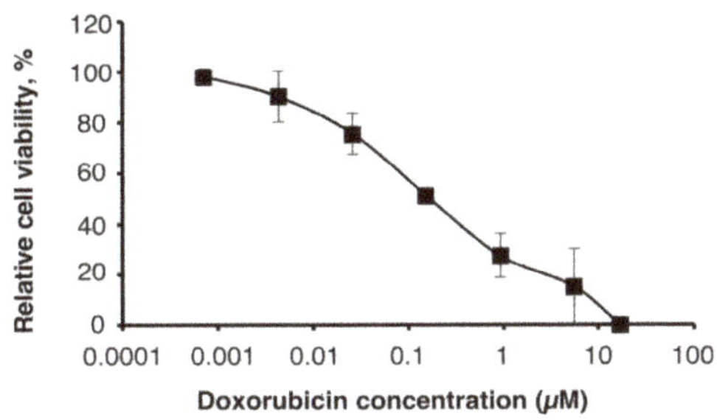

Figure 1. The U251MG-P1 cells are sensitive to doxorubicin. Cytotoxicity of doxorubicin was assessed on U251MG-P1 cells by MTT assay. The data presented as the mean ± SD (n = 3) from independent experiment.

However, cardiotoxicity is the well-known side effect of doxorubicin, so that the amount of administration through the lifetime is critically limited. If doxorubicin is one of the rare candidates of effective agents to treat such as cancer stem cells, the development of drug delivery systems to avoid the side effects should seriously be important.

2.2. Expression of MMP-2 in U251MG-P1 Cells

We assessed the expression of MMP-2 in U251MG-P1 cells to confirm that MMP-2 could be a sufficient marker to target the cells. A172 (MMP-2-positive) and human breast cancer SK-BR-3 (MMP-2-negative) cell line by both Western blot and reverse transcription-polymerase chain reaction (qRT-PCR) as shown in Figure 2A–C. The 72-kDa protein treated with anti-MMP-2 antibody corresponding to proMMP-2 was observed in U251MG-P1 and A172 but less (or not seen) in SK-BR-3. Since U251MG-P1 cells were confirmed to express MMP-2, we decided to employ CTX to target U251MG-P1 cells.

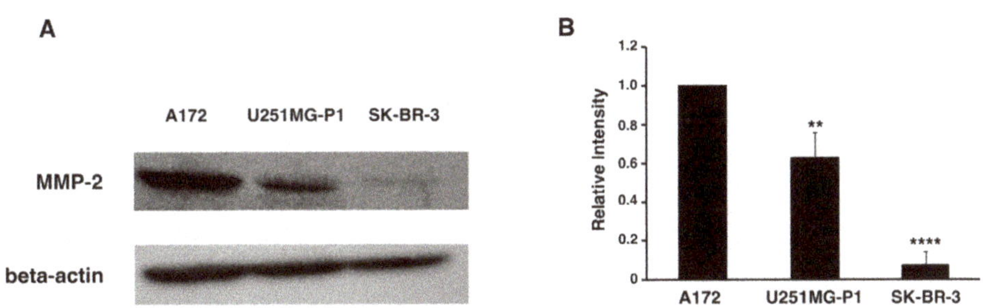

Figure 2. *Cont.*

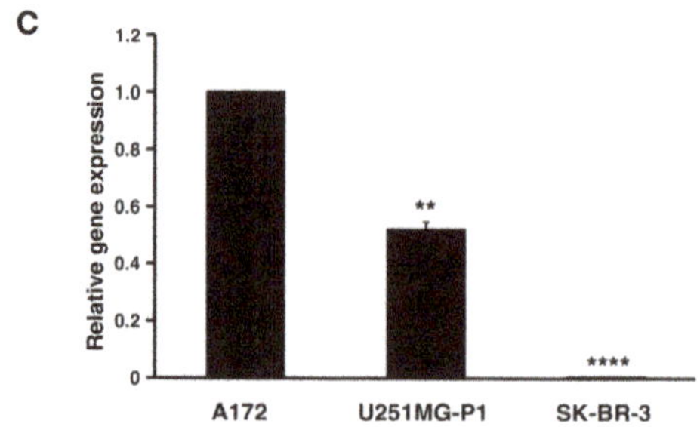

Figure 2. The U251MG-P1 cells are expressing MMP-2. (**A**) Western blot of cells with anti-MMP-2 and anti-beta-actin antibodies. (**B**) Relative intensity of the bands in Western blot densitometrically analyzed by ImageJ. (**C**) Relative gene expression analyzed by reverse transcription quantitative PCR. The data presented as the mean ± SD (n = 3) from three independent experiments. The data were analyzed by two-tailed students t-test using A172 cells as a control; **, $p < 0.01$; ****, $p < 0.001$.

2.3. *Characterization of M-CTX-Fc*

Peptide ligands specific to cell surface molecules have extensively been used for various drug delivery targeting in cancer cells. However, they are often labile and degraded resulting in short half-life due to their antigenicity and reticuloendothelial system (RES) [23]. One of the approaches to overcome those problems is to fuse the ligand with human IgG Fc domain. The Fc domain provides significant advantages such as improving the solubility and stability of partner molecules and to prolong the half-life in plasma [24,25].

After preparation of M-CTX-Fc from *E. coli*, the ability to inhibit the gelatinase activity of secreted MMP-2 in the condition medium of U251MG-P1 cells was observed by gelatin zymography (Figure 3A). The intensity of gelatinase activity decreased in the presence of M-CTX-Fc in a dose-dependent manner. This confirmed the interaction of M-CTX-Fc with MMP-2 even though the exact molecular target of CTX is still unknown [26]. We then evaluated the activity of M-CTX-Fc on the proliferation and viability of U251MG-P1 cells. M-CTX-Fc suppressed the growth of the cells as described in other cells such as the glioma cell line A172 and pancreatic carcinoma cell line PANC-1 cells [16,17] (Figure 3B,C). This cell growth inhibition does not appear to induce cell death because the cell viability recovered after removal of the protein from the culture medium while CTX alone did not affect the cell growth of U251MG-P1. Collectively, we concluded that M-CTX-Fc is folded properly. Hence, we chose M-CTX-Fc instead of CTX as the ligand to target U251MG-P1 cells considering that the Fc moiety would help the conformation of intact CTX when conjugated on the surface of liposomes.

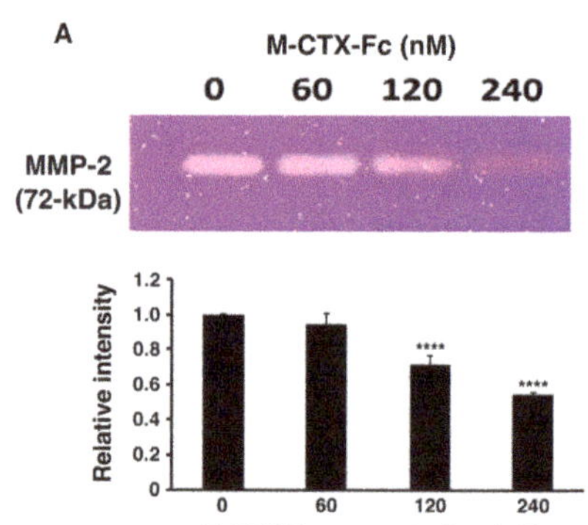

Figure 3. *Cont.*

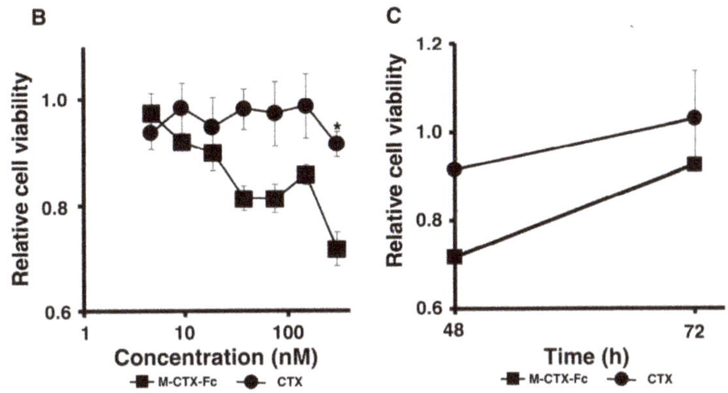

Figure 3. M-CTX-Fc inhibits gelatinase activity in the condition medium of U251MG-P1 cells (**A**) and cell growth of U251MG-P1 cells (**B**,**C**). The M-CTX-Fc inhibit gelatinase activity. (**A**) MMP-2 activity in the condition medium of U251MG-P1 cells was monitored by zymography in the presence of 0, 60, 120, and 240 nM of M-CTX-Fc. The data presented as the mean ± SD (n = 3) from technical replicates. The statistical significance in mean values of more than two groups was determined using one-way analysis of variance (ANOVA) and post hoc Tukey HSD were applied using CM without M-CTX-Fc as control *, $p < 0.05$; ****, $p < 0.001$. (**B**) The inhibition of cell growth in the presence of M-CTX-Fc and CTX after 48 h. (**C**) The viable cells at 48 h were kept cultured without M-CTX-Fcs or CTX up to 72 h. Cell numbers in each well were assessed by MTT assay. The absorbance at 570 nm corresponding to the initial number of the cells was defined as 1. The data presented as the mean ± SD (n = 3) from three independent experiments. The data were analyzed by 2-tailed students t-test using M-CTX-Fc as a control; *, $p < 0.05$; ****, $p < 0.001$.

2.4. Characterization and Optimization of M-CTX-Fc Conjugated to Liposome

Prior to further investigation of their cytotoxicity in vitro and in vivo, we optimized the amount of M-CTX-Fc (nmol) conjugated to liposomes. The preparation of liposomes conjugated with M-CTX-Fc encapsulating doxorubicin (M-CTX-Fc-L-Dox) is summarized in Figure 4A. Various amounts of M-CTX-Fc such as 5, 10, 15 or 20 nmol conjugated to liposome encapsulating doxorubicin was prepared, respectively. The mean particle size of these liposomes was almost 150 nm, which was not significantly affected by the amount of ligand as previously described [27,28]. The optimal amount of M-CTX-Fc conjugated to liposomes was determined by the IC_{50} (Figure 4B). When 10 nmol of M-CTX-Fc were used to conjugate to liposome, the cytotoxicity of doxorubicin was at a maximum whereas 5 nmol was not enough for binding to the receptor. Similarly, M-CTX-Fc at 15 nmol and 20 nmol did not improve the cytotoxic effect of liposome encapsulating doxorubicin. We further investigated the capability of specific targeting of liposomes by using M-CTX-Fc at 10 nmol.

The characteristics of the formulations of liposomes encapsulating doxorubicin used in this study are summarized in Table 1. The liposomes conjugated with human IgG (hIgG-L-Dox) or without ligands (L-Dox) were prepared as references of nonspecific targeting. All the prepared liposomes showed diameters of approximately 150 nm with a low polydispersity index less than 0.1 indicating homogeneous uniformity. The uniformity of particle size is considered important to obtain stable receptor-mediated endocytosis in the intracellular delivery by the nano-carrier system [29]. A particle size between 50 and 200 nm is considered sufficient to accumulate the drug in the tumor via enhanced permeability and retention (EPR) effect since particles with larger sizes are generally trapped by the RES resulting in a short half-life by rapid clearance from blood flow [30].

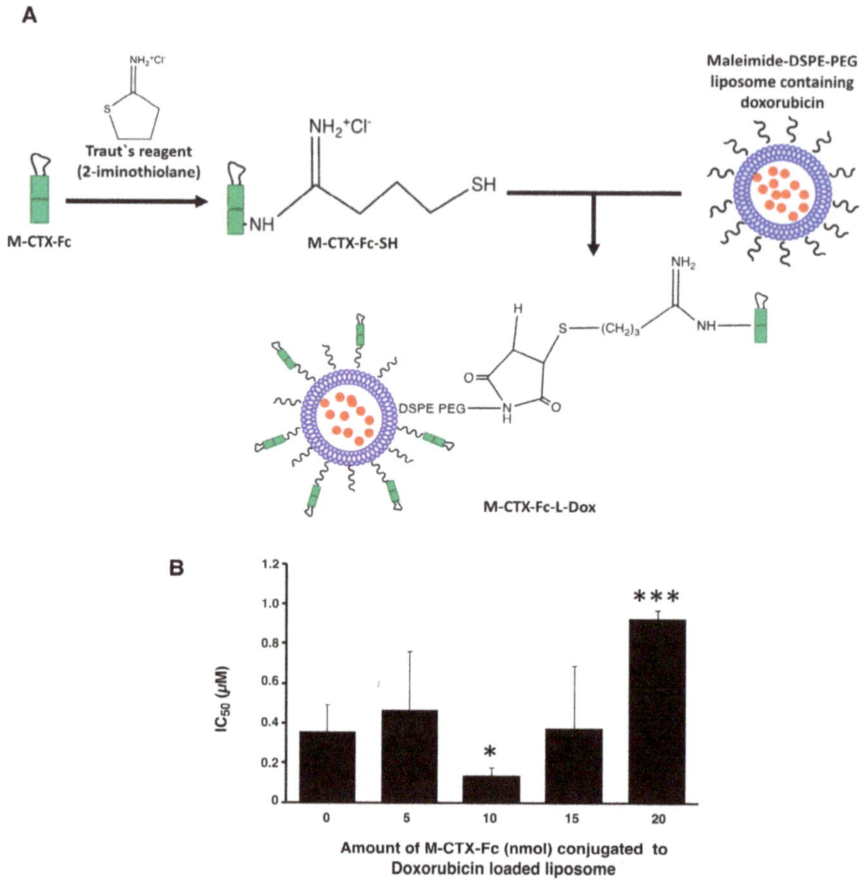

Figure 4. The amount of M-CTX-Fc conjugated to liposomes encapsulating doxorubicin was optimal at 10 nmol/48 μmol DPPC. (**A**) Conjugation procedure of M-CTX-Fc to liposomes encapsulating doxorubicin. (**B**) IC_{50} of doxorubicin encapsulated in liposomes conjugating various amount of M-CTX-Fc) against U251MG-P1 cells. The data presented as the mean ± SD (n = 3) from independent experiments. The statistical significance in mean values of more than two groups was determined using one-way analysis of variance (ANOVA) and post hoc Tukey HSD were applied using no M-CTX-Fc (0 mol) as control. *, $p < 0.05$; ***, $p < 0.005$.

Table 1. Characteristics of the formulations of liposomes encapsulating doxorubicin.

Formulations	Diameter (nm)	Polydispersity Index	Zeta Potential (−mV)	Encapsulation Efficiency (%)	Loading Efficiency (%)
L-Dox	133.4 ± 12.7	0.09 ± 0.03	8.13 ± 2.32	97.5 ± 3.1	3.4 ± 0.1
M-CTX-Fc-L-Dox	148.3 ± 3.0	0.05 ± 0.03	7.86 ± 1.19	98.2 ± 1.3	4.5 ± 0.4
hIgG-L-Dox	151.3 ± 4.3	0.07 ± 0.02	6.66 ± 3.78	94.4 ± 7.2	4.1 ± 0.4

Each experiment was performed in triplicate and the values are given as mean ± SD.

Transmission electron microscopy (TEM) revealed that all the formulations of liposome encapsulating doxorubicin exhibited precipitates of fibrous-bundle aggregates when doxorubicin was encapsulated into the inner core of liposomes due to the loading method of pH gradient (Figure 5) [19].

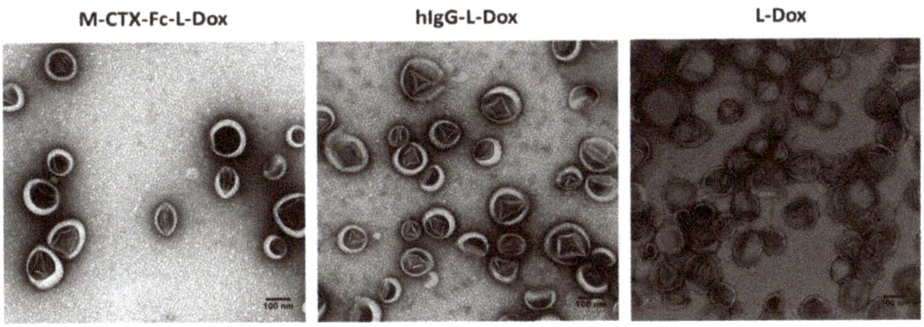

Figure 5. M-CTX-Fc-L-Dox showed unilamellar vesicles with diameter of approximately 100 nm. TEM images of liposome formulation encapsulating doxorubicin were compared between M-CTX-Fc-L-Dox, hIgG-L-Dox and L-Dox. Each scale bar shows 100 nm.

2.5. Cellular Uptake of Liposomes

The cellular uptake of M-CTX-Fc-L-Dox and L-Dox into U251MG-P1 and SK-BR-3 cells was evaluated under a confocal microscope after 1 h incubation at 37 °C (Figure 6). It is worth noting that the strong fluorescence of doxorubicin was observed in U251MG-P1 cells treated with M-CTX-Fc-L-Dox, especially in the nuclei. This observation might be attributed to the specific interaction of M-CTX-Fc-L-Dox with the cell surface molecule on U251MG-P1 cells by receptor-mediated endocytosis. On the other hand, when cells were treated with L-Dox, the fluorescence from L -Dox was reduced compared to that from M-CTX-Fc-L-Dox. It is important to mention that receptor-mediated endocytosis achieved by M-CTX-Fc-L-Dox might be faster than endocytosis gained by L-Dox only. This explanation agrees with our result for IT_{50} in Figure 7. Oppositely, almost no signal for doxorubicin uptake was observed in the SK-BR-3 cells, which showed low expression of MMP-2 compared to U251MG-P1 cells. Collectively, M-CTX-Fc may have the potential to target MMP-2-expressing cancer cells and internalize into the cells while further investigation is required to identify the molecule on the cell surface directly binding to M-CTX-Fc.

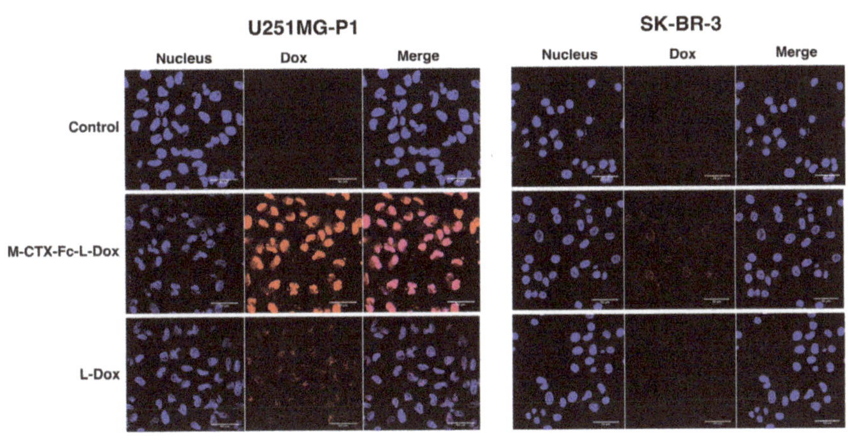

Figure 6. Cellular uptake of Doxorubicin in U251MG-P1 cells was enhanced through M-CTX-Fc-L-Dox. U251MG-P1 cells and SK-BR-3 cells were evaluated for the cellular uptake of doxorubicin under a confocal microscope. Cell nuclei were stained with DAPI (blue). Red color arises from natural fluorescence properties of doxorubicin. Each scale bar shows 50 μm.

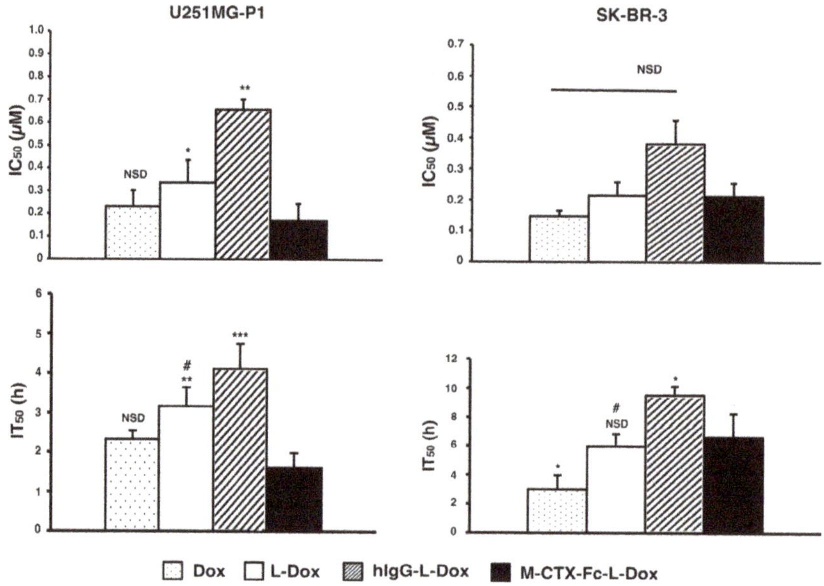

Figure 7. M-CTX-Fc-L-Dox exhibited the lowest inhibition concentration (IC_{50}) and shortest exposure time (IT_{50}) in U251MG-P1 cells. In vitro cytotoxicity IC_{50} of doxorubicin in different formulations after 72 h of exposure to U251MG-P1 and SK-BR-3 cells was evaluated and compared (top). IT_{50} with doxorubicin at IC_{100} were evaluated and compared (bottom). The data presented as the mean ± SD (n = 3). The statistical significance in mean values of more than two groups was determined using one-way analysis of variance (ANOVA) and post hoc Tukey HSD were applied using M-CTX-Fc-L-Dox liposome as control. *, $p < 0.05$; **, $p < 0.01$; ***, $p < 0.005$; NSD, no significant difference; #, $p < 0.05$ versus Dox.

2.6. Cytotoxicity In Vitro

The IC_{50}s of doxorubicin were assessed when U251MG-P1 and SK-BR-3 cells were treated for 72 h with each formulation of naked doxorubicin, L-Dox, M-CTX-Fc-L-Dox and hIgG-L-Dox (Figure 7). Among the liposomal formulations, M-CTX-Fc-L-Dox showed the highest cytotoxicity with the lowest IC_{50} of 0.17 μM in U251MG-P1 cells. However, the cytotoxicity of M-CTX-Fc-L-Dox showed no significant different with naked doxorubicin. In SKBR-3 cells without MMP-2 expression, M-CTX-Fc-L-Dox appeared almost equally effective with hIgG-L-Dox and L-Dox. Collectively, the results appear consistent with the dependency of MMP-2 expression. As described previously, we thought the time of exposure allowing the cellular uptake should also be important to determine the effectiveness [31]. To make this point clearer, we evaluated the IT_{50}.

In U251MG-P1 cells, M-CTX-Fc-L-Dox showed significantly rapid exposure time of IT_{50} at around 1.6 h. This is the shortest time when compared with those by other formulations. Meanwhile, in both cells, naked doxorubicin had rapid exposure time compared to L-Dox. This result could be explained by the difference of cellular mechanism of internalization. The cellular uptake of liposomes is mediated by endocytosis, whereas the naked doxorubicin molecules internalized into the cell via passive diffusion. However, in the case of M-CTX-Fc-L-Dox, the conjugated ligand specific to MMP-2 receptor, exhibited shorter time for liposomes internalized into the cells and reaches IT_{50} comparable to the naked doxorubicin. On the other hand, in SK-BR-3 cells, no significant difference was found between L-Dox and M-CTX-Fc-L-Dox. Thus, M-CTX-Fc-L-Dox successfully demonstrated the specific targeting of U251MG-P1 cells in vitro.

2.7. Suppression of Tumor Growth In Vivo

The suppression of tumor growth by M-CTX-Fc-L-Dox was evaluated in BALB/c mice bearing tumors of transplanted U251MG-P1 cells (Figure 8). We found that the tumor latency of U251MG-P1 is so rapid that it should not be comparable with that of U251MG cells. Meanwhile, the tumor latency of U251MG is not stable. This means that the targeting effect of our liposomal formulation is difficult to be demonstrated on the tumors from U251MG cells. First of all, the effect of doxorubicin on the body weight was assessed by the three-time injections of 10 mg/kg (Figure 8A). As a result, the loss of body weight was less than 20% even when the naked doxorubicin was injected. The liposomal formulations were less toxic than naked doxorubicin as they showed body weight loss less than 10%. After three times of injection in seven-day intervals, the tumor growth was observed for 20 days, and the efficacy of the suppression of tumor growth was calculated as a relative tumor volume normalized to the initial tumor volume before the treatment. The tumor volume in the PBS group increased aggressively, whereas M-CTX-Fc-L-Dox slowed the tumor growth more significant than naked Dox and hIgG-L-Dox at day 20 ($p < 0.001$) (Figure 8B). M-CTX-Fc-L-Dox appeared slightly more effective than L-Dox at day 20 ($p = 0.043 < 0.05$). The representative tumors excised from mice treated with five different formulations at day 20 demonstrated tumor suppression effect of M-CTX-Fc-L-Dox (Figure 8C). In this context, M-CTX-Fc-Dox exhibited an inhibitory effect on tumor growth, which could be attributed to the combined action of the passive targeting via the EPR effect and active targeting via receptor-mediated endocytosis. However, a larger cohort-evaluation is needed to confirm the effect of M-CTX-Fc-L-Dox more precisely.

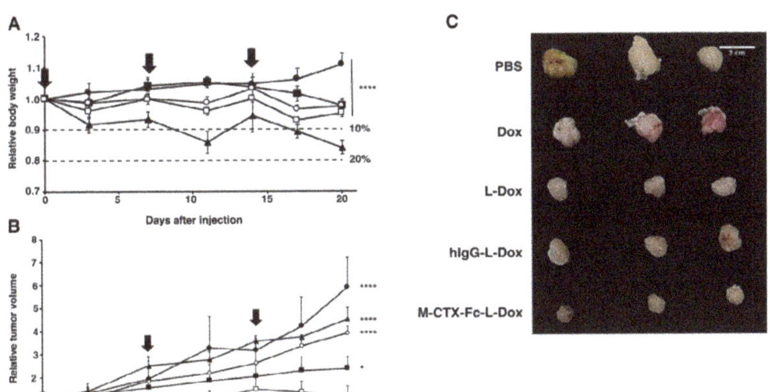

Figure 8. M-CTX-Fc-L-Dox suppressed tumor growth in the most effective manner in vivo. (**A**) The relative body weight of mice bearing tumors during the treatment. M-CTX-FC-L-Dox and other liposome formulations were less toxic than naked doxorubicin. The statistical significance in mean values of more than two groups was determined using one-way analysis of variance (ANOVA) and post hoc Tukey HSD were applied using relatives body weight of Dox treatment as control. ****, $p < 0.001$. (**B**) The effect of different formulations of doxorubicin on the volume of tumors. M-CTX-Fc-L-Dox was the most effective formulation to suppress the growth of tumor. Doxorubicin in each formulation was administered at 7-day intervals (indicated by vertical arrows). The statistical significance in mean values of more than two groups was determined using one-way analysis of variance (ANOVA) and post hoc Tukey HSD were applied using relatives tumor volume of M-CTX-Fc-L-Dox treatment as control, *, $p < 0.05$; ****, $p < 0.001$ (**C**) The tumors from the experiment (**B**) representing each group were displayed exhibiting the effect of each formulation of doxorubicin. Data are expressed as the mean with ±SD where $n = 3$.

3. Materials and Methods

3.1. Materials

Dipalmitoylphosphatidylcholine (DPPC), 1,2-distearoyl-sn-glycerol-3 phosphoethanolamine-*N* -[methoxy (polyethylene glycol)-2000] (mPEG–DSPE) and 1,2-distearoyl-sn-glycerol-3-phosphoethanolamine-*N*-[maleimide (polyethylene glycol)-2000] (Mal–PEG–DSPE) were obtained from NOF Co. (Tokyo, Japan). Cholesterol (Chol) was purchased from Kanto Chemical Co., Inc. (Tokyo, Japan). Thiazolyl blue tetrazolium bromide (MTT), 2-iminothiolane hydrochloride, human IgG, RPMI 1640 medium, and DMEM were from Sigma-Aldrich (St Louis, MO, USA). Doxorubicin hydrochloride was purchased from Wako Pure Chemical (Osaka, Japan). Amicon Ultra filters were from Merck Millipore Ltd. (Billerica, MA, USA). PD-10 columns were from GE Healthcare (Carlsbad, CA, USA).

3.2. Cell Cultures and Experimental Animals

The human glioblastoma cell line A172 cells and human breast cancer cell line SK-BR-3 cells were obtained from ATCC (Manassas, VA, USA). U251MG-P1 was isolated from a xenograft tumor of human glioblastoma cell line U251MG cells in mouse [18]. A172 and U251MG-P1 were maintained in DMEM medium supplemented with 10% fetal bovine serum (FBS) (PAA Laboratories, Pasching, Austria) in the presence of penicillin (100 IU/mL) and streptomycin (100 μg/mL) (Nacalai Tesque, Kyoto, Japan). SK-BR-3 cells were cultured in RPMI 1640 medium supplemented with 10% FBS in the presence of 100 IU/mL penicillin and 100 μg/mL streptomycin. Cells were cultured in a humidified incubator at 37 °C with the atmosphere of 5% CO_2.

Four-week-old female BALB/c nude mice from Charles River (Kanagawa, Japan) were bred at 23 °C and fed with sterilized food and water during the experiments. All animal experimental protocols were reviewed and approved by the ethics committee (Animal Care and Use Committee) of Okayama University under the project identification code IDs OKU-2016078 (Date of approval: 1 April 2016).

3.3. Preparation of M-CTX-Fc

M-CTX-Fc fusion protein was produced in our laboratory by recombinant expression in *E. coli* [14,15]. Briefly, M-CTX-Fc was expressed in *E. coli* as inclusion body and was refolded (Figure S1 in Supplementary Material).

3.4. Gelatin Zymography for MMP-2 Activity

MMP-2 gelatinolytic activity in the CM was determined by zymography. Fifteen microliter aliquots of CM with or without M-CTX-Fc were subjected to 10% SDS-PAGE containing 0.05% gelatine. The samples were prepared without reducing reagent and boiling prior to electrophoresis. After electrophoresis, the gel was washed twice in 2.5% Triton X-100 for 30 min and once in 50 mM Tris-HCl, pH 7.4, 10 mM $CaCl_2$ and 0.02% NaN_3. After incubation, gel was stained with Coomassie brilliant blue in 50% methanol and 10% acetic acid and destained in 10% methanol and 10% acetic acid.

3.5. Preparation of Liposomes Encapsulating Doxorubicin

3.5.1. Encapsulation of Doxorubicin into Liposomes

Liposomes composed of DPPC and Chol with 5 mol % mPEG–DSPE were prepared by the thin-film hydration method followed by the transmembrane pH gradient method [19,32]. In brief, DPPC, Chol, and mPEG–DSPE were dissolved in an organic solvent mixture consisting of chloroform and methanol (9:1, *v*/*v*) in a round-bottom flask equipped with rotary evaporator at 50 °C under aspirator vacuum. The resulting lipid film was left overnight under vacuum to ensure that all traces of organic solvent are removed from the film. Then, the film was hydrated with 300 mM citrate buffer, pH 4.0, by gentle mixing, resulting in spontaneously organized multilamellar vesicles (MLVs). MLVs were freeze-thawed five times and passed through a Whatman polycarbonate membrane with a pore size of 100 nm (GE Healthcare, Carlsbad, CA, USA) ten times using an extruder (Avanti Polar Lipids, Inc., Alabaster, AL, USA) to form small, unilamellar vesicles. The liposome suspension was eluted using Sephadex G-25 (PD-10 desalting column) pre-equilibrated with PBS, pH 7.4, to form a pH gradient. Dox at pH 7.4 was introduced into the liposome suspension, and an excess of free Dox was removed by washing with PBS followed by ultrafiltration using a 100K-membrane filter.

3.5.2. Preparation of M-CTX-Fc-L-Dox or hIgG-L-Dox

The M-CTX-Fc or human IgG were coupling to liposome using classical method by employing the maleimide-thiol addition reaction [33]. Briefly, 2 mol % of Mal–PEG–DSPE were incubated with Dox loaded liposome at 50 °C for 10 min. Simultaneously, thiol group (-SH) were introduced into M-CTX-Fc and human IgG by incubation with Traut's reagent (2-iminothiolane) at molar ratio of 1:10 and 1:50 in 25 mM HEPES, pH 8.0, containing 140 mM NaCl respectively. The reaction occurred under gentle stirring for 1 h in the dark at room temperature. Unreacted 2-IT reagent was removed by using gel chromatography G25 PD-10 column (GE Healthcare, Carlsbad, CA, USA). Thiolated M-CTX-Fc or human IgG was then coupled to Mal–PEG–DSPE by thioether linkage. The coupling reaction was performed overnight in the dark with gentle stirring at 4 °C. Free M-CTX-Fc or human hIgG was removed by ultrafiltration with 100K and 300K membrane filter respectively (Sartorius Stedim Biotech GmbH, Goettingen, Germany) (Figure S2).

3.6. Characterization of Liposomes

3.6.1. Size Distribution and Zeta Potential

The size and zeta potential of liposomes were determined by dynamic and electrophoretic light scattering using an ELS-8000 (Photal Otsuka Electronics, Osaka, Japan).

3.6.2. Encapsulation Efficiency (EE) and Loading Efficiency (LE)

The concentration of doxorubicin was quantified by microplate reader SH-9000 (Corona Electric, Ibaraki, Japan) at 490 nm. EE was then calculated as the amount of drug loaded in the liposomes divided by initial amount of the drug. LE was calculated as the molar ratio of the drug loaded into liposomes to the total of lipid and Chol.

3.7. Evaluation of Cellular Uptake of Liposomes

The cellular uptake of liposomes was evaluated according to the previously described method [15]. U251MG-P1 and SK-BR-3 cells were cultured on gelatin-coated glass coverslips in a 12-well plate at a density of 3×10^5 cells per well for 24 h. Then, M-CTX-Fc and L-Dox (containing Dox at 30 μg/mL) were added to a serum-free medium and incubated at 37 °C in the darkness for 1 h. The cells treated with a medium were used as a negative control. After the incubation, cells were washed three times with cold PBS and fixed with 4% paraformaldehyde for 20 min. After nuclear staining with DAPI (Vector Laboratories, Burlingame, CA, USA), the fluorescent signal was imaged using a laser scanning confocal microscope (FV-1000, Olympus, Tokyo, Japan).

3.8. Cytotoxic Assay

To investigate the in vitro cytotoxicity of various Dox-loaded liposomes in U251MG-P1 and SK-BR-3 cells, tetrazolium reduction assay was employed as previously described [31]. Briefly, 5000 cells/well were seeded onto a 96-well plate and cultured for 24 h. Then, the cells were incubated with 50 μL of a culture medium containing free doxorubicin or different liposome formulations with various doxorubicin concentrations for 72 h at 37 °C and 5% CO_2. Afterward, the cells were exposed to 5 mg/mL of MTT in PBS (the final concentration of 1 mg/mL) for 4 h. Formazan crystals formed during the incubation period were dissolved overnight at 37 °C by adding 10% SDS containing 20 mM HCl. The absorbance was measured at 570 nm using 96-well microplate reader SH-9000 (Corona Electric, Ibaraki, Japan). The exposure time required to kill 50% of the cells was evaluated as IT_{50}, which should be obtained at the minimum concentration of doxorubicin killing 100% of the cells after 72 h of exposure defined as IC_{100}.

3.9. Time-Dependent Cytotoxic Effects

Time-Dependent cellular cytotoxicity [31] was evaluated by the MTT assay with drugs at IC_{100} (Supplementary Table S1). Briefly, U251MG-P1 and SK-BR-3 were seeded in a 96-well plate at 5×10^3 cell/well. After incubation at 37 °C under 5% CO_2 for 24 h, drugs at each IC_{100} were added to each well and incubated for 1, 2, 6, 12, 24, 48, and 72 h. After each round of drug exposure, the medium was replaced with fresh medium without drugs and the incubation was continued until 72 h. Cell viability was determined by MTT assay. The time required for 50% growth inhibition (IT_{50}) was estimated from the survival curve.

3.10. Anti-Tumor Study In Vivo

The xenograft of U251MG-P1 cells in mice was prepared by a subcutaneous injection of 1×10^6 cells/mouse. Tumor volume was measured by a vernier caliper and calculated as [length × $(\text{width})^2$]/2. Anti-tumor effect of each formulation was evaluated when the tumor volume reached 100–200 mm^3. Mice were randomly assigned to five groups (n = 3); group 1 for saline, group 2 for naked Dox, group 3 for L-Dox, group 4 for hIgG-L-Dox and group 5 for M-CTX-Fc-L-Dox. Ten mg of doxorubicin per kg body weight was injected three times via tail vein at the intervals of 7-day. Tumor volume was measured every three days.

3.11. Statistical Analysis

All the experiments were at least three-time repetition. Data were depicted as means ± standard deviation. The statistical significance in mean values between two groups was determined by 2-tailed student's t-test. The statistical significance between the mean values of more than two groups was determined using one-way analysis of variance (ANOVA) and post hoc Tukey HSD. A p-value less than 0.05 was considered to be statistically significant, and a p-value less than 0.01 was regarded as highly significant.

4. Conclusions

In our previous study, U251MG-P1 cells had shown CD44 expression was enriched when compared to the parental U251MG cells, along with aberrant activation of principal stemness marker genes *OCT3/4*, *SOX2*, *KLF4* and *Nanog* and with less expression of glial fibrillary acidic protein. This meant that U251MG-P1 cells consist of rather undifferentiated cells than the parental U251MG cells. Additionally, U251MG-P1 cells are highly tumorigenic exhibiting rapid tumor growth in vivo, when compared to U251MG cells. This high tumorigenicity enabled us to study the drug delivery in vivo.

Drug delivery targeting cancer cells derived from glioblastoma was successfully demonstrated in vitro and in vivo with chlorotoxin fused to human IgG Fc domain in this study. The results showed that chlorotoxin fusion protein shortened the IT_{50} of doxorubicin encapsulated in liposomes in vitro and suppressed the growth of tumor in vivo when compared with the liposomes without chlorotoxin ligand. Although we could not show the direct binding of M-CTX-Fc to MMP-2, we could show significant difference in the relative tumor growth between the treatment with M-CTX-Fc-L-Dox and that with L-Dox at day 20 ($p < 0.05$). This difference appeared to depend on the expression of MMP-2 according to a previous report, which demonstrated the specific interaction of chlorotoxin and MMP-2 [10]. Thus, the combination of doxorubicin and chlorotoxin is proposed in this paper as a successful candidate of liposomal DDS formulation to target glioblastoma cells and derivative tumors.

Supplementary Materials: The following figures are available online at www.mdpi.com/1422-0067/19/3/659/s1.

Acknowledgments: This research was partly supported by the Grant-in-Aid for Scientific Research (A) No. 25242045 (Masaharu Seno); the Grant-in-Aid for Challenging Exploratory Research No. 26640079 (Masaharu Seno); the Japan Science and Technology Agency, Matching Planner Program-Tansaku Shiken-Grant (Tomonari Kasai) and Wesco Scientific Promotion Foundation (Tomonari Kasai).

Author Contributions: Masaharu Seno directed the whole study. Hafizah Mahmud and Tomonari Kasai designed the experiments. Hafizah Mahmud, Tomonari Kasai, Apriliana Cahya Khayrani, and Mami Asakura performed the experiments. Hafizah Mahmud, Apriliana Cahya Khayrani and Tomonari Kasai analyzed the data. Aung Ko Ko Oo, Juan Du, Samah El-Ghlban, and Akifumi Mizutani contributed supplying reagents and materials and supported analysis. Hafizah Mahmud, Tomonari Kasai, Arun Vaidyanath, Akimasa Seno, Akifumi Mizutani, Junko Masuda, Hiroshi Murakami and Masaharu Seno participated in the discussion. Hafizah Mahmud wrote the manuscript.

Conflicts of Interest: The authors declare no conflict of interest.

Abbreviations

BBB	Blood brain barrier
MMP-2	Matrix metalloproteinase-2
CTX	Chlorotoxin
DPPC	Dipalmitoylphosphatidylcholine
MTT	Thiazolyl blue tetrazolium bromide
RES	Reticuloendothelial system
EPR	Enhanced permeability retention
TEM	Transmission electron microscopy
E. coli	*Escherichia coli*
FBS	Fetal bovine serum
M-CTX-Fc-L-Dox	Liposomes conjugated with M-CTX-Fc encapsulating doxorubicin
hIgG-L-Dox	Liposome conjugated with human IgG encapsulating doxorubicin
L-Dox	Liposome encapsulating doxorubicin without ligand

References

1. Van Tellingen, O.; Yetkin-Arik, B.; de Gooijer, M.C.; Wesseling, P.; Wurdinger, T.; de Vries, H.E. Overcoming the blood–brain tumor barrier for effective glioblastoma treatment. *Drug Resist. Updat.* **2015**, *19*, 1–12. [CrossRef] [PubMed]
2. Davis, M.E. Glioblastoma: Overview of disease and treatment. *Clin. J. Oncol. Nurs.* **2016**, *20*, S2–S8. [CrossRef] [PubMed]
3. Wang, S.; Meng, Y.; Li, C.; Qian, M.; Huang, R. Receptor-mediated drug delivery systems targeting to glioma. *Nanomaterials* **2016**, *6*, 3. [CrossRef] [PubMed]
4. Chen, J.; Li, Y.; Yu, T.-S.; McKay, R.M.; Burns, D.K.; Kernie, S.G.; Parada, L.F. A restricted cell population propagates glioblastoma growth following chemotherapy. *Nature* **2012**, *488*, 522–526. [CrossRef] [PubMed]
5. Jhanwar-Uniyal, M.; Labagnara, M.; Friedman, M.; Kwasnicki, A.; Murali, R. Glioblastoma: Molecular pathways, stem cells and therapeutic targets. *Cancers* **2015**, *7*, 538–555. [CrossRef] [PubMed]
6. Ramachandran, R.K.; Sørensen, M.D.; Aaberg-Jessen, C.; Hermansen, S.K.; Kristensen, B.W. Expression and prognostic impact of matrix metalloproteinase-2 (mmp-2) in astrocytomas. *PLoS ONE* **2017**, *12*, e0172234. [CrossRef] [PubMed]
7. Sato, H.; Takino, T. Coordinate action of membrane-type matrix metalloproteinase-1 (mt1-mmp) and mmp-2 enhances pericellular proteolysis and invasion. *Cancer Sci.* **2010**, *101*, 843–847. [CrossRef] [PubMed]
8. Bauvois, B. New facets of matrix metalloproteinases mmp-2 and mmp-9 as cell surface transducers: Outside-in signaling and relationship to tumor progression. *BBA Rev. Cancer* **2012**, *1825*, 29–36. [CrossRef] [PubMed]
9. DeBin, J.A.; Strichartz, G.R. Chloride channel inhibition by the venom of the scorpion leiurus quinquestriatus. *Toxicon* **1991**, *29*, 1403–1408. [CrossRef]
10. Deshane, J.; Garner, C.C.; Sontheimer, H. Chlorotoxin inhibits glioma cell invasion via matrix metalloproteinase-2. *J. Biolog. Chem.* **2003**, *278*, 4135–4144. [CrossRef] [PubMed]

11. Lyons, S.A.; O'Neal, J.; Sontheimer, H. Chlorotoxin, a scorpion-derived peptide, specifically binds to gliomas and tumors of neuroectodermal origin. *Glia* **2002**, *39*, 162–173. [CrossRef] [PubMed]
12. Graf, N.; Mokhtari, T.E.; Papayannopoulos, I.A.; Lippard, S.J. Platinum(iv)-chlorotoxin (ctx) conjugates for targeting cancer cells. *J. Inorg. Biochem.* **2012**, *110*, 58–63. [CrossRef] [PubMed]
13. Fang, C.; Wang, K.; Stephen, Z.R.; Mu, Q.; Kievit, F.M.; Chiu, D.T.; Press, O.W.; Zhang, M. Temozolomide nanoparticles for targeted glioblastoma therapy. *ACS Appl. Mater. Interfaces* **2015**, *7*, 6674–6682. [CrossRef] [PubMed]
14. Soroceanu, L.; Gillespie, Y.; Khazaeli, M.B.; Sontheimer, H. Use of chlorotoxin for targeting of primary brain tumors. *Cancer Res.* **1998**, *58*, 4871–4879. [PubMed]
15. Qin, C.; He, B.; Dai, W.; Lin, Z.; Zhang, H.; Wang, X.; Wang, J.; Zhang, X.; Wang, G.; Yin, L.; et al. The impact of a chlorotoxin-modified liposome system on receptor mmp-2 and the receptor-associated protein clc-3. *Biomaterials* **2014**, *35*, 5908–5920. [CrossRef] [PubMed]
16. Kasai, T.; Nakamura, K.; Vaidyanath, A.; Chen, L.; Sekhar, S.; El-Ghlban, S.; Okada, M.; Mizutani, A.; Kudoh, T.; Murakami, H.; et al. Chlorotoxin fused to igg-fc inhibits glioblastoma cell motility via receptor-mediated endocytosis. *J. Drug Deliv.* **2012**, *2012*, 975763. [CrossRef] [PubMed]
17. El-Ghlban, S.; Kasai, T.; Shigehiro, T.; Yin, H.X.; Sekhar, S.; Ida, M.; Sanchez, A.; Mizutani, A.; Kudoh, T.; Murakami, H.; et al. Chlorotoxin-fc fusion inhibits release of mmp-2 from pancreatic cancer cells. *BioMed Res. Int.* **2014**, *2014*, 10. [CrossRef] [PubMed]
18. Vaidyanath, A.; Mahmud, H.; Apriliana, C.K.; Oo, A.K.K.; Seno, A.; Asakura, M.; Kasai, T.; Seno, M. Hyaluronic acid mediated enrichment of cd44 expressing glioblastoma stem cells in u251mg xenograft mouse model. *J. Stem Cell Res. Ther.* **2017**, *7*, 384. [CrossRef]
19. Abraham, S.A.; Waterhouse, D.N.; Mayer, L.D.; Cullis, P.R.; Madden, T.D.; Bally, M.B. The liposomal formulation of doxorubicin. In *Methods in Enzymology*; Academic Press: Cambridge, MA, USA, 2005; Volume 391, pp. 71–97.
20. Haran, G.; Cohen, R.; Bar, L.K.; Barenholz, Y. Transmembrane ammonium sulfate gradients in liposomes produce efficient and stable entrapment of amphipathic weak bases. *Biochim. Biophys. Acta (BBA) Biomembr.* **1993**, *1151*, 201–215. [CrossRef]
21. Fabel, K.; Dietrich, J.; Hau, P.; Wismeth, C.; Winner, B.; Przywara, S.; Steinbrecher, A.; Ullrich, W.; Bogdahn, U. Long-term stabilization in patients with malignant glioma after treatment with liposomal doxorubicin. *Cancer* **2001**, *92*, 1936–1942. [CrossRef]
22. Hau, P.; Fabel, K.; Baumgart, U.; Rümmele, P.; Grauer, O.; Bock, A.; Dietmaier, C.; Dietmaier, W.; Dietrich, J.; Dudel, C.; et al. Pegylated liposomal doxorubicin-efficacy in patients with recurrent high-grade glioma. *Cancer* **2004**, *100*, 1199–1207. [CrossRef] [PubMed]
23. Fosgerau, K.; Hoffmann, T. Peptide therapeutics: Current status and future directions. *Drug Discov. Today* **2015**, *20*, 122–128. [CrossRef] [PubMed]
24. Czajkowsky, D.M.; Hu, J.; Shao, Z.; Pleass, R.J. Fc-fusion proteins: New developments and future perspectives. *EMBO Mol. Med.* **2012**, *4*, 1015–1028. [CrossRef] [PubMed]
25. Sekhar, S.C.; Kasai, T.; Satoh, A.; Shigehiro, T.; Mizutani, A.; Murakami, H.; El-Aarag, B.Y.A.; Salomon, D.S.; Massaguer, A.; de Llorens, R.; et al. Identification of caveolin-1 as a potential causative factor in the generation of trastuzumab resistance in breast cancer cells. *J. Cancer* **2013**, *4*, 391–401. [CrossRef] [PubMed]
26. Veiseh, M.; Gabikian, P.; Bahrami, S.-B.; Veiseh, O.; Zhang, M.; Hackman, R.C.; Ravanpay, A.C.; Stroud, M.R.; Kusuma, Y.; Hansen, S.J.; et al. Tumor paint: A chlorotoxin:Cy5.5 bioconjugate for intraoperative visualization of cancer foci. *Cancer Res.* **2007**, *67*, 6882–6888. [CrossRef] [PubMed]
27. Li, H.; Yuan, D.; Sun, M.; Ping, Q. Effect of ligand density and peg modification on octreotide-targeted liposome via somatostatin receptor in vitro and in vivo. *Drug Deliv.* **2016**, *23*, 3562–3572. [CrossRef] [PubMed]
28. Chu, C.; Xu, P.; Zhao, H.; Chen, Q.; Chen, D.; Hu, H.; Zhao, X.; Qiao, M. Effect of surface ligand density on cytotoxicity and pharmacokinetic profile of docetaxel loaded liposomes. *Asian J. Pharm. Sci.* **2016**, *11*, 655–661. [CrossRef]
29. Düzgüneş, N.; Nir, S. Mechanisms and kinetics of liposome–cell interactions. *Adv. Drug Deliv. Rev.* **1999**, *40*, 3–18. [CrossRef]
30. Drummond, D.C.; Meyer, O.; Hong, K.; Kirpotin, D.B.; Papahadjopoulos, D. Optimizing liposomes for delivery of chemotherapeutic agents to solid tumors. *Pharmacol. Rev.* **1999**, *51*, 691–744. [PubMed]

31. Shigehiro, T.; Kasai, T.; Murakami, M.; Sekhar, S.C.; Tominaga, Y.; Okada, M.; Kudoh, T.; Mizutani, A.; Murakami, H.; Salomon, D.S.; et al. Efficient drug delivery of paclitaxel glycoside: A novel solubility gradient encapsulation into liposomes coupled with immunoliposomes preparation. *PLoS ONE* **2014**, *9*, e107976. [CrossRef] [PubMed]
32. Alyane, M.; Barratt, G.; Lahouel, M. Remote loading of doxorubicin into liposomes by transmembrane ph gradient to reduce toxicity toward h9c2 cells. *SPJ* **2016**, *24*, 165–175. [CrossRef] [PubMed]
33. Shigehiro, T.; Masuda, J.; Saito, S.; Khayrani, A.; Jinno, K.; Seno, A.; Vaidyanath, A.; Mizutani, A.; Kasai, T.; Murakami, H.; et al. Practical liposomal formulation for taxanes with polyethoxylated castor oil and ethanol with complete encapsulation efficiency and high loading efficiency. *Nanomaterials* **2017**, *7*, 290. [CrossRef] [PubMed]

International Journal of
Molecular Sciences

MDPI

Review

Multifunctional Cargo-Free Nanomedicine for Cancer Therapy

Ying Wang [1], Pengfei Yang [1], Xinrui Zhao [1], Di Gao [1], Na Sun [1], Zhongmin Tian [1], Tianyou Ma [2,*] and Zhe Yang [1,3,*]

1 The Key Laboratory of Biomedical Information Engineering of Ministry of Education, School of Life Science and Technology, Xi'an Jiaotong University, Xi'an 710049, China; wy1127@stu.xjtu.edu.cn (Y.W.); pengfeiyang323@stu.xjtu.edu.cn (P.Y.); zhaoxinrui0103@stu.xjtu.edu.cn (X.Z.); digao5-c@my.cityu.edu.hk (D.G.); shanqilingyun@stu.xjtu.edu.cn (N.S.); zmtian@mail.xjtu.edu.cn (Z.T.)

2 Institute of Endemic Diseases, Environment and Diseases-Related Gene of Key Laboratory of Education Ministry, Medical School of Xi'an Jiaotong University, Xi'an 710061, China

3 Department of Ophthalmology, University of California, San Diego, La Jolla, CA 92093, USA

* Correspondence: maty@xjtu.edu.cn (T.M.); yangzhe@xjtu.edu.cn (Z.Y.); Tel.: +86-29-82-655-193 (T.M.); +86-29-82-663-926-93 (Z.Y.)

Received: 22 August 2018; Accepted: 22 September 2018; Published: 28 September 2018

Abstract: Nanocarriers encapsulating multiple chemotherapeutics are a promising strategy to achieve combinational chemotherapy for cancer therapy; however, they generally use exotic new carriers without therapeutic effect, which usually suffer from carrier-related toxicity issues, as well as having to pass extensive clinical trials to be drug excipients before any clinical applications. Cargo-free nanomedicines, which are fabricated by drugs themselves without new excipients and possess nanoscale characteristics to realize favorable pharmacokinetics and intracellular delivery, have been rapidly developed and drawn much attention to cancer treatment. Herein, we discuss recent advances of cargo-free nanomedicines for cancer treatment. After a brief introduction to the major types of carrier-free nanomedicine, some representative applications of these cargo-free nanomedicines are discussed, including combination therapy, immunotherapy, as well as self-monitoring of drug release. More importantly, this review draws a brief conclusion and discusses the future challenges of cargo-free nanomedicines from our perspective.

Keywords: cargo-free; nanomedicine; combination therapy; cancer therapy

1. Introduction

Cancer is one of the most serious life-threatening diseases [1] and the World Health Organization (WHO) predicted that the number of new cancer cases would increase continuously up to 19.3 million by 2025. Among a series of cancer treatments currently applied in clinical practice, including conventional (surgery, chemotherapy and radiotherapy) and innovative modalities (e.g., phototherapy, immunotherapy and gene therapy), chemotherapy still occupies a crucial position [2,3]. As systemic treatment, chemotherapy can not only shrink the cancer before radiation therapy and surgery but also eliminate metastatic cancer cells due to the widespread body distribution of anticancer drugs in the bloodstream [4,5]. However, there are still many obvious drawbacks to chemotherapy. It has been widely recognized that severe side effects could be caused during treatment due to the nonspecific in vivo distribution of traditional chemotherapeutics. Additionally, the occurrence of multidrug resistance (MDR) after repeated administration of same chemotherapeutic agents could also compromise the chemotherapeutic effect [6]. Thus, it is necessary to develop more effective ways to improve the performance of chemotherapy by fostering its strengths and circumventing its weaknesses.

The delivery of therapeutics using drug carriers appears as a promising and innovative approach to improve cancer treatment. It is largely attributed to the unique benefits of carriers in improving the circulation and therapeutic index of therapeutics and reducing the adverse effects of them on normal tissue [6]. With the development in nanotechnology over past few decades, various nanoformulations have been studied and applied in cancer therapy, including liposomes, micelles, albumin nanoparticles, polymeric nanoparticles and inorganic nanoparticles [7,8]. More importantly, several nanomedicines have been approved for clinical application (Table 1). For example, Doxil® (doxorubicin-loaded PEGylated liposomes) was the first nanomedicine approved by the U.S. Food and Drug Administration (FDA), which is commercially available in the United States. It had shown an enhanced therapeutic effect against acquired immune-deficiency syndrome (AIDS)-related Kaposi's sarcoma and ovarian cancers than standard therapies [9]. In addition, Abraxane® (an albumin-bound paclitaxel nanoparticle) could even generate $967 million in annual revenue due to its impressive therapeutic efficiency by improving its pharmacokinetic properties of paclitaxel compared to Taxol® [10,11].

Table 1. Clinical-stage nanomedicines for cancer treatment [7].

Trade Name	Nanotechnology Platform	Cancer Type	Status
Doxil	Liposome	Ovarian cancer, HIV-related Kaposi sarcoma and multiple myeloma	Approved by the U.S. Food and Drug Administration (FDA)
Genexol-PM	Micelle	Breast cancer and non-small-cell lung cancer	Approved in Korea
Abraxane	Albumin NP	Lung, breast and pancreatic cancer	Approved by FDA
CRLX-101	Polymeric NP	Metastatic renal-cell carcinoma, non-small-cell lung cancer and recurrent tubal, ovarian, or peritoneal cancer	Phase II
CYT-6091	Colloidal gold NP	Advanced solid tumors	Phase I
Kadcyla	Antibody-drug conjugate	HER-2 positive breast cancer	Approved by FDA

Although these carrier-assistant nanomedicine present advantages in combating cancers, there are still some limitations. Firstly, the process of fabricating nanocarriers is extremely complicated in most cases. The challenges that were confronted during materials synthesis and purification, as well as nanoparticle preparation, could prevent their production on a large scale [12–14]. Secondly, almost all nanocarriers for therapeutics delivery are used as excipients without any therapeutic effect. For some particular carriers, their degradations can even induce some adverse effects (e.g., mitochondrial damage and cardiovascular effects) and immune reactions [14–16]. For instance, although poly(lactic acid) (PLA) as a biomaterial has been approved by FDA, it can induce a local inflammatory response following in vivo application [17,18]. Mesoporous silica nanoparticles (MSNs), an excellent candidate among drug carriers, present high drug-loading capacity due to their large surface area [19,20]. However, the long-term toxicity and ineffective biodegradability of MSNs may limit their further application as a drug carrier [15]. Thus, it is necessary to design and prepare innovative nanomedicine platforms to take advantage of nanoformulation without using any unnecessary material.

Cargo-free nanomedicine was formed through self-assembly of therapeutics in the absence of extraneous nanocarriers. This novel nanomedicine has been considered as an excellent paradigm for effective cancer therapy for the following reasons: (1) Cargo-free nanomedicine could prevent drugs from rapid blood/renal clearance by altering their aggregation state. It can also facilitate drugs to accumulate at tumor tissue through the enhanced permeability and retention (EPR) effect due to their reserved nanoscale features; (2) circumvention of additional carrier materials when preparing cargo-free nanomedicine could not only improve drug-loading efficiency (even up to 100% for drug nanocrystals) but also allay concerns about the biosafety of these carrier materials, such as their toxicity and immunogenicity; (3) compared with conventional drug-delivery systems, cargo-free nanomedicine is composed of small molecule units, usually resulting in smaller size and higher penetrability in tumor tissue. These superiorities promote cargo-free nanomedicine to overcome multilayered stromal-cell barriers for deeper tumor penetration and drug perfusion; (4) large-scale fabrication of cargo-free nanomedicine is easily achievable due to their comparatively simple structure [21–24]. Due to the aforementioned advantages, cargo-free nanomedicine has been considered a promising strategy for future clinical applications. Therefore, the objectives of this review are to present an overview of cargo-free nanomedicine, summarize their classification and describe representative applications of these carrier-free nanomedicines for different purposes, including combinational therapy, immunotherapy and self-monitoring (Figure 1). Moreover, this review presents a brief conclusion and discusses the future challenges of cargo-free nanomedicine from our perspective.

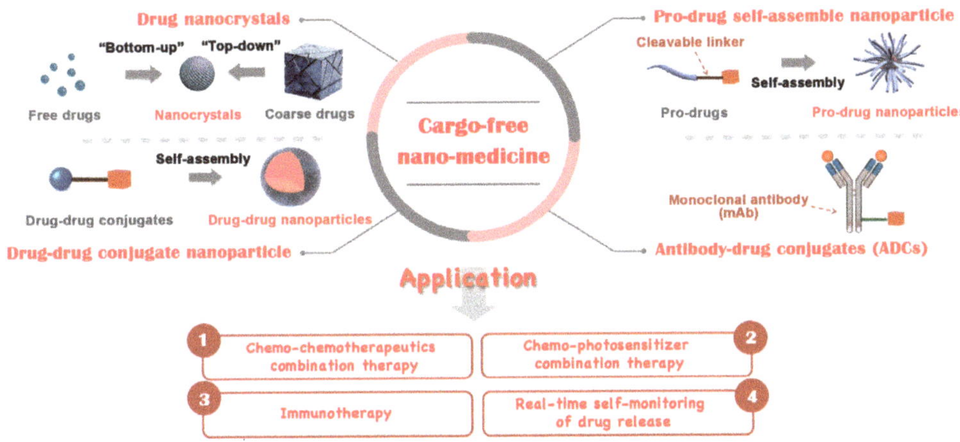

Figure 1. Overview of classification and representative applications of cargo-free nanomedicine.

2. Classification of Cargo-Free Nanomedicine

Cargo-free nanomedicine refers to drug-delivery nanosystems, with no nontherapeutic ingredients, which are used in cancer treatment. According to the difference in the building blocks of these nanosystems, cargo-free nanomedicine could be classified into drug nanocrystals, prodrug self-assembled nanoparticles, drug–drug conjugate nanoparticles and antibody-drug conjugates (ADCs). Drug nanocrystals are a representative example of a self-delivery system aggregated by pure drugs. These nanocrystals could be formulated by one drug alone or multiple drugs and 100% drug-loading capacity is achievable with them. Compared with nanocrystals that consist of disordered, kinetically trapped chains, self-assembled nanomedicine is in a dynamic equilibrium, showing a high level of internal order as well as enhanced stability. However, the hydrophobicity of anticancer drugs usually limits their self-assembly in aqueous environments. Upon introducing hydrophilic groups into drug molecules, the formation of nanoparticles via self-assembly could easily be achieved. Prodrug self-assembled nanoparticles are a type of nanoparticle assembled from drugs predecorated with small groups via susceptible linkers. Amphiphilic drug–drug conjugates consisting of hydrophobic and hydrophilic anticancer drugs were also designed for self-assembling nanoparticles for drug delivery. ADCs, containing three crucial components, that is, monoclonal antibodies (mAb), cytotoxins and chemical linkers, are also cargo-free nanomedicine, showing potential in enhancing cancer-therapy efficacy. The following sections discuss these types of cargo-free nanomedicine in detail with the assistance of several representative examples.

2.1. Drug Nanocrystals

Drug nanocrystals are drug crystals with a nanoscale particle size. They are usually composed of almost 100% hydrophobic drugs and small amounts of ionic or nonionic surfactants to maintain their stability [25–27]. Currently, there are three major approaches for fabricating drug nanocrystals, "top–down," "bottom–up" and combination methods [25,28]. In the "top–down" method, high-pressure homogenization (HPH) [29] or wet-ball milling [30] is adopted to decrease the size of coarse drugs down to nanocrystals; conversely, the "bottom–up" method refers to the self-assembly of drugs into nanoscale particles by nanoprecipitation [31]. Some thermolabile drugs could also experience controlled crystallization during freeze-drying with this method [28,32]. For combination techniques, the "top–down" and "bottom–up" methods are well-integrated with the purpose of fabricating a nanoscale and narrowing size distribution of drug nanocrystals [33–35]. So far, a series of nanocrystal formulations have been prepared and studied for cancer therapy by using the aforementioned methods. Excitingly, six therapeutic nanocrystal products have been commercially available since the 1990s and many others are undergoing clinical trials [25,28]. For example, Theralux®, which is fabricated by Thymectacin, has been studied in a Phase I/II trial for colon-cancer treatment [28]. Panzem®-NCD, a composite of 2-methoxyestradiol nanocrystals (a natural metabolite of estradiol), can efficiently inhibit angiogenesis. For this reason, the combinational treatment of Panzem®-NCD and bevacizumab or sunitinib are used to treat metastatic carcinoid tumors in Phase II clinical trials [36].

The fabrication of two drugs into single-drug nanocrystals has also proven to be another promising strategy to achieve enhanced therapy efficiency through the synergistic effect of these drugs. For instance, Liang's group used a simple reprecipitation method to successfully mix doxorubicin (DOX) and 10-hydroxycamptothecin (HCPT) into a single nanoparticle in the absence of any nanocarrier or solubilizing surfactants [37]. This dual-drug nanocrystal showed enhanced synergistic cytotoxicity to suppress breast cancer. Additionally, Shen's group prepared an amphiphilic antitumor drug (irinotecan hydrochloride, CPT11) as surfactant to emulsify hydrophobic drugs such as camptothecin (CPT) or paclitaxel (PTX) into nanodispersions [13]. Based on their results, both drug bioavailability and anticancer ability improved significantly by employing cargo-free nanomedicine. More importantly, ready clinic translation is more achievable by avoiding additional carriers.

2.2. Prodrug Self-Assembly Nanoparticles

Amphiphilic prodrugs could self-assemble into nanoparticles through intermolecular interaction between their hydrophobic moieties [38,39]. Through using insoluble drugs as these hydrophobic segments, amphiphilic prodrugs with low molecular weight have been successfully synthesized and formulated into a series of cargo-free prodrug self-assembly nanoparticles. Furthermore, due to the presence of some stimuli-responsive cleavable linkers in prodrugs, such as enzyme- [40], acid- [41] and redox-sensitive linkers [42], prodrug nanoparticles can achieve controlled release of the drugs under a special tumor environment or external stimuli [43]. Attributed to overexpressed enzymes/specific secretions, abnormal pH and redox potential in tumor tissue or cells, biologically inactive drugs could readily transform into pharmacologically active ones, leading to the effective inhibition of tumor growth. Chen's group synthesized the acid-sensitive PEGylated doxorubicin (mPEG-CAD) by conjugation of PEG with *cis*-aconitic anhydride-modified DOX (CAD) [41]. The resultant mPEG-CAD was able to self-assemble into micelles in PBS at pH 7.4, which exhibited the clear spherical morphology with the diameter of 100 nm. Moreover, due to its pH-sensitive release, this prodrug nanoparticle possessed enhanced antitumor efficacy as well as upregulated security in vivo compared to free DOX.

Matrix metalloproteinases (MMPs) are overexpressed in a tumor extracellular matrix [44]. Thus, activation of prodrugs by MMPs is another smart way to induce the disassembly of prodrug self-assembly nanoparticles and the controlled release of drugs, leading to enhanced therapeutic efficiency [40,45]. Maruyama's group prepared an MMP7-sensitive precursor of hydrogelation [46]. In the absence of MMP7, the hydrogelation of the precursor solution was inhibited due to the conjugation of the gelator with gelation-preventing moiety. Once taken up by cancer cells, the MMP7-sensitive peptide linkers between these two parts could be cleaved, finally converting into a hydrogel inside the cancer cell. This process resulted in vital stress on cancer cells, initiating cancer-cell necrosis. Sui's group coupled two CPT molecules with an oligomer chain of ethylene glycol (OEG), synthesizing amphiphilic phospholipid-mimicking prodrugs [47]. These obtained prodrugs could form stable liposome-like nanocapsules with high drug-loading efficacy, which achieved higher antitumor activity in vitro and in vivo.

To further endow prodrug nanoparticles with active targeting capacity, some peptides or other molecules were integrated into the prodrugs [48]. These prodrug nanoparticles could specifically recognize receptors on the surface of the cancer-cell membrane. Arginine-glycine-aspartic acid (RGD) peptide is a typical targeting peptide to recognize the integrin $\alpha v\beta 3$ receptor [49]. Zhang's group conjugated CPT and a hydrophobic tail to the end of the Arg-Gly-Asp-Ser (RGDS) peptide, which could self-assemble into fibrillary nanoarchitectures and effectively target integrin-overexpressed cancer cells. Based on their results, this nanomedicine could significantly suppress tumor growth via RGD-mediated specific targeting [50]. Yan's group utilized lactose (Lac) as the targeting ligand to modify the DOX. The Lac-DOX conjugate could self-assemble into nanoparticles in aqueous solution [14]. Moreover, both in vitro and in vivo evaluation demonstrated that the Lac-DOX nanomedicine showed improved antitumor activity and alleviative side effects.

2.3. Drug–Drug Conjugate Nanoparticles

Drug–drug conjugates are a particular kind of amphiphilic prodrug, deserving of a separate discussion. Different from polymer-drug or peptide-drug nanoparticles, amphiphilic drug–drug conjugates are composed of hydrophobic and hydrophilic antitumor drugs. They can also self-assemble into nanoparticles without additional stabilizers, yielding relatively stable drug-delivery systems. In recent years, Yan's group prepared a series of amphiphilic drug–drug pairs by conjugating water-soluble antitumor drugs to poor water-soluble drugs via a hydrolysable ester linker such as irinotecan-chlorambucil (Ir-Cb) [51], irinotecan-doxorubicin (Ir-DOX) [52], floxuridine-bendamustine (FdU-BdM) [53] and floxuridine-CPT (FdU-CPT) [54]. From there, Ir-DOX conjugate nanoparticles were capable of overcoming multidrug resistance (MDR) on breast cancer. Due to aggregation-caused quenching (ACQ) of both Ir and DOX in these nanoparticles, the blue fluorescence of Ir and red fluorescence of DOX were quenched significantly. However, after the Ir-DOX linkage broke and Ir and DOX were released, dual-color fluorescence was recovered, which could be used to real-time track two drugs during cancer therapy. To endow drug–drug conjugate nanoparticles with environment-sensitive release behavior, some sensitive linkers, represented by an ester bond and disulfide bond, were adopted into drug–drug conjugates. For example, Chen's group utilized redox-responsive disulfide linkers to connect evans blue (EB) with CPT, which could self-assemble into nanoformulation in water [55]. Meanwhile, the redox-responsive drug-release manner in cancer cells was also accordingly proved. This cargo-free nanomedicine showed 130-fold longer blood circulation and more efficient tumor accumulation than free drugs, which were beneficial to achieve excellent therapeutic efficacy.

Instead of using covalent bonds to form drug–drug conjugates, Zhu's group employed noncovalent molecular interaction to link hydrophobic (quinazoline-based folate analogue, RT) and hydrophilic drugs (purine nucleoside analog, CA) together [56]. Through computer simulation and transmission electron microscope (TEM) studies, it was demonstrated that the nanomedicine formed by the resultant amphiphilic drug–drug conjugates exhibited stable nanostructures with higher drug loading. Additionally, these nanodrugs were able to simultaneously inhibit ribonucleotide reductase, DNA polymerase and thymidylate synthase through the synergetic effect of CA and RT both in vitro and in vivo.

Moreover, anticancer drugs could be used as a monomer to synthesize polymeric drugs. These polymeric drugs could self-assemble into nanoparticles and present impressive pharmacological activity. For example, Zhou's group selected the antitumor drugs, including cisplatin (Pt(IV)) and demethylcantharidin (DMC, protein phosphatase 2A (PP2A) inhibitor), as monomers for polymerization with ethylenediamine (EDA) via *N*-hydroxysuccinimide (NHS)/1-ethyl-3-(3-dimethylaminopropyl) carbodiimide hydrochloride (EDC), eventually forming the polymeric drugs (Figure 2) [57]. Without inducing any additional carriers, the nanomedicine prepared by the resultant polymeric drugs possessed several excellent advantages: (1) The ratio of Pt(IV) and

DMC in polymeric drugs can be precisely controlled; (2) due to the existence of a heavy metal (Pt) in polymeric drugs, this nanomedicine could also be utilized as a computer tomography (CT) imaging agent to monitor the tumor's status during the therapy; (3) this nanomedicine can effectively suppress the tumor burden on a patient-derived lung-cancer (PDLC) model.

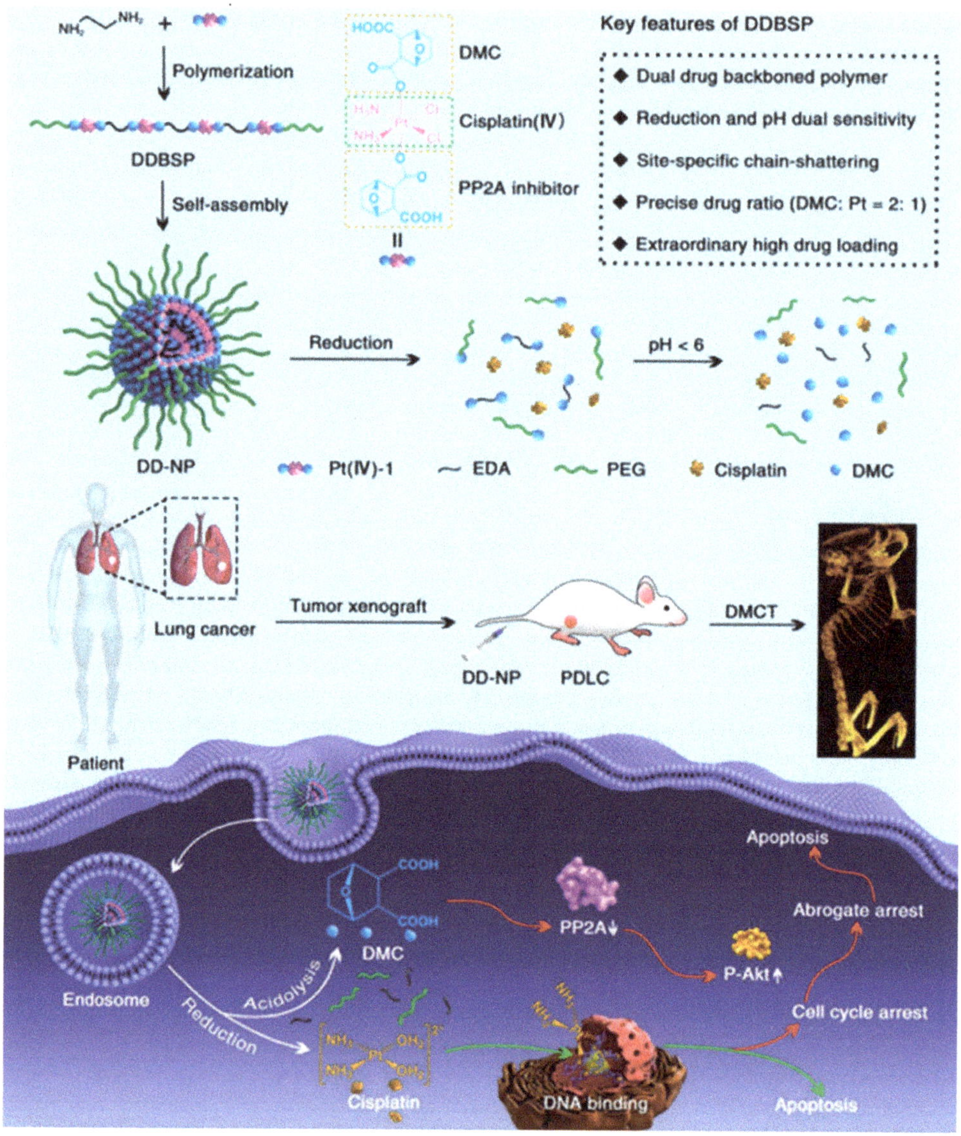

Figure 2. Schematic illustration of a polymeric drug consisting of cisplatin and demethylcantharidin for computer-tomography (CT) imaging and cancer therapy [57].

2.4. ADCs

In recent decades, ADCs have also been recognized as an attractive solution to overcome challenges in cancer therapy. By exploiting the specific binding ability of monoclonal antibodies (mAb) to cell-surface target antigens, ADCs could selectively deliver cytotoxic drugs to tumor cells.

All three components of ADCs, mAb, cytotoxin and the chemical linker that connects them together, are crucial for developing highly efficient ADCs, making ADCs an important type of cargo-free nanomedicine for cancer therapy. The first action of ADCs after being administered intravenously is to recognize the specific antigen on the surface of target cells by its antibody component. Then, the ADC/antigen complex could be internalized by cancer cells through receptor-mediated endocytosis. After transporting the payloads to endosomes and lysosomes, the complex is degraded and the drugs are further released into the cytoplasm to induce cell death with various mechanisms, mainly depending on the type of the toxic drugs. Currently, several ADCs have been approved by the FDA for clinical cancer treatment and over 50 ADCs are in the process of clinical trials [58,59]. In 2000, Gemtuzumab ozogamicin (GO), as the first FDA-approved ADC and the first-generation ADC, was used for monotherapy for patients with acute myeloid leukemia (AML), who were not suitable for conventional chemotherapy. This ADC contains a humanized IgG4 mAb against CD33, a surface antigen that presents in 85–90% of AML patients, which is conjugated to calicheamicin through a cleavable hydrazone linker. The outcomes of GO in the initial single-agent Phase II trials were positive, with a remission rate of 30% [60]. Nevertheless, the required post approval study demonstrated no improvement in survival rate and even higher fatal toxicity in patients treated with GO and chemotherapy compared with patients treated with chemotherapy alone, resulting in the voluntary withdrawal of GO in 2010 [61].

In 2013, one of the second-generation ADCs, ado-trastuzumab emtansine (T-DM1), was approved by the FDA for the treatment of non-hematological malignancies. In T-DM1, a non-cleavable linker connects the well-known antibody trastuzumab (humanized IgG1 anti-HER-2 antibody) and drug maytansinoid DM1. Trastuzumab could not only play the role of active targeting ligand for binding to the HER-2/neu receptor on target cancer cells but also prevent homodimerization or heterodimerization (Her2/Her3) of receptors, finally inhibiting the activation of MAPK and PI3K/AKT cellular signaling pathways to stop the growth of cancer cells. Meanwhile, DM1 are internalized into cancer cells followed by binding to tubulin to induce cell death [62,63]. However, there is a black-box warning of serious side effects, such as hepatotoxicity, embryofetal and cardiac toxicity for T-DM1, creating a need for third-generation ADCs [63].

Vadastuximab talirine, also known as SGN-CD33A, is representative of third-generation ADCs, containing a novel synthetic pyrrolobenzodiazepine (PBD) dimer, which is structurally similar to anthramycin. A humanized anti-CD33 IgG1 antibody is coupled to PBD through a maleimidocaproyl valine-alanine dipeptide linker. Vadastuximab talirine showed robust activity in a series of AML animal models, including those in which GO had minimal effect [64,65]. Moreover, CD33-directed delivery of PBD dimers could overcome transporter-mediated multidrug resistance. Currently, vadastuximab talirine, in combination with azacitidine or decitabine, is undergoing evaluation via a Phase III trial in older patients with newly diagnosed AML.

Apart from ADCs under clinical trials, there is a large number of ADCs in the research phase that are also showing potential for improving cancer therapy [66–68]. Through the exquisite design in the components of mAb, linker and cytotoxic drugs, these ADCs could be considered in the development of potent antitumor therapeutic agents in the future.

3. Representative Applications of Cargo-Free Nanomedicine

Due to tumor heterogeneity [69], single-drug treatment is usually not adequate to eliminate cancer. Additionally, MDR acquired after repeated treatment with single chemotherapeutics could enhance the possibility of tumor metastasis or recurrence after traditional treatments [70]. Thus, combinational treatment with multiple chemotherapeutics or therapeutic modalities is considered a more promising way of cancer therapy. As described above, cargo-free nanomedicine has the potential advantage of incorporating multiple therapeutics into a single nanoplatform in an easy manner, which is beneficial for achieving combination-cancer therapy through a synergetic effect. In the following section, we present some representative applications of cargo-free nanomedicine for combined cancer therapy.

3.1. Chemo–Chemotherapeutics Combination Therapy

The basic principle of chemo–chemotherapy combination therapy is to use drugs with different therapeutic mechanisms and deliver multiple drugs at the maximum tolerated dose. There is a growing awareness that synergistic cytotoxicity could be achieved both in vitro and in vivo with the appropriate combination of drugs. However, the administration of a drug combination through the conventional free "cocktail" method has shown unsatisfactory therapeutic outcomes in vivo due to the different distribution and metabolism of individual drugs. Additionally, a drug combination may result in unexpected systemic toxicity. Therefore, it is necessary to employ an effective system to achieve chemo–chemotherapeutics combination therapy. Due to excellent features, cargo-free nanomedicine is one candidate to achieve this goal. For example, Wu's group prepared a dual-drug nanohybrid that consisted of 10-HCPT nanocrystals integrated with an exterior layer of the methotrexate (MTX)-chitosan prodrug [71]. Since MTX could simultaneously be used as a chemotherapeutic agent for killing cancer cells and a ligand for tumor targeting, this HCPT- and MTX-based dual-drug nanoformulation possessed enhanced cellular internalization and antitumor capacity in Hela cells when compared to either individual drug or their physical mixture. Similarly, Zhang's group employed three poor water-soluble drugs, namely, HCPT, MTX and paclitaxel (PTX), to fabricate cargo-free multidrug nanocrystals (MDNCs) for combination chemotherapy [72]. Through the synergetic effect of these three drugs, this MDNC could effectively overcome MDR to achieve significantly superior anticancer efficacy. MDNCs could also be used as an imaging agent by loading some imaging probes, which render the multidrug nanorods as an all-in-one processing system for cancer diagnosis and treatment.

In recent years, it has been found that there exists a very small population of cancer cells (cancer stem cells, CSCs) in tumors that possess tumor-growth, metastasis and recurrence abilities [73–75]. Thus, our group designed and prepared a pH-responsive cargo-free nanomedicine consisting of pH-responsive prodrug (PEG-C = N-DOX) and 7-ethyl-10-hydroxyl camptothecin (SN38) through their self-assembling. The synergistic effect of the two drugs could simultaneously eradicate breast-cancer cells and cancer stem cells (Figure 3) [76]. In this system, because of the pH-responsive imine linker between DOX and PEG, this cargo-free nanomedicine could achieve pH-controlled release and avoid unexpected drug release during circulation. More importantly, because topoisomerases in CSCs play a key role in facilitating transcription rate, enhancing the expression of antiapoptotic proteins and repairing DNA, the inhibition of both TOP I and TOP II using DOX and SN38 can effectively eliminate non-CSCs and CSCs, improving in vivo therapy effect with fewer adverse effects [77].

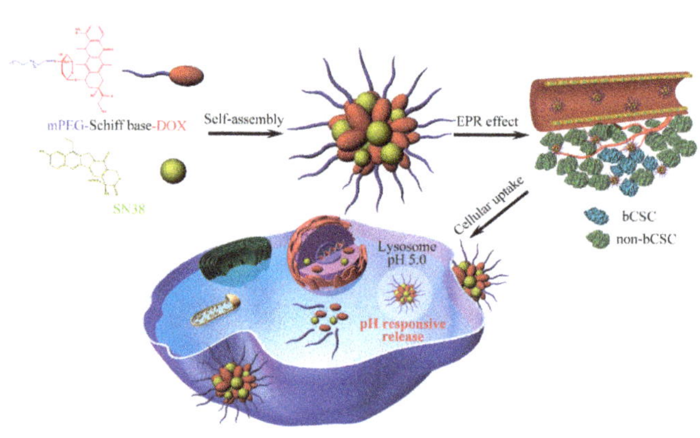

Figure 3. Schematic illustration of pH-responsive cargo-free nanomedicine with improved cellular uptake and controlled drug release for the elimination of bCSCs (breast cancer stem cell) and non-bCSC [76].

3.2. Chemo–Photosensitizer Combination Therapy

3.2.1. Chemo–Photodynamic Combination Therapy

Photodynamic therapy (PDT) is increasingly recognized as an alternative to treating various cancers in clinical trials due to its simplicity, efficiency and fewer side effects [78–80]. PDT uses a light source at appropriate wavelengths to activate tumor local photosensitizers, producing highly toxic reactive oxygen species (ROS), such as singlet oxygen ($^1O_2$) and free radicals. Then, ROS can directly induce cytotoxic effects in tumor cells, damage to the tumor vasculature and a strong inflammatory response that can lead to systemic immunity [78]. The combination of chemotherapy and PDT has been widely studied in recent years, the results of which found that chemo–photodynamic combination therapy can significantly maximize therapeutic efficiency against cancer [81]. As a multifunctional delivery platform for cancer therapy, cargo-free nanomedicine containing several chemotherapeutics and photosensitizers were employed for combined chemo–PDT.

Chlorin e6 (Ce6) is a second-generation photosensitizer, synthesized from chlorophyll. Upon light irradiation by light in the red spectrum (655 nm), the activated Ce6 could efficiently generate $^1O_2$ [82]. Moreover, there exist obvious clinical benefits achieved by the Ce6-mediated PDT on various cancer treatments, such as bladder cancer, melanoma and nasopharyngeal cancer. Therefore, various kinds of cargo-free nanomedicines composed of Ce6 have been developed for chemo–photodynamic combination therapy. For example, Liang's group employed a green and simple reprecipitation method to form stable carrier-free pure nanodrugs (PNDs) consisting of Ce6 and HCPT [82]. Based on their results, it has been found that the ROS-generation efficiency of Ce6 and the cytotoxicity of HCPT were not influenced during the co-self-assembly process. In addition, HCPT/Ce6 PNDs were able to effectively accumulate in tumor tissue due to their nanoscale features. The results of in vitro and in vivo antitumor studies exhibited that chemo–photodynamic combination therapy possessed obviously enhanced anticancer efficacy when compared to single chemotherapy or PDT. Tumor growth was also remarkably suppressed when treated with HCPT/Ce6 PNDs upon light irradiation. It is well known that Ce6 and some chemotherapeutic agents that exist in the aromatic structure can self-assemble into nanoformulation while excluding unnecessary materials.

Yoon's group used a hydrophilic photosensitizer (zinc(II) phthalocyanine tetrasubstituted with 6,8-disulfonate-2-naphthyloxy groups (PcS)) and mitoxantrone (MA) as the host and guest molecules for supramolecular assembly [83]. The resultant nanomedicine displayed enhanced PDT by increasing intratumoral blood flow and relieving tumor hypoxia. Meanwhile, the concomitant chemotherapeutic effect due to the formation of free MA proved to eliminate deeply located tumor cells out of the range of PDT. Similarly, Yin's group also developed a targeted phototheranostic nanodrug (PTN) by covalently conjugating CPT (chemotherapeutic agents) with cyclic RGD (targeting peptides) and pentamethine indocyanine (ICy5, photosensitizers) for self-assembly (Figure 4) [80]. Due ICy5 aggregation, $^1O_2$ generation efficiency was significantly enhanced upon red-light irradiation. Meanwhile, particle size decreased from 90 nm to 10 nm, which was beneficial for deeper tumor penetration as well as CPT release. Thus, this targeted PTN obtained 77.5% tumor-inhibition efficiency through chemo–photodynamic combination therapy. In contrast, the value was only 33.1% and 20% for the single PDT or chemotherapy, respectively. It was indicated that this multifunctional nanodrug would be a promising platform to improve antitumor performance.

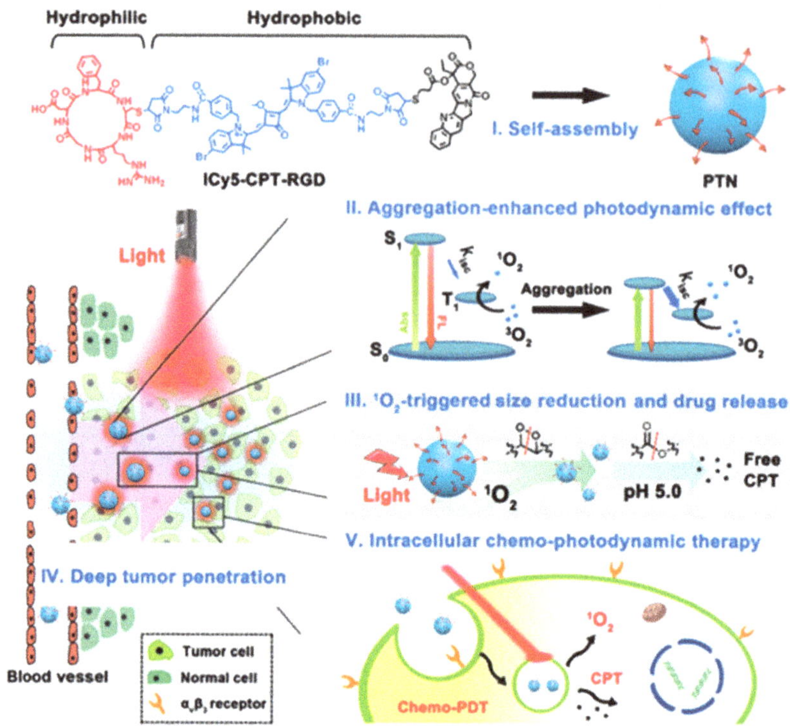

Figure 4. Schematic illustration of cargo-free nanomedicine formed by targeted drug–drug conjugates (ICy5-CPT-RGD) for chemo–photodynamic therapy [80].

3.2.2. Chemo–Photothermal Combination Therapy

In recent years, photothermal therapy (PTT), an extension of photodynamic therapy, has attracted great attention for cancer therapy. Photosensitizers in an excited state activated by near-infrared (NIR) radiation can release energy via non-radiative vibrational relaxation to induce local hyperthermia and cell death, which is the basal principle for organic photosensitizer-mediated PTT. There are numerous advantages for this therapeutic modality, such as increased preservation of surrounding tissue, minimized invasiveness, short recovery time and reduced complication rates [84]. More importantly, it was found that the local hyperthermia effect can assist in the cellular uptake of chemotherapeutics and facilitate the therapeutic efficiency of chemotherapeutic agents both in vitro and in vivo [85]. Thus, a growing number of nanomedicines in combination of chemotherapy and PTT have been fabricated and used to suppress tumor growth through their synergetic effect.

During PTT, photothermal agents play an important role in improving the chemo–photothermal combination therapeutic effect. Various inorganic and organic photothermal agents have been prepared. For inorganic photothermal agents, different carbon nanoformulations [86], gold nanostructures [87], copper sulfide nanoparticles [88], palladium nanosheets [89] and a few others were included [90]. Although these inorganic photothermal agents can effectively transform absorbed NIR optical energy into heat, they still show some disadvantages, such as nonbiodegradability and long-term retention in the body. Due to the potential toxicity caused by inorganic PTT agents to the human body, their further application in the clinic has been limited. In contrast, organic photothermal agents not only possess high NIR absorption and photothermal conversion efficiency but also good biodegradability and structural modifiability. Therefore, organic photothermal agents are usually combined with chemotherapeutics and exploited to construct cargo-free nanomedicine for chemo–photothermal combination therapy.

Indocyanine green (ICG) is a cyanine dye (fluorescent dye) that has been approved by the FDA for medical diagnostics in clinics [91]. It was demonstrated that ICG has better capacity of transforming NIR light energy into heat to induce hyperthermia [92]. Thus, it has also been applied for PTT and photoacoustic (PA) imaging for cancer therapy and diagnosis simultaneously. Hou's group reported a chemotherapeutic drug-photothermal agent, which was composed of ICG and another FDA-approved antitumor drug, epirubicin (EPI), by co-self-assembling into nanoformulation through collaborative interactions (π–π stacking and electrostatic and hydrophobic interactions) in the absence of a molecular precursor and excipient [93]. Similar to other cargo-free nanomedicine, these ICG-EPI nanodrugs has high drug loading (~92%) and excellent stability in physiological conditions. Moreover, due to their preferable photothermal conversion efficacy, these nanodrugs exhibited enhanced synergistic chemo–PTT efficiency and inappreciable toxicity compared to free ICG or EPI. More importantly, as described above, ICG also was a near-infrared fluorescent (NIRF) and PA imaging agent; ICG-EPI nanodrugs also showed high spatial resolution and deep penetration via in vivo NIRF/PA dual-modal imaging. Thus, a multifunctional nanotheranostic platform was able to obtain the targeting multinadal imaging-guided cancer therapy and real-time self-monitoring of drug release. Similarly, Shao's group prepared a targeted nanodrug formed by the co-self-assembling of ursolic acid (UA) as antitumor drug, ICG as PTT and PDT agent and lactobionic acid (LA) as targeting ligand [94]. This nanodrug was then successfully used for cancer imaging and chemo–photo combination therapy on asialoglycoprotein receptor (ASGPR)-overexpressing human hepatocellular carcinoma (HepG2). It is worth mentioning that this multifunctional nanoparticle can simultaneously combine chemotherapy, PTT and PDT to treat cancer under the guidance of NIR-fluorescence imaging, which is essential for a precise and synergistic antitumor effect.

Additionally, Jing's group developed cyanine–curcumin self-assembling nanoformulations (CCNPs) through a single-step reprecipitation method (Figure 5) [95]. The resultant nanomedicine not only showed in vivo photothermal tumor ablation under NIR laser irradiation but also had excellent NIR-imaging ability, indicating that this multifunctional cargo-free nanomedicine would be a promising strategy for drug delivery and cancer treatment.

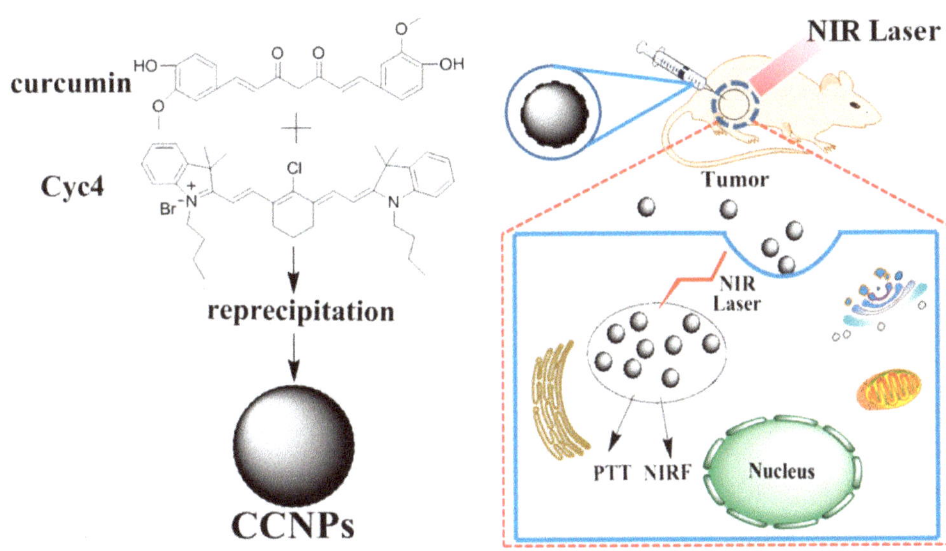

Figure 5. Schematic illustration of self-assembly of CUR and Cyc4 for chemo–photothermal therapy [95].

3.3. Other Applications

3.3.1. Immunotherapy

As an effective way for cancer therapy, cancer immunotherapy has had much focus in recent years, as it can eliminate both metastatic cancer cells and primary cancer cells [96–98]. In addition, with the development of nanotechnology, a growing number of nanoparticles were employed for cancer immunotherapy, including vaccines, immune checkpoint blockade therapy and adoptive T-cell transfer [99,100]. Liu's group prepared cancer-cell membrane-modified adjuvant nanoparticles as a vaccine for cancer therapy [101]; You's group developed hollow gold nanoshell (HAuNS)-loaded PLGA nanoparticles to encapsulate antiprogrammed death 1 (PD-1) peptides, which achieved a preferable anticancer effect through the combination of PTT and immunotherapy [102].

Cargo-free nanomedicines can also be applied for immunotherapy. Ammonio methacrylate copolymers (AMC), the components of controlled-release oral formulations, have been well studied in the pharmaceutical field [103]. Lamprecht's group prepared cargo-free particles of AMC in small (130 nm) and large (500 nm) sizes for immunotherapy [103]. Since immunomodulation can be regulated by different factors of particles, including size, surface charges, composition and morphology, AMC nanoparticles with different size can activate varying immune responses. Based on their results, the level of IL-6, TNF-α, G-CSF and RANTES (CCL5) on immune cells treated with AMC nanoparticles increased with the size of nanoparticles. Moreover, the secretion of other cytokines, like IL-12p40 and IFN-γ, could be facilitated upon cellular exposure to quaternary ammonium groups, all of which induced the apoptosis of cancer cells and immunological memory.

As shown in Figure 6, Shao's group also prepared a carrier-free nanodrug by self-assembly of a pure drug (ursolic acid, UA), which is a food plant-derived natural product possessing good capacity for cancer therapy. As shown in Figure 6, this cargo-free nanomedicine with a diameter of 150 nm could not only induce antiproliferative activity but also inhibit the expression of COX2/VEGFR2/VEGFA and facilitate the immunostimulatory activity of IL-6, IFN-β and TNF-α. Thus, it significantly caused the apoptosis of A549 cells in vitro and suppressed tumor growth in vivo. Additionally, compared to the free drug, this UA carrier-free nanodrug further improved immune function and anticancer efficacy [104].

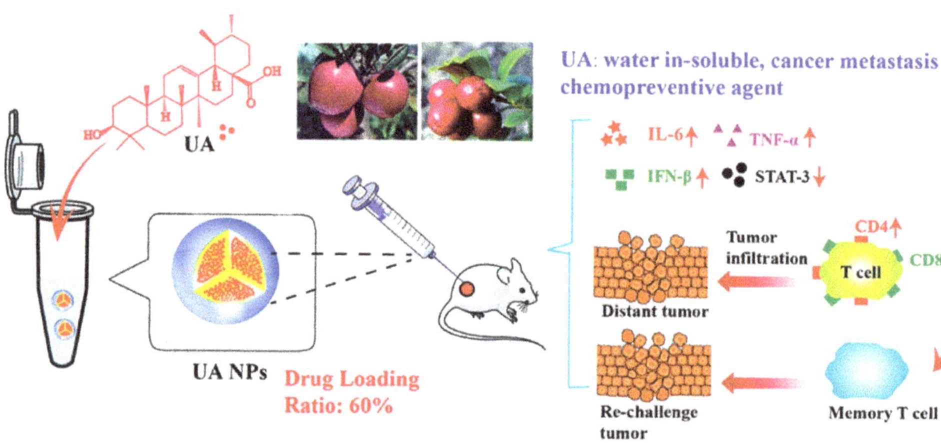

Figure 6. Cargo-free self-delivery system developed by natural anticancer drug ursolic acid for immunotherapy [104].

3.3.2. Real-Time Self-Monitoring of Drug Release

Through an elaborate design by researchers, cargo-free nanomedicines have also been used to real-time self-monitor drug release or track cancer cells as an extension for the application of these nanoparticles for cancer diagnosis. Liang's group employed anticancer drug curcumin (CUR) to construct cargo-free medicine with the reprecipitation method, followed by surface modification with poly(maleic anhydride-alt-1-octadecene)-polyethylene glycol (C18PMH-PEG) through hydrophobic interactions to facilitate further in vivo applications [105]. The green fluorescence of CUR was quenched significantly in the nanomedicine due to serious aggregation and fluorescence would not be restored unless CUR could be released from the nanoparticles. Through their studies by confocal laser-scanning microscope and flow cytometry, the green fluorescence of CUR could not be detected in cancer cells at the beginning of this nanomedicine treatment. However, after longer incubation, CUR's green-fluorescence intensity was gradually enhanced, indicating that this cargo-free nanomedicine exhibited time-dependent cellular release capacity, which could be monitored by the nanomedicines themselves. Similarly, Lee's group developed another self-monitored drug-delivery system, which was composed of CUR, perylene and 5,10,15,20-tetro (4-pyridyl) porphyrin (H2TPyP) (Figure 7) [106]. In this system, CUR could also inhibit cancer-cell growth and self-monitor release by observing their green "ON–OFF" fluorescence. It is worth mentioning that $^1O_2$ generation efficiency could be enhanced via Förster Resonance Energy Transfer (FRET) from perylene under NIR laser irradiation. Therefore, this self-assembled nanomedicine was regarded as a real-time self-monitored and self-delivered drug-delivery system, capable of multiple-fluorescent imaging and chemo–photodynamic therapy.

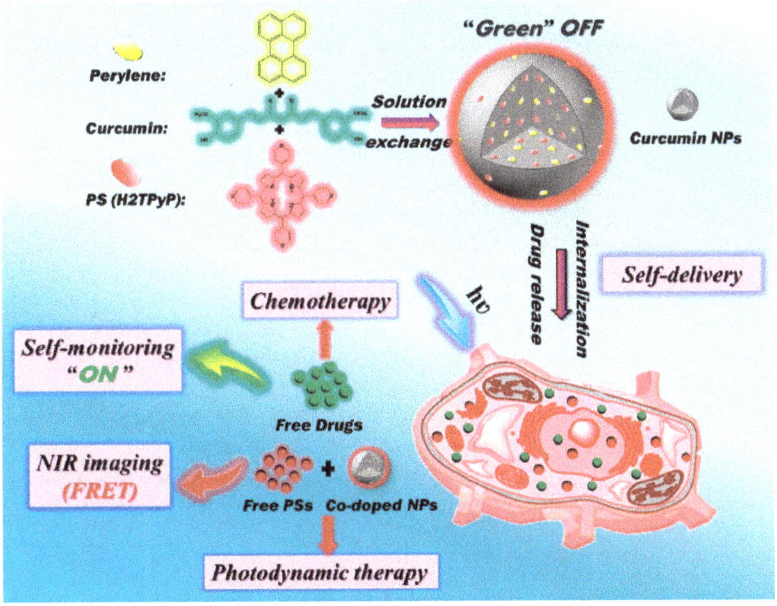

Figure 7. Schematic illustration of a self-monitored drug-delivery system composed of curcumin (CUR), perylene and 5,10,15,20-tetro (4-pyridyl) porphyrin (H2TPyP) for self-monitoring drug release, NIR imaging and cancer therapy [106].

Zhang's group developed surface-functionalized small-molecule organic-dye nanoparticles for cell imaging [107]. Based on their design, poly(maleic anhydride-alt-1-octadecene)-polyethylene glycol (C18PMH-PEG) was selected as the surfactant to surface-modify the preformulated pure organic-dye (2-tert-butyl-9,10-di(naphthalen-2-yl)anthracene, TBADN) nanoparticles. Due to the absence of additional inert materials as carriers, the dye-loading capacity of the resulting TBADN NPs was much higher than those reported carrier-based nanoparticles, leading to significantly enhanced photoluminescence intensity, which was even higher than quantum dots (CdSe/ZnS) under the same conditions. Especially when this pure dye NPs is modified with the targeting ligands (Folic acid), it could be an efficient targeted fluorescence probe for tracking cancer cells.

4. Challenges and Future Perspectives of Cargo-Free Nanomedicine

Although cargo-free nanomedicine can avoid carrier toxicity, effectively deliver antitumor agents to tumor sites and exhibit excellent therapeutic efficiency, it still has a long way to go to be further developed for clinical applications. To further improve the performance of cargo-free nanomedicine, the following two challenges should be carefully considered.

The first challenge is the enhancement of the EPR effect. It is well known that a large number of nanomedicines have been developed for cancer therapy, the effective performance of which was largely depended on the EPR effect. Unfortunately, few of them have accomplished clinical transformation, which is not only attributed to their complicated fabrication processes but also the poor EPR effect they have on humans [11]. In mice, the tumor grows rapidly and blood vessels in tumors tend be much leakier, achieving an effective EFR effect. However, not all tumor vessels are leaky in human tumors; the heterogeneous distribution of pore sizes in human tumors usually compromises the intratumoral accumulation of nanomedicine, resulting in inadequate therapeutic efficacy. To solve this problem, it is necessary to select the most appropriate tumors for cargo-free nanomedicines to exert therapeutic action. In clinics, several kinds of tumors, such as head and neck tumors and Kaposi sarcoma, have been proven to possess an excellent EPR effect and they could be selected as the first series of treatment objects [11]. Using auxiliary approaches to enhance the EPR effect in human tumors could also be another promising way for successful cancer treatment with cargo-free nanomedicine. For example, Jaffray's group found that heat and radiation could obviously facilitate the EPR effect and delivery efficiency of nanomedicine by adjusting human intratumoral-fluid dynamics [108]. Thus, combination of radiation or heat with cargo-free nanomedicine may also be a potential way to improve the therapeutic effect of nanomedicine on human cancer.

The second challenge for cargo-free nanomedicine is to precisely control drug ratios in multidrug nanoparticles. It is demonstrated that combination therapy using more than two kinds of drugs could achieve a better therapeutic effect. Moreover, to maximize efficiency, it is necessary to optimize the ratios of different drugs in nanomedicine or results backfire due to additional side effects of more chemotherapeutics. However, for most prodrugs or drug–drug conjugates, it is hard to precisely control their ratios. Thus, it is urgent to design and develop more reasonable ways to construct cargo-free nanomedicine, like polymeric drugs.

Through the discussion of their advantages, applications and challenges, it is believed that cargo-free nanomedicines are an effective and promising drug-delivery system for cancer treatment. After considerable progress in clinical applications, it is also expected that multifunctional cargo-free nanomedicine would be a milestone in the near future for cancer therapy using nanotechnology.

Author Contributions: Y.W. wrote this manuscript; Y.W., P.Y., X.Z., D.G. and N.S. collected and classified the references; review was performed based on the direction of Z.T., T.M. and Z.Y.

Funding: This work was supported by the National Natural Science Foundation of China (51703178, 81770728, 81570655), the Postdoctoral Science Foundation of Shaanxi Province (2016BSHEDZZ100), the China Postdoctoral Science Foundation (2015M580855, 2018T111057), the Natural Science Foundation of Shaanxi Province (2018JQ5053) and the Fundamental Research Fund for the Central Universities (xjj2016084).

Conflicts of Interest: The authors declare no conflict of interest.

References

1. Siegel, R.L.; Miller, K.D.; Jemal, A. Cancer statistics, 2018. *CA Cancer J. Clin.* **2018**, 68, 7–30. [CrossRef] [PubMed]

2. Chen, Q.; Ke, H.; Dai, Z.; Liu, Z. Nanoscale theranostics for physical stimulus-responsive cancer therapies. *Biomaterials* **2015**, *73*, 214–230. [CrossRef] [PubMed]
3. Zhao, C.-Y.; Cheng, R.; Yang, Z.; Tian, Z.-M. Nanotechnology for cancer therapy based on chemotherapy. *Molecules* **2018**, *23*, 826. [CrossRef] [PubMed]
4. Wu, Q.; Yang, Z.; Nie, Y.; Shi, Y.; Fan, D. Multi-drug resistance in cancer chemotherapeutics: Mechanisms and lab approaches. *Cancer Lett.* **2014**, *347*, 159–166. [CrossRef] [PubMed]
5. Perez-Herrero, E.; Fernandez-Medarde, A. Advanced targeted therapies in cancer: Drug nanocarriers, the future of chemotherapy. *Eur. J. Pharm. Biopharm.* **2015**, *93*, 52–79. [CrossRef] [PubMed]
6. Dong, X.; Mumper, R.J. Nanomedicinal strategies to treat multidrug-resistant tumors: Current progress. *Nanomedicine* **2010**, *5*, 597–615. [CrossRef] [PubMed]
7. Shi, J.; Kantoff, P.W.; Wooster, R.; Farokhzad, O.C. Cancer nanomedicine: Progress, challenges and opportunities. *Nat. Rev. Cancer* **2017**, *17*, 20–37. [CrossRef] [PubMed]
8. Wang, Y.; Sun, S.; Zhang, Z.; Shi, D. Nanomaterials for cancer precision medicine. *Adv. Mater.* **2018**, *30*, e1705660. [CrossRef] [PubMed]
9. Barenholz, Y.C. Doxil®—The first fda-approved nano-drug: Lessons learned. *J. Control. Release* **2012**, *160*, 117–134. [CrossRef] [PubMed]
10. Sofias, A.M.; Dunne, M.; Storm, G.; Allen, C. The battle of "nano" paclitaxel. *Adv. Drug Deliv. Rev.* **2017**, *122*, 20–30. [CrossRef] [PubMed]
11. Danhier, F. To exploit the tumor microenvironment: Since the EPR effect fails in the clinic, what is the future of nanomedicine? *J. Control. Release* **2016**, *244*, 108–121. [CrossRef] [PubMed]
12. Sun, Q.; Radosz, M.; Shen, Y. Challenges in design of translational nanocarriers. *J. Control. Release* **2012**, *164*, 156–169. [CrossRef] [PubMed]
13. Hu, S.; Lee, E.; Wang, C.; Wang, J.; Zhou, Z.; Li, Y.; Li, X.; Tang, J.; Lee, D.H.; Liu, X. Amphiphilic drugs as surfactants to fabricate excipient-free stable nanodispersions of hydrophobic drugs for cancer chemotherapy. *J. Control. Release* **2015**, *220*, 175–179. [CrossRef] [PubMed]
14. Mou, Q.; Ma, Y.; Zhu, X.; Yan, D. A small molecule nanodrug consisting of amphiphilic targeting ligand–chemotherapy drug conjugate for targeted cancer therapy. *J. Control. Release* **2016**, *230*, 34–44. [CrossRef] [PubMed]
15. De Jong, W.H.; Borm, P.J. Drug delivery and nanoparticles: Applications and hazards. *Int. J. Nanomed.* **2008**, *3*, 133–149. [CrossRef]
16. Franz, S.; Rammelt, S.; Scharnweber, D.; Simon, J.C. Immune responses to implants–a review of the implications for the design of immunomodulatory biomaterials. *Biomaterials* **2011**, *32*, 6692–6709. [CrossRef] [PubMed]
17. Athanasiou, K.A.; Niederauer, G.G.; Agrawal, C.M. Sterilization, toxicity, biocompatibility and clinical applications of polylactic acid/polyglycolic acid copolymers. *Biomaterials* **1996**, *17*, 93–102. [CrossRef]
18. Meinel, L.; Hofmann, S.; Karageorgiou, V.; Kirker-Head, C.; McCool, J.; Gronowicz, G.; Zichner, L.; Langer, R.; Vunjak-Novakovic, G.; Kaplan, D.L. The inflammatory responses to silk films in vitro and in vivo. *Biomaterials* **2005**, *26*, 147–155. [CrossRef] [PubMed]
19. Wang, Y.; Zhao, Q.; Han, N.; Bai, L.; Li, J.; Liu, J.; Che, E.; Hu, L.; Zhang, Q.; Jiang, T. Mesoporous silica nanoparticles in drug delivery and biomedical applications. *Nanomed. Nanotechnol.* **2015**, *11*, 313–327. [CrossRef] [PubMed]
20. Yang, P.; Gai, S.; Lin, J. Functionalized mesoporous silica materials for controlled drug delivery. *Chem. Soc. Rev.* **2012**, *41*, 3679–3698. [CrossRef] [PubMed]
21. Qin, S.-Y.; Zhang, A.-Q.; Cheng, S.-X.; Rong, L.; Zhang, X.-Z. Drug self-delivery systems for cancer therapy. *Biomaterials* **2017**, *112*, 234–247. [CrossRef] [PubMed]
22. Ma, W.; Cheetham, A.G.; Cui, H. Building nanostructures with drugs. *Nano Today* **2016**, *11*, 13–30. [CrossRef] [PubMed]
23. Jia, J.; Zhu, F.; Ma, X.; Cao, Z.W.; Li, Y.X.; Chen, Y.Z. Mechanisms of drug combinations: Interaction and network perspectives. *Nat. Rev. Drug Discov.* **2009**, *8*, 111. [CrossRef] [PubMed]
24. Ma, Y.; Mou, Q.; Zhu, X.; Yan, D. Small molecule nanodrugs for cancer therapy. *Mater. Today Chem.* **2017**, *4*, 26–39. [CrossRef]
25. Möschwitzer, J.P. Drug nanocrystals in the commercial pharmaceutical development process. *Int. J. Pharm.* **2013**, *453*, 142–156. [CrossRef] [PubMed]

26. Gao, L.; Zhang, D.; Chen, M. Drug nanocrystals for the formulation of poorly soluble drugs and its application as a potential drug delivery system. *J. Nanopart. Res.* **2008**, *10*, 845–862. [CrossRef]
27. Guo, S.; Huang, L. Nanoparticles containing insoluble drug for cancer therapy. *Biotechnol. Adv.* **2014**, *32*, 778–788. [CrossRef] [PubMed]
28. Miao, X.; Yang, W.; Feng, T.; Lin, J.; Huang, P. Drug nanocrystals for cancer therapy. *Wires. Nanomed. Nanobiotechnol.* **2018**, *10*, e1499. [CrossRef] [PubMed]
29. Keck, C.M.; Müller, R.H. Drug nanocrystals of poorly soluble drugs produced by high pressure homogenisation. *Eur. J. Pharm. Biopharm.* **2006**, *62*, 3–16. [CrossRef] [PubMed]
30. Juhnke, M.; Märtin, D.; John, E. Generation of wear during the production of drug nanosuspensions by wet media milling. *Eur. J. Pharm. Biopharm.* **2012**, *81*, 214–222. [CrossRef] [PubMed]
31. Chan, H.-K.; Kwok, P.C.L. Production methods for nanodrug particles using the bottom-up approach. *Adv. Drug Deliv. Rev.* **2011**, *63*, 406–416. [CrossRef] [PubMed]
32. Van Eerdenbrugh, B.; Van den Mooter, G.; Augustijns, P. Top-down production of drug nanocrystals: Nanosuspension stabilization, miniaturization and transformation into solid products. *Int. J. Pharm.* **2008**, *364*, 64–75. [CrossRef] [PubMed]
33. Rabinow, B.E.; Gupta, P.; Wong, J.; Papadopoulos, P.; Chaubal, M. Formulation of water-insoluble antineoplastic agents as nanosuspensions using nanoedge formulation technology. *Cancer Res.* **2004**, *64*, 146.
34. Salazar, J.; Müller, R.H.; Möschwitzer, J.P. Combinative particle size reduction technologies for the production of drug nanocrystals. *J. Pharm.* **2014**, *2014*, 265754. [CrossRef] [PubMed]
35. Van Eerdenbrugh, B.; Froyen, L.; Van Humbeeck, J.; Martens, J.A.; Augustijns, P.; Van den Mooter, G. Drying of crystalline drug nanosuspensions—The importance of surface hydrophobicity on dissolution behavior upon redispersion. *Eur. J. Pharm. Sci.* **2008**, *35*, 127–135. [CrossRef] [PubMed]
36. Kulke, M.H.; Chan, J.A.; Meyerhardt, J.A.; Zhu, A.X.; Abrams, T.A.; Blaszkowsky, L.S.; Regan, E.; Sidor, C.; Fuchs, C.S. A prospective phase ii study of 2-methoxyestradiol administered in combination with bevacizumab in patients with metastatic carcinoid tumors. *Cancer Chemoth. Pharm.* **2011**, *68*, 293–300. [CrossRef] [PubMed]
37. Chen, F.; Zhao, Y.; Pan, Y.; Xue, X.; Zhang, X.; Kumar, A.; Liang, X.-J. Synergistically enhanced therapeutic effect of a carrier-free hcpt/dox nanodrug on breast cancer cells through improved cellular drug accumulation. *Mol. Pharm.* **2015**, *12*, 2237–2244. [CrossRef] [PubMed]
38. Pastorino, F.; Loi, M.; Sapra, P.; Becherini, P.; Cilli, M.; Emionite, L.; Ribatti, D.; Greenberger, L.M.; Horak, I.D.; Ponzoni, M. Tumor regression and curability of preclinical neuroblastoma models by pegylated sn38 (ezn-2208), a novel topoisomerase I inhibitor. *Clin. Cancer Res.* **2010**, *16*, 4809–4821. [CrossRef] [PubMed]
39. Koseki, Y.; Ikuta, Y.; Kamishima, T.; Onodera, T.; Oikawa, H.; Kasai, H. Drug release is determined by the chain length of fatty acid-conjugated anticancer agent as one component of nano-prodrug. *Bull. Chem. Soc. JPN* **2016**, *89*, 540–545. [CrossRef]
40. Yao, Q.; Kou, L.; Tu, Y.; Zhu, L. Mmp-responsive 'smart' drug delivery and tumor targeting. *Trends Pharmacol. Sci.* **2018**, *39*, 766–781. [CrossRef] [PubMed]
41. Sun, D.; Ding, J.; Xiao, C.; Chen, J.; Zhuang, X.; Chen, X. Preclinical evaluation of antitumor activity of acid-sensitive pegylated doxorubicin. *ACS Appl. Mater. Interfaces* **2014**, *6*, 21202–21214. [CrossRef] [PubMed]
42. Zhang, Q.; Ko, N.R.; Oh, J.K. Recent advances in stimuli-responsive degradable block copolymer micelles: Synthesis and controlled drug delivery applications. *Chem. Commun.* **2012**, *48*, 7542–7552. [CrossRef] [PubMed]
43. Torchilin, V.P. Multifunctional, stimuli-sensitive nanoparticulate systems for drug delivery. *Nat. Rev. Drug Discov.* **2014**, *13*, 813–827. [CrossRef] [PubMed]
44. Egeblad, M.; Werb, Z. New functions for the matrix metalloproteinases in cancer progression. *Nat. Rev. Cancer* **2002**, *2*, 161–174. [CrossRef] [PubMed]
45. Sun, Z.; Li, R.; Sun, J.; Peng, Y.; Xiao, L.; Zhang, X.; Xu, Y.; Wang, M. Matrix metalloproteinase cleavable nanoparticles for tumor microenvironment and tumor cell dual-targeting drug delivery. *ACS Appl. Mater. Interfaces* **2017**, *9*, 40614–40627. [CrossRef] [PubMed]
46. Tanaka, A.; Fukuoka, Y.; Morimoto, Y.; Honjo, T.; Koda, D.; Goto, M.; Maruyama, T. Cancer cell death induced by the intracellular self-assembly of an enzyme-responsive supramolecular gelator. *J. Am. Chem. Soc.* **2015**, *137*, 770–775. [CrossRef] [PubMed]

47. Zhang, H.; Wang, J.; Mao, W.; Huang, J.; Wu, X.; Shen, Y.; Sui, M. Novel sn38 conjugate-forming nanoparticles as anticancer prodrug: In vitro and in vivo studies. *J. Control. Release* **2013**, *166*, 147–158. [CrossRef] [PubMed]
48. Wang, Y.; Cheetham, A.G.; Angacian, G.; Su, H.; Xie, L.; Cui, H. Peptide–drug conjugates as effective prodrug strategies for targeted delivery. *Adv. Drug Deliv. Rev.* **2017**, *110*, 112–126. [CrossRef] [PubMed]
49. Yang, Z.; Luo, X.; Zhang, X.; Liu, J.; Jiang, Q. Targeted delivery of 10-hydroxycamptothecin to human breast cancers by cyclic rgd-modified lipid–polymer hybrid nanoparticles. *Biomed. Mater.* **2013**, *8*, 025012. [CrossRef] [PubMed]
50. Peng, M.; Qin, S.; Jia, H.; Zheng, D.; Rong, L.; Zhang, X. Self-delivery of a peptide-based prodrug for tumor-targeting therapy. *Nano Res.* **2016**, *9*, 663–673. [CrossRef]
51. Huang, P.; Wang, D.; Su, Y.; Huang, W.; Zhou, Y.; Cui, D.; Zhu, X.; Yan, D. Combination of small molecule prodrug and nanodrug delivery: Amphiphilic drug–drug conjugate for cancer therapy. *J. Am. Chem. Soc.* **2014**, *136*, 11748–11756. [CrossRef] [PubMed]
52. Wang, S.; Deng, H.; Huang, P.; Sun, P.; Huang, X.; Su, Y.; Zhu, X.; Shen, J.; Yan, D. Real-time self-tracking of an anticancer small molecule nanodrug based on colorful fluorescence variations. *RSC Adv.* **2016**, *6*, 12472–12478. [CrossRef]
53. Zhang, T.; Huang, P.; Shi, L.; Su, Y.; Zhou, L.; Zhu, X.; Yan, D. Self-assembled nanoparticles of amphiphilic twin drug from floxuridine and bendamustine for cancer therapy. *Mol. Pharm.* **2015**, *12*, 2328–2336. [CrossRef] [PubMed]
54. Hu, M.; Huang, P.; Wang, Y.; Su, Y.; Zhou, L.; Zhu, X.; Yan, D. Synergistic combination chemotherapy of camptothecin and floxuridine through self-assembly of amphiphilic drug–drug conjugate. *Bioconjug. Chem.* **2015**, *26*, 2497–2506. [CrossRef] [PubMed]
55. Zhang, F.; Zhu, G.; Jacobson, O.; Liu, Y.; Chen, K.; Yu, G.; Ni, Q.; Fan, J.; Yang, Z.; Xu, F. Transformative nanomedicine of an amphiphilic camptothecin prodrug for long circulation and high tumor uptake in cancer therapy. *ACS Nano* **2017**, *11*, 8838–8848. [CrossRef] [PubMed]
56. Wang, D.; Yu, C.; Xu, L.; Shi, L.; Tong, G.; Wu, J.; Liu, H.; Yan, D.; Zhu, X. Nucleoside analogue-based supramolecular nanodrugs driven by molecular recognition for synergistic cancer therapy. *J. Am. Chem. Soc.* **2018**, *140*, 8797–8806. [CrossRef] [PubMed]
57. Cong, Y.; Xiao, H.; Xiong, H.; Wang, Z.; Ding, J.; Li, C.; Chen, X.; Liang, X.J.; Zhou, D.; Huang, Y. Dual drug backboned shattering polymeric theranostic nanomedicine for synergistic eradication of patient-derived lung cancer. *Adv. Mater.* **2018**, *30*, 1706220. [CrossRef] [PubMed]
58. Teicher, B.A.; Chari, R.V.J. Antibody conjugate therapeutics: Challenges and potential. *Clin. Cancer Res.* **2011**, *17*, 6389–6397. [CrossRef] [PubMed]
59. Nasiri, H.; Valedkarimi, Z.; Aghebati-Maleki, L.; Majidi, J. Antibody-drug conjugates: Promising and efficient tools for targeted cancer therapy. *J. Cell Physiol.* **2018**, *233*, 6441–6457. [CrossRef] [PubMed]
60. Tsimberidou, A.-M.; Giles, F.J.; Estey, E.; O'Brien, S.; Keating, M.J.; Kantarjian, H.M. The role of gemtuzumab ozogamicin in acute leukaemia therapy. *Br. J. Haematol.* **2006**, *132*, 398–409. [CrossRef] [PubMed]
61. Petersdorf, S.H.; Kopecky, K.J.; Slovak, M.; Willman, C.; Nevill, T.; Brandwein, J.; Larson, R.A.; Erba, H.P.; Stiff, P.J.; Stuart, R.K.; et al. A phase 3 study of gemtuzumab ozogamicin during induction and postconsolidation therapy in younger patients with acute myeloid leukemia. *Blood* **2013**, *121*, 4854–4860. [CrossRef] [PubMed]
62. LoRusso, P.M.; Weiss, D.; Guardino, E.; Girish, S.; Sliwkowski, M.X. Trastuzumab emtansine: A unique antibody-drug conjugate in development for human epidermal growth factor receptor 2–positive cancer. *Clin. Cancer Res.* **2011**, *17*, 6437. [CrossRef] [PubMed]
63. Verma, S.; Miles, D.; Gianni, L.; Krop, I.E.; Welslau, M.; Baselga, J.; Pegram, M.; Oh, D.-Y.; Diéras, V.; Guardino, E.; et al. Trastuzumab emtansine for her2-positive advanced breast cancer. *N. Engl. J. Med.* **2012**, *367*, 1783–1791. [CrossRef] [PubMed]
64. Kung Sutherland, M.S.; Walter, R.B.; Jeffrey, S.C.; Burke, P.J.; Yu, C.; Kostner, H.; Stone, I.; Ryan, M.C.; Sussman, D.; Lyon, R.P.; et al. Sgn-cd33a: A novel cd33-targeting antibody–drug conjugate using a pyrrolobenzodiazepine dimer is active in models of drug-resistant aml. *Blood* **2013**, *122*, 1455. [CrossRef] [PubMed]

65. Kennedy, D.A.; Alley, S.C.; Zhao, B.; Feldman, E.J.; O'Meara, M.; Sutherland, M. Abstract ddt02-04: Sgn-cd33a: Preclinical and phase 1 interim clinical trial results of a cd33-directed pbd dimer antibody-drug conjugate for the treatment of acute myeloid leukemia (AML). *Cancer Res.* **2015**, *75*. [CrossRef]
66. Rudin, C.M.; Pietanza, M.C.; Bauer, T.M.; Ready, N.; Morgensztern, D.; Glisson, B.S.; Byers, L.A.; Johnson, M.L.; Burris, H.A.; Robert, F.; et al. Rovalpituzumab tesirine, a dll3-targeted antibody-drug conjugate, in recurrent small-cell lung cancer: A first-in-human, first-in-class, open-label, phase 1 study. *Lancet Oncol.* **2017**, *18*, 42–51. [CrossRef]
67. Hamblett, K.J.; Hammond, P.W.; Barnscher, S.D.; Fung, V.K.; Davies, R.H.; Wickman, G.R.; Hernandez, A.; Ding, T.; Galey, A.S.; Winters, G.C.; et al. Abstract 3914: Zw49, a her2-targeted biparatopic antibody-drug conjugate for the treatment of her2-expressing cancers. *Cancer Res.* **2018**, *78*, 3914. [CrossRef]
68. Koopman, L.A.; Janmaat, M.L.; Jacobsen, K.; Terp, M.G.; Heuvel, E.G.-V.D.; Forssman, U.; Lingnau, A.; Parren, P.W.; Ditzel, H.; Breij, E.C. Abstract 832: An axl-specific antibody-drug conjugate shows preclinical anti-tumor activity in non-small cell lung cancer, including egfr-inhibitor resistant nsclc. *Cancer Res.* **2018**, *78*, 832. [CrossRef]
69. Marusyk, A.; Almendro, V.; Polyak, K. Intra-tumour heterogeneity: A looking glass for cancer? *Nat. Rev. Cancer* **2012**, *12*, 323–334. [CrossRef] [PubMed]
70. Housman, G.; Byler, S.; Heerboth, S.; Lapinska, K.; Longacre, M.; Snyder, N.; Sarkar, S. Drug Resistance in Cancer: An Overview. *Cancers* **2014**, *6*, 1769–1792. [CrossRef] [PubMed]
71. Wu, S.C.; Yang, X.R.; Lu, Y.; Fan, Z.X.; Li, Y.; Jiang, Y.; Hou, Z.Q. A green approach to dual-drug nanoformulations with targeting and synergistic effects for cancer therapy. *Drug Deliv.* **2017**, *24*, 51–60. [CrossRef] [PubMed]
72. Zhou, M.; Zhang, X.; Yang, Y.; Liu, Z.; Tian, B.; Jie, J.; Zhang, X. Carrier-free functionalized multidrug nanorods for synergistic cancer therapy. *Biomaterials* **2013**, *34*, 8960–8967. [CrossRef] [PubMed]
73. Meacham, C.E.; Morrison, S.J. Tumour heterogeneity and cancer cell plasticity. *Nature* **2013**, *501*, 328–337. [CrossRef] [PubMed]
74. Yang, Z.; Sun, N.; Cheng, R.; Zhao, C.; Liu, Z.; Li, X.; Liu, J.; Tian, Z. Ph multistage responsive micellar system with charge-switch and peg layer detachment for co-delivery of paclitaxel and curcumin to synergistically eliminate breast cancer stem cells. *Biomaterials* **2017**, *147*, 53–67. [CrossRef] [PubMed]
75. Yang, Z.; Sun, N.; Cheng, R.; Zhao, C.; Liu, J.; Tian, Z. Hybrid nanoparticles coated with hyaluronic acid lipoid for targeted co-delivery of paclitaxel and curcumin to synergistically eliminate breast cancer stem cells. *J. Mater. Chem. B* **2017**, *5*, 6762–6775. [CrossRef]
76. Sun, N.; Zhao, C.; Cheng, R.; Liu, Z.; Li, X.; Lu, A.; Tian, Z.; Yang, Z. Cargo-free nanomedicine with ph-sensitivity for co-delivery of dox conjugated prodrug with sn38 to synergistically eradicate breast cancer stem cells. *Mol. Pharm.* **2018**, *15*, 3343–3355. [CrossRef] [PubMed]
77. Wang, H.; Agarwal, P.; Zhao, S.; Xu, R.X.; Yu, J.; Lu, X.; He, X. Hyaluronic acid-decorated dual responsive nanoparticles of pluronic f127, plga, and chitosan for targeted co-delivery of doxorubicin and irinotecan to eliminate cancer stem-like cells. *Biomaterials* **2015**, *72*, 74–89. [CrossRef] [PubMed]
78. Castano, A.P.; Mroz, P.; Hamblin, M.R. Photodynamic therapy and anti-tumour immunity. *Nat. Rev. Cancer* **2006**, *6*, 535–545. [CrossRef] [PubMed]
79. Moan, J.; Peng, Q. An outline of the hundred-year history of pdt. *Anticancer Res.* **2003**, *23*, 3591–3600. [PubMed]
80. Ji, C.; Gao, Q.; Dong, X.; Yin, W.; Gu, Z.; Gan, Z.; Zhao, Y.; Yin, M. Size-reducible nanodrug with aggregation-enhanced photodynamic effect for deep chemo-photodynamic therapy. *Angew. Chem. Int. Ed.* **2018**. [CrossRef]
81. Tian, G.; Zhang, X.; Gu, Z.; Zhao, Y. Recent advances in upconversion nanoparticles-based multifunctional nanocomposites for combined cancer therapy. *Adv. Mater.* **2015**, *27*, 7692–7712. [CrossRef] [PubMed]
82. Wen, Y.; Zhang, W.; Gong, N.; Wang, Y.-F.; Guo, H.-B.; Guo, W.; Wang, P.C.; Liang, X.-J. Carrier-free, self-assembled pure drug nanorods composed of 10-hydroxycamptothecin and chlorin e6 for combinatorial chemo-photodynamic antitumor therapy in vivo. *Nanoscale* **2017**, *9*, 14347–14356. [CrossRef] [PubMed]
83. Li, X.; Yu, S.; Lee, D.; Kim, G.; Lee, B.; Cho, Y.; Zheng, B.-Y.; Ke, M.-R.; Huang, J.-D.; Nam, K.T. Facile supramolecular approach to nucleic-acid-driven activatable nanotheranostics that overcome drawbacks of photodynamic therapy. *ACS Nano* **2017**, *12*, 681–688. [CrossRef] [PubMed]

84. Yang, Z.; Cheng, R.; Zhao, C.; Sun, N.; Luo, H.; Chen, Y.; Liu, Z.; Li, X.; Liu, J.; Tian, Z. Thermo-and ph-dual responsive polymeric micelles with upper critical solution temperature behavior for photoacoustic imaging-guided synergistic chemo-photothermal therapy against subcutaneous and metastatic breast tumors. *Theranostics* **2018**, *8*, 4097–4115. [CrossRef] [PubMed]
85. Li, Z.; Wang, H.; Chen, Y.; Wang, Y.; Li, H.; Han, H.; Chen, T.; Jin, Q.; Ji, J. Ph-and nir light-responsive polymeric prodrug micelles for hyperthermia-assisted site-specific chemotherapy to reverse drug resistance in cancer treatment. *Small* **2016**, *12*, 2731–2740. [CrossRef] [PubMed]
86. Yang, K.; Feng, L.; Shi, X.; Liu, Z. Nano-graphene in biomedicine: Theranostic applications. *Chem. Soc. Rev.* **2013**, *42*, 530–547. [CrossRef] [PubMed]
87. Terentyuk, G.; Panfilova, E.; Khanadeev, V.; Chumakov, D.; Genina, E.; Bashkatov, A.; Tuchin, V.; Bucharskaya, A.; Maslyakova, G.; Khlebtsov, N. Gold nanorods with a hematoporphyrin-loaded silica shell for dual-modality photodynamic and photothermal treatment of tumors in vivo. *Nano Res.* **2014**, *7*, 325–337. [CrossRef]
88. Tian, Q.; Tang, M.; Sun, Y.; Zou, R.; Chen, Z.; Zhu, M.; Yang, S.; Wang, J.; Wang, J.; Hu, J. Hydrophilic flower-like cus superstructures as an efficient 980 nm laser-driven photothermal agent for ablation of cancer cells. *Adv. Mater.* **2011**, *23*, 3542–3547. [CrossRef] [PubMed]
89. Huang, X.; Tang, S.; Mu, X.; Dai, Y.; Chen, G.; Zhou, Z.; Ruan, F.; Yang, Z.; Zheng, N. Freestanding palladium nanosheets with plasmonic and catalytic properties. *Nat. Nanotechnol.* **2011**, *6*, 28–32. [CrossRef] [PubMed]
90. Chou, S.S.; Kaehr, B.; Kim, J.; Foley, B.M.; De, M.; Hopkins, P.E.; Huang, J.; Brinker, C.J.; Dravid, V.P. Chemically exfoliated mos2 as near-infrared photothermal agents. *Angew. Chem. Int. Ed.* **2013**, *125*, 4254–4258. [CrossRef]
91. Kobayashi, H.; Ogawa, M.; Alford, R.; Choyke, P.L.; Urano, Y. New strategies for fluorescent probe design in medical diagnostic imaging. *Chem. Rev.* **2009**, *110*, 2620–2640. [CrossRef] [PubMed]
92. Zheng, M.; Yue, C.; Ma, Y.; Gong, P.; Zhao, P.; Zheng, C.; Sheng, Z.; Zhang, P.; Wang, Z.; Cai, L. Single-step assembly of dox/icg loaded lipid–polymer nanoparticles for highly effective chemo-photothermal combination therapy. *ACS Nano* **2013**, *7*, 2056–2067. [CrossRef] [PubMed]
93. Li, Y.; Liu, G.; Ma, J.; Lin, J.; Lin, H.; Su, G.; Chen, D.; Ye, S.; Chen, X.; Zhu, X. Chemotherapeutic drug-photothermal agent co-self-assembling nanoparticles for near-infrared fluorescence and photoacoustic dual-modal imaging-guided chemo-photothermal synergistic therapy. *J. Control. Release* **2017**, *258*, 95–107. [CrossRef] [PubMed]
94. Zhao, R.; Zheng, G.; Fan, L.; Shen, Z.; Jiang, K.; Guo, Y.; Shao, J.-W. Carrier-free nanodrug by co-assembly of chemotherapeutic agent and photosensitizer for cancer imaging and chemo-photo combination therapy. *Acta Biomater.* **2018**, *70*, 197–210. [CrossRef] [PubMed]
95. Zhang, J.; Liu, S.; Hu, X.; Xie, Z.; Jing, X. Cyanine-curcumin assembling nanoparticles for near-infrared imaging and photothermal therapy. *ACS Biomater. Sci. Eng.* **2016**, *2*, 1942–1950. [CrossRef]
96. Galluzzi, L.; Buque, A.; Kepp, O.; Zitvogel, L.; Kroemer, G. Immunological effects of conventional chemotherapy and targeted anticancer agents. *Cancer cell* **2015**, *28*, 690–714. [CrossRef] [PubMed]
97. Patel, S.A.; Minn, A.J. Combination cancer therapy with immune checkpoint blockade: Mechanisms and strategies. *Immunity* **2018**, *48*, 417–433. [CrossRef] [PubMed]
98. Sharma, P.; Allison, J.P. Immune checkpoint targeting in cancer therapy: Toward combination strategies with curative potential. *Cell* **2015**, *161*, 205–214. [CrossRef] [PubMed]
99. Luo, L.; Shu, R.; Wu, A. Nanomaterial-based cancer immunotherapy. *J. Mater. Chem. B* **2017**, *5*, 5517–5531. [CrossRef]
100. Fan, Q.; Chen, Z.; Wang, C.; Liu, Z. Toward biomaterials for enhancing immune checkpoint blockade therapy. *Adv. Funct. Mater.* **2018**. [CrossRef]
101. Yang, R.; Xu, J.; Xu, L.; Sun, X.; Chen, Q.; Zhao, Y.; Peng, R.; Liu, Z. Cancer cell membrane-coated adjuvant nanoparticles with mannose modification for effective anticancer vaccination. *ACS Nano* **2018**, *12*, 5121–5129. [CrossRef] [PubMed]
102. Luo, L.; Zhu, C.; Yin, H.; Jiang, M.; Zhang, J.; Qin, B.; Luo, Z.; Yuan, X.; Yang, J.; Li, W. Laser immunotherapy in combination with perdurable pd-1 blocking for treatment of metastatic tumor. *ACS Nano* **2018**. [CrossRef] [PubMed]

103. Boushehri, M.A.S.; Stein, V.; Lamprecht, A. Cargo-free particles of ammonio methacrylate copolymers: From pharmaceutical inactive ingredients to effective anticancer immunotherapeutics. *Biomaterials* **2018**, *166*, 1–12. [CrossRef] [PubMed]
104. Fan, L.; Zhang, B.; Xu, A.; Shen, Z.; Guo, Y.; Zhao, R.; Yao, H.; Shao, J.W. Carrier-free, pure nanodrug formed by the self-assembly of an anticancer drug for cancer immune therapy. *Mol. Pharm.* **2018**, *15*, 2466–2478. [CrossRef] [PubMed]
105. Zhang, J.; Li, S.; An, F.-F.; Liu, J.; Jin, S.; Zhang, J.-C.; Wang, P.C.; Zhang, X.; Lee, C.-S.; Liang, X.-J. Self-carried curcumin nanoparticles for in vitro and in vivo cancer therapy with real-time monitoring of drug release. *Nanoscale* **2015**, *7*, 13503–13510. [CrossRef] [PubMed]
106. Zhang, J.; Liang, Y.-C.; Lin, X.; Zhu, X.; Yan, L.; Li, S.; Yang, X.; Zhu, G.; Rogach, A.L.; Yu, P.K. Self-monitoring and self-delivery of photosensitizer-doped nanoparticles for highly effective combination cancer therapy in vitro and in vivo. *ACS Nano* **2015**, *9*, 9741–9756. [CrossRef] [PubMed]
107. Diao, X.; Li, W.; Yu, J.; Wang, X.; Zhang, X.; Yang, Y.; An, F.; Liu, Z.; Zhang, X. Carrier-free, water dispersible and highly luminescent dye nanoparticles for targeted cell imaging. *Nanoscale* **2012**, *4*, 5373–5377. [CrossRef] [PubMed]
108. Stapleton, S.; Dunne, M.; Milosevic, M.; Tran, C.W.; Gold, M.J.; Vedadi, A.; McKee, T.; Ohashi, P.S.; Allen, C.; Jaffray, D.A. Radiation and heat improve the delivery and efficacy of nanotherapeutics by modulating intra-tumoral fluid dynamics. *ACS Nano* **2018**. [CrossRef] [PubMed]

International Journal of
Molecular Sciences

MDPI

Article

Probing Interactions between AuNPs/AgNPs and Giant Unilamellar Vesicles (GUVs) Using Hyperspectral Dark-field Microscopy

Anupama Bhat [1], Kewei Huan [2], Tiana Cooks [1], Hacene Boukari [1] and Qi Lu [1,*]

[1] Department of Physics and Engineering, Delaware State University, 1200 N. DuPont Hwy, Dover, DE 19977, USA; abhat13@students.desu.edu (A.B.); tdcooks15@students.desu.edu (T.C.); hboukari@desu.edu (H.B.)

[2] School of Science, Changchun University of Science and Technology, 7089 Weixing Rd, Changchun 130022, China; huankewei@126.com

* Correspondence: qilu@desu.edu; Tel.: +1-302-857-6806

Received: 16 February 2018; Accepted: 20 March 2018; Published: 28 March 2018

Abstract: Noble metallic nanoparticles (NPs) such as gold and silver nanoparticles (AuNPs and AgNPs) have been shown to exhibit anti-tumor effect in anti-angiogenesis, photothermal and radio therapeutics. On the other hand, cell membranes are critical locales for specific targeting of cancerous cells. Therefore, NP-membrane interactions need be studied at molecular level to help better understand the underlying physicochemical mechanisms for future applications in cancer nanotechnology. Herein, we report our study on the interactions between citrate stabilized colloidal AuNPs/AgNPs (10 nm in size) and giant unilamellar vesicles (GUVs) using hyperspectral dark-field microscopy. GUVs are large model vesicle systems well established for the study of membrane dynamics. GUVs used in this study were prepared with dimyristoyl phosphatidylcholine (DMPC) and doped with cholesterol at various molar concentrations. Both imaging and spectral results support that AuNPs and AgNPs interact very differently with GUVs, i.e., AuNPs tend to integrate in between the lipid bilayer and form a uniform golden-brown crust on vesicles, whereas AgNPs are bejeweled on the vesicle surface as isolated particles or clusters with much varied configurations. The more disruptive capability of AuNPs is hypothesized to be responsible for the formation of golden brown crusts in AuNP-GUV interaction. GUVs of 20 mol% CHOL:DMPC were found to be a most economical concentration for GUVs to achieve the best integrity and the least permeability, consistent with the finding from other phase studies of lipid mixture that the liquid-ordered domains have the largest area fraction of the entire membrane at around 20 mol% of cholesterol.

Keywords: gold nanoparticles; silver nanoparticles; giant unilamellar vesicles; dark-field microscopy; hyperspectral imaging; DMPC; cholesterol

---

1. Introduction

Noble metallic nanoparticles (NPs) such as gold and silver nanoparticles (AuNPs and AgNPs) are clusters of tens to thousands of gold or silver atoms with sizes ranging from 1 to 100 nm. They have been increasingly integrated into a wide array of biomedical applications owing to their unique optoelectronic characteristics and surface chemistry as well as the possibility of well-controlled synthesis. Being biocompatible and inert [1,2], AuNPs are very attractive for biomedical and pharmaceutical applications [3,4], such as drug and gene delivery [5], medical diagnostics [6,7], and therapeutics [8,9]. On the other hand, silver nanoparticles (AgNPs) have long been known for their potent antimicrobial and anti-inflammatory effects for such applications as wound dressing and biomedical implants [10]. Both gold and silver NPs have been shown to exhibit anti-tumor effect through inhibiting the inherent function of heparin binding growth factors thereby suppressing the abnormal growth of blood vessels [11]. Multiple animal studies have shown that gold or silver NPs can significantly reduce the tumor size and enhance the survival rate of animals in either photothermal or radio therapy [11]. Given the promise of gold or silver NPs in targeted elimination of cancerous cells while sparing the normal tissue [11–13], more research is needed at molecular levels for the development of safer, robust, effective and efficient gold and silver NP-based therapeutic systems.

On the other hand, cell membranes are critical locales for the specific targeting of cancerous cells, yet the fundamental mechanisms that govern the interactions of gold or silver NPs and cell membranes remain largely inconclusive. The inevitable adsorption of nanomaterials on the surface of membranes modifies the physicochemical properties of the membrane [14]. Therefore, it is imperative to conduct studies based on simplified model systems for the physicochemical factors to be revealed on interactions occurring between nanoparticles and biological membranes.

Giant unilamellar vesicles (GUVs) are the model membrane system used in this study, because they provide a cell-sized confined volume for the study of biochemical reactions as well as self-assembly processes that allow for tunable lipid composition. The size of GUVs ranges from 1–100 μm, making them easily distinguishable using light and fluorescence microscopy. The simplest GUV consists of a spherical lipid bilayer enclosing a buffer. This membrane bound entity mimics three important features of a cell, namely: compartmentalization, finite surface area and bending elasticity of the cellular membranes. GUVs allow optical examinations at the single-vesicle level thus offer a facile model system to study fundamental membrane thermodynamics, membrane domains and mechanical properties such as membrane curvature, membrane morphology and shape changes [15]. Since phosphatidylcholine is the most abundant lipid in the membranes of eukaryotic cells, the base composition of the GUVs used in this study was chosen to be dimyristoyl phosphatidylcholine (DMPC). DMPC is a widely used zwitterionic phospholipid molecule in model membranes with 14 carbon atoms in the alkyl chain and a melting temperature at 24 °C.

The composition of lipid membranes is an important factor which influences the interaction of NPs with the membrane. Cholesterol (CHOL) is a dominant sterol component in mammalian cell membranes that regulates the fluidity of the latter. Cholesterol behaves as a spacer molecule or dynamic "glue" which fills in between sphingolipids and phospholipids that enables the tight packing to form the floating membrane microdomains known as lipid rafts [16,17]. Lipid rafts are dynamic liquid-ordered platforms that can include or exclude proteins to various extents [18], thus providing anchorage for receptors, coupling factors, effector enzymes, and substrates for the orchestrated function of cell signaling [17]. The modulation of malignant phenotype of cancer cells in terms of cancer cell adhesion, aggressive invasion and metastatic spread occurs at the surface, to a large extent, mediated by lipid rafts [19]. For example, CD44 is a marker molecule expressed in cancer cells and has been associated with cancer cell adhesion, migration, and metastasis. Its abundance in lipid rafts has been noted in several reports [20–22]. Cholesterol depletion treatment has been shown to trigger the shedding of CD44 and hence suppress tumor cell migration [23]. Considering the important role cholesterol plays in regulating the fluidity of cell membranes hence the ability to hold key proteins in cancer therapy, we tested GUVs at various molar concentration of DMPC:CHOL to gain understanding of the effects of cholesterol on GUV-NP interactions.

The major observation technique employed in this work is the dark-field microscopy (DFM), which works by a special optical design that allows only the light scattered from an unstained specimen to be collected by the objective lens while rejecting the illumination. As a result, a brightly-lit image appears against a dark, almost black background with much enhanced signal to noise ratio. Dark-field microscopy is an enabling optical technique for the observation of nanoparticles as small as a few to tens of nanometers under native conditions [24], as the tiny size of nanoparticles is way beyond the resolution limit around 200 nm for regular light microscopes. The other distinctive advantage of DFM, especially in contrast to the fluorescence microscopy, is to enable the observation of biological molecules without conjugation of fluorescence tags. The label-free feature in DMF is advantageous in the investigation of biological systems in a native state, which is of better biomedical relevance as compared to those that are conjugated with fluorescence tags sizable enough to alter the biological condition.

An added benefit of the dark-field microscopy system (CytoViva, Inc., Auburn, AL, USA) used in this work is the formation of hyperspectral imaging (HSI), a technique combining spectrophotometry and imaging [25]. The advanced optics and algorithms built into the system allows for the simultaneous acquisition of the spatial images and the reflectance spectral response at every pixel, nondestructively. Instead of taking a single photograph with only the dominant wavelength, HSI contains the complete spectral response at each pixel which enables the quantitative characterization of NPs and their interactions with biomolecules. The scattering spectra arising from the light-matter interactions encompass implications from the electrical, optical, and plasmonic properties of NPs as well as their local environments. The spectral response can range from the visible near infrared (VNIR) to short wave infrared (SWIR) dependent on the setup of CCD camera. A recent study shows that HSI can enhance the cytologic diagnosis of cancer cells [26].

In this paper, we report our major findings made with hyperspectral dark-field microscopy on the interactions between citrate stabilized colloidal AuNPs or AgNPs and GUVs of varied composition including pure DMPC to different percent molar concentration of CHOL:DMPC (10, 20, 30 and 40 mol%). The size chosen for both AuNPs and AgNPs was 10 nm, because in our earlier work 10-nm AuNPs were found to be most capable in inducing the phase and shape changes in lipid vesicles [27]. The findings reported in this work are intended to provide better understanding regarding the mechanisms of AuNP/AgNP and membrane interactions, which are fundamental and critical for their future applications in both therapeutics and diagnostics of cancer.

2. Results

2.1. Dark-Field Microscopy (DFM) Images and Spectral Profiles of AuNPs and AgNPs

The dark-field images of AuNPs or AgNPs of 10 nm in size appear as bright and shiny dots against a dark background as shown in Figures 1a and 2a. AuNPs exhibit a yellowish-brown hue whereas AgNPs exhibit a greenish hue. Three representative NPs are highlighted in each image acquired with ENVI hyperspectral imaging. The spectral profiles averaged for each NP are displayed in Figures 1b and 2b, respectively. The spectral profiles from AuNPs peak around 590 nm whereas those from AgNPs peak around 550 nm. The spectral profiles from different NPs vary slightly in shape and width but hold the same peak for both AuNPs and AgNPs (Figures 1b and 2b). The spectral profiles from AuNPs are broader than those from AgNPs. It is also found that the spectral profiles from AuNPs (Figure 1b) are more symmetric than those from AgNPs (Figure 2b), with the spectral profiles from AgNPs leaning slightly toward red.

2.2. DFM Images and Spectral Profiles of GUVs Interacting with AuNPs or AgNPs

The GUVs made up of DMPC alone or DMPC doped with varied percent molar concentration of CHOL (10, 20, 30 and 40 mol%) were imaged with or without AuNPs and AgNPs. As found in the analysis of spectral profiles to follow that GUVs of 20 mol% CHOL:DMPC tend to have most prominent shifts in peak wavelength and most broadening in reflectance spectral width, therefore we display the DFM images as well as the spectral profiles obtained from the batch that GUVs were prepared with 20 mol% CHOL:DMPC.

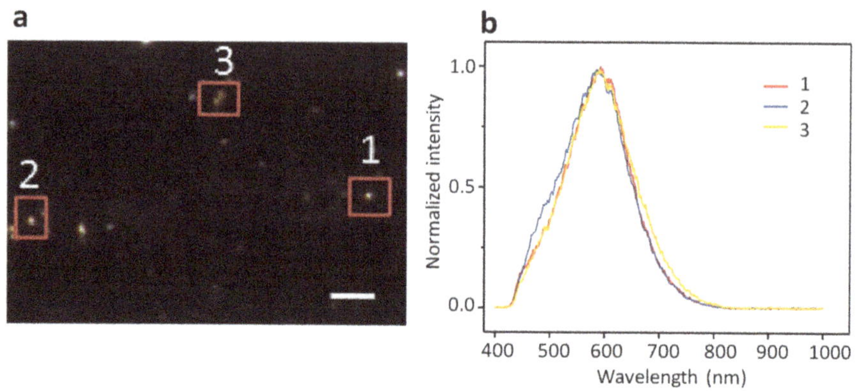

Figure 1. (**a**) A dark-field microscopy (DFM) image of 10-nm AuNPs dispersed in PBS buffer and (**b**) the normalized spectral profiles collected from three different particles specified in (**a**). The three spectral profiles in (**b**) peak are around 590 nm. Scale bar is 5 μm.

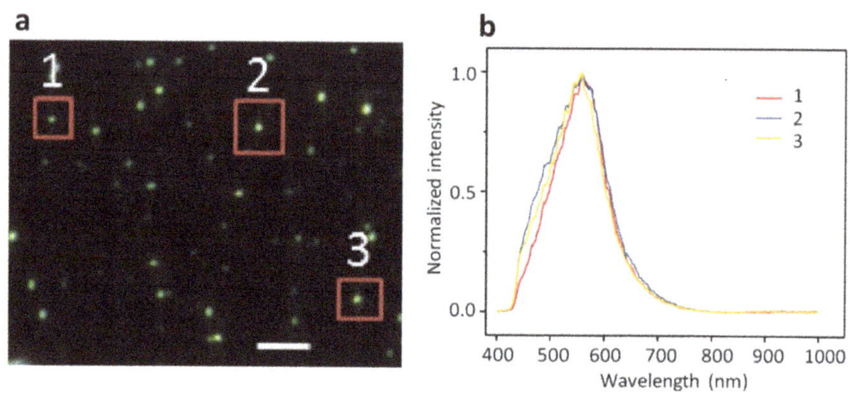

Figure 2. (**a**) A DFM image of 10-nm AgNPs dispersed in PBS buffer and (**b**) the normalized spectral profiles collected from three different particles specified in (**a**). The three spectral profiles in (**b**) peak are around 550 nm. Scale bar is 5 μm.

Shown in Figure 3 are DFM images for (a) GUVs alone, (b) GUVs with AuNPs and (c) GUVs with AgNPs. The spectral profiles acquired inside the squared regions on each image are shown right below in (d), (e) and (f), respectively. The normalized spectral profiles (Figure 3d) from four different lipid bilayer patch regions of the imaged GUV (Figure 3a, 20 mol% CHOL:DMPC in composition) are identical in shape and all peak around 550 nm. The spectral profiles from GUV+AuNPs (Figure 3e) are broadened and shifted in peak from ~590 nm to ~620 nm as compared to those from AuNPs alone (Figure 1b). Also, the spectral profiles collected at different regions show more variations in shape and width as compared to those from GUV alone (Figure 3d). The variations occur mostly on the left halves of the spectral profiles. The spectral profiles from GUV+AuNPs lose the symmetry as found with AuNPs alone (Figure 1b). Similar spectral broadening is observed for GUV+AgNPs (Figure 3f) as well as the peak shift from ~550 nm to ~590 nm as compared to those from AgNPs alone (Figure 2b). Variations in spectral shape and width are also observed for GUV+AgNPs from different regions. Ironically, spectral profiles from GUV+AgNPs are more symmetric than those from AgNPs alone (Figure 2b).

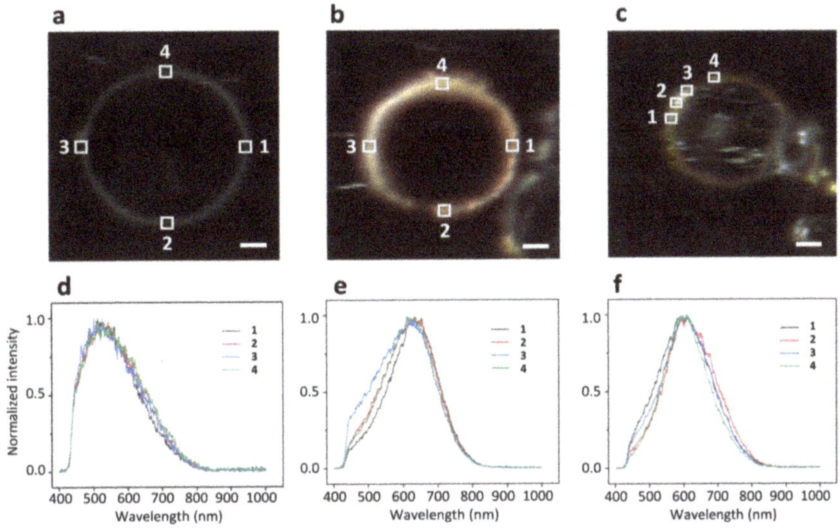

Figure 3. DFM images (**a**) GUV alone, (**b**) AuNPs and GUVs, as well as (**c**) AgNPs and GUVs. The GUVs shown here are made of 20 mol% CHOL:DMPC. (**d**–**f**) are the normalized spectral profiles averaged from inside of the squared regions in images (**a**–**c**), respectively. Scale bars are 5 μm.

Shown in Figure 4 are the normalized spectral profiles averaged from the highlighted regions in Figures 1–3. Five plots are from AuNPs, AgNPs, GUVs, GUV+AuNPs, and GUV+AgNPs, respectively. The GUVs shown in this graph were made of 20 mol% CHOL:DMPC. The red shifts occur for both AuNPs and AgNPs upon interactions with GUVs. The spectral broadening also occurs for both AuNPs and AgNPs upon interactions with GUVs. The spectral profile of GUV alone is more asymmetric than the other four, leaning toward blue and tapering off toward red. The spectral profiles of AgNPs and GUVs are both asymmetric yet resulting in a symmetric profile upon interactions of AgNPs and GUVs. The spectral profile of AuNPs is relatively symmetric yet resulting in a slightly asymmetric profile upon interacting with GUVs.

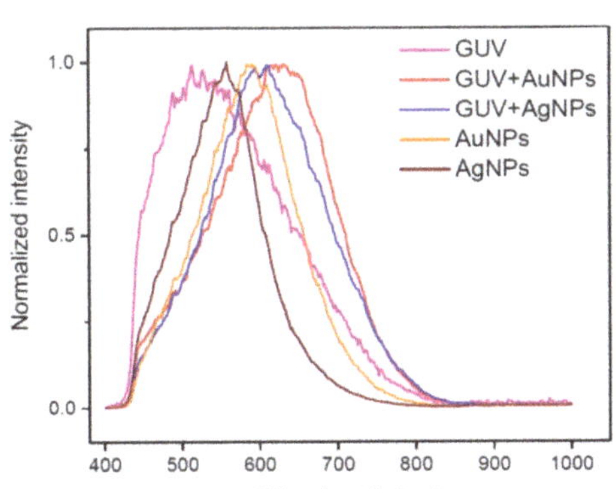

Figure 4. The normalized spectral profiles averaged from the highlighted regions in previous figures for AuNPs, AgNPs, GUVs, GUV+AuNPs, and GUV+AgNPs, respectively. The GUVs shown in this graph were made of 20 mol% CHOL:DMPC.

Figure 5 displays the peak wavelengths and FWHM (full width at half maximum) analyzed from spectral profiles collected from a wide range of samples including AuNPs alone, AgNPs alone, GUVs alone, GUV+AuNPs, and GUV+AgNPs, noting that GUVs used were of varied composition including DMPC alone and 10, 20, 30, 40 mol% of CHOL:DMPC. These two graphs are the most comprehensive exhibition of the experiments conducted in this study. Quite a few data points here have been shown in Figures 1–4 as spectral profiles. Each data point in the graph is the mean and the standard deviation calculated from the spectral profiles of 15 regions of interest selected from that sample.

It is found in Figure 5a and our data chart that the mean peak wavelength of AuNPs is 591 nm and the mean peak wavelength of AgNPs is 548 nm. The peak wavelengths for GUVs of varied composition are around 528 nm except for GUVs of 20 mol% CHOL:DMPC which peaks at 549 nm, coinciding with the peak wavelength of AgNPs. The red shifts of peak wavelength are observed for all cases of GUV+AuNPs and GUV+AgNPs from those of NPs alone. The peak wavelength of GUV+AuNPs shifts to ~616 nm from 591 nm for AuNPs alone, an increase of 25 nm. The peak wavelength of GUV+AgNPs shifts to ~579 nm from 548 nm for AgNPs alone, an increase of 31 nm. In general, the peak wavelengths AuNPs and AgNPs are both of small errors, as expected from the uniformity of particles in size and shape. However, even smaller errors are found for GUVs alone at 40 mol% CHOL:DMPC and GUV+AuNPs at 30 and 40 mol% CHOL:DMPC, suggesting a very stable peak wavelength for high concentration of CHOL especially when interacting with AuNPs. In contrast, much greater errors are noticed for peak wavelengths of GUV+AgNPs for all different compositions of GUVs, indicating that the peak wavelengths shift widely when GUVs interact with AgNPs. This stark contrast in peak wavelength shifts between AuNPs and AgNPs upon interacting with GUVs may suggest a fundamental difference of the underlying mechanism governing the NP-membrane interaction. Also worth noting is the GUV composition of 20 mol% CHOL:DMPC, which not only sees the greatest peak wavelength among all GUV compositions but also the greatest errors for GUVs only and GUV+AgNPs.

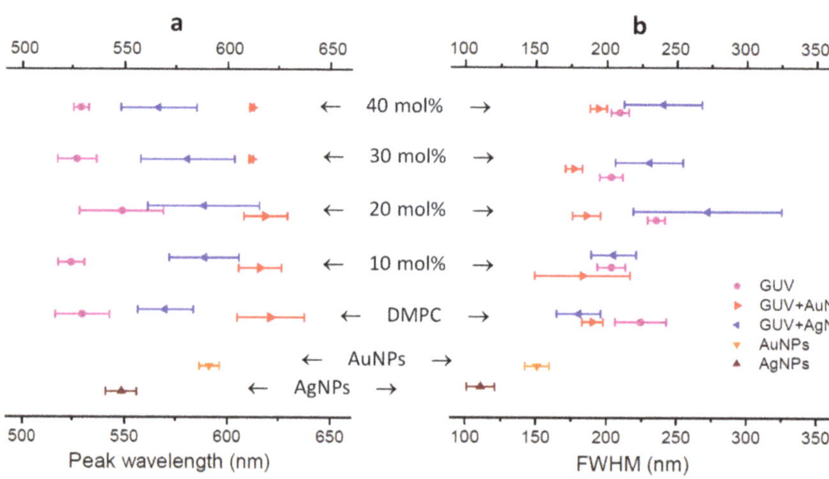

Figure 5. (**a**) The peak wavelengths and (**b**) the FWHM (full width at half maximum) of the spectral profiles from AuNPs alone, AgNPs alone, GUVs alone, GUV+AuNPs, and GUV+AgNPs. The GUV composition was varied from DMPC only to 10, 20, 30 and 40 mol% CHOL:DMPC. The error bars are based on the standard deviations calculated from 15 regions of interest for each sample.

It is found in Figure 5b and our data chart that the mean FWHM of AuNPs is 151 nm and the mean FWHM of AgNPs is 111 nm. The FWHM for GUVs of varied composition are around 210 nm except for GUVs of 20 mol% CHOL:DMPC whose FWHM is 235 nm. The broadening of FWHM is noted for all cases of GUV+AuNPs and GUV+AgNPs from those of NPs alone. The FWHM of GUV+AuNPs increases to ~186 nm from 151 nm for AuNPs alone, a broadening of 35 nm. The FWHM of GUV+AgNPs increases to ~226 nm from 111 nm for AgNPs alone, an astounding broadening of 115 nm, more than doubled. This stark contrast in FWHM broadening between AuNPs and AgNPs upon interacting with GUVs further suggests a difference underlying NP-membrane interactions. Again, at GUV composition of 20 mol% CHOL:DMPC, the greatest error is found for GUV+AgNPs as is the case for peak wavelength of GUV+AgNPs at this GUV composition. GUV+AgNPs at GUV composition of 20 mol% CHOL:DMPC is the most interesting case among all plotted as it shows the greatest error for both the peak wavelength and the FWHM.

Another noticeable general trend observed in Figure 5b is that the mean FWHM of GUVs of all different compositions is ~215 nm, 29 nm broader than that of GUV+AuNPs with FWHM at ~186 nm whereas 11 nm narrower than that of GUV+AgNPs with FWHM at ~226 nm. It is somewhat counterintuitive that AuNPs decrease the spectral width of GUVs upon NP-membrane interactions whereas AgNPs increase the spectral width of GUVs upon NP-membrane interactions.

2.3. DFM Images of GUVs of Varied Composition Interacting with AuNPs or AgNPs

Figure 6 displays an array of DFM images of GUV+AuNPs and Figure 7 displays an array of DFM images of GUV+AgNPs, where GUVs were made of varied composition including DMPC alone and 10, 20, 30, or 40 mol% CHOL:DMPC. When AuNPs interact with GUVs, they tend to form a continuous coverage on the vesicle, forming a golden-brown crust with rarely distinguishable isolated NPs (Figure 6). On the other hand, when AgNPs interact with GUVs, they tend to anchor on the lipid membranes bejeweled on the vesicle surface with much distinguishable isolated NPs (Figure 7).

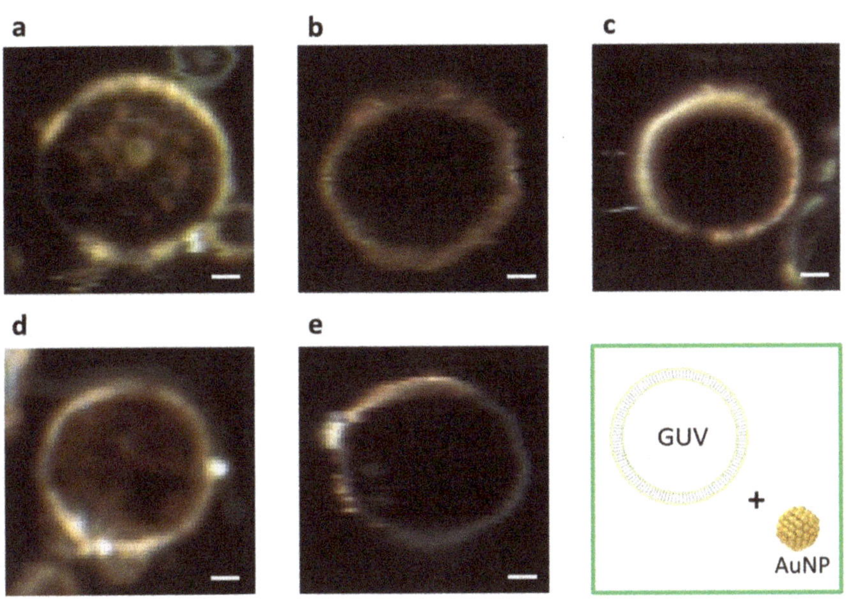

Figure 6. DFM images of GUV+AuNPs where GUVs are of varied composition including (**a**) DMPC only, (**b**) 10 mol%, (**c**) 20 mol%, (**d**) 30 mol%, and (**e**) 40 mol% CHOL:DMPC, respectively. Scale bars are 5 μm.

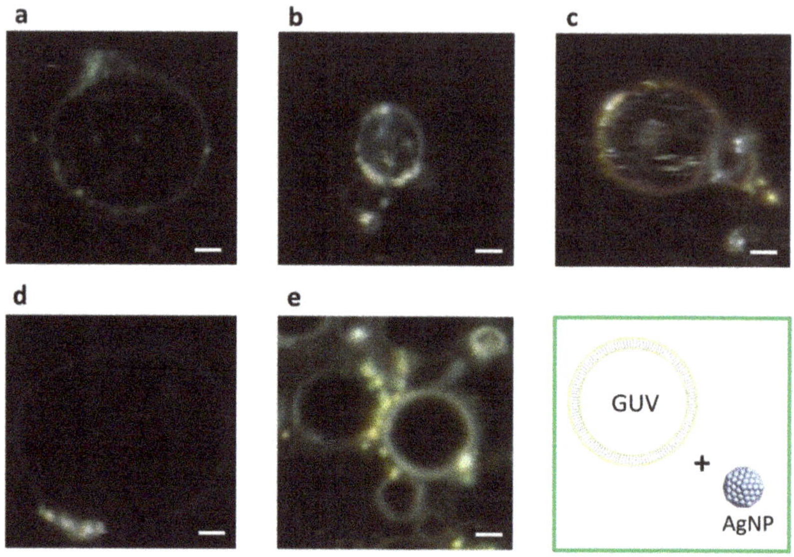

Figure 7. DFM images of GUV+AgNPs where GUVs are of varied composition including (**a**) DMPC only, (**b**) 10 mol%, (**c**) 20 mol%, (**d**) 30 mol%, and (**e**) 40 mol% CHOL:DMPC, respectively. Scale bars are 5 μm.

It is also noticed that among all images displayed, GUVs made of DMPC alone are most permeable, allowing the translocation of noticeable amount of AuNPs or AgNPs inside the vesicles (Figures 6a and 7a). However, AuNPs found inside the vesicles (Figure 6a) are fuzzy, bulky and agglomerated whereas AgNPs found inside the vesicles (Figure 7a) are well-defined, isolated and dispersed. This suggest that AuNPs are more integrated in the lipid bilayers than AgNPs, therefore upon entry into the vesicles lipid coating was carried with AuNPs but rarely with AgNPs. By comparing the NPs inside the vesicles in the image array of Figures 6 and 7, it can be found that the permeability of GUVs in general decreases with the increasing molar concentration of CHOL. This makes sense because the addition of CHOL in GUVs modulates the membrane phase and stiffens the bilayer. AgNPs are more capable of penetrating across the lipid membrane than AuNPs, as AgNPs are still spotted inside GUVs when the concentration of CHOL increased to 10 and 20 mol% (Figure 7b,c). However, the leakage is not observed for AuNPs inside CHOL-doped GUVs (Figure 6b–e).

3. Discussion

The distinct spectral shape and peak wavelength as seen for the spectral profiles from AuNPs vs. AgNPs (Figure 1b vs. Figure 2b) demonstrate that the hyperspectral function of DFM is capable of characterizing NPs regarding their types, shapes, and sizes. The spectral profiles as a result of scattering and reflectance from plasmonic NPs are sensitive to the dielectric medium in the surrounding, thus can be further characterized upon adsorption to lipid membranes to understand the interactions. Red shifts of 25 nm and 31 nm are observed for AuNPs and AgNPs respectively, upon interacting with GUVs (Figure 5). Previously, large red shifts were observed for AuNPs as attributed to NP cluster formation [28]. Shifts in the optical spectra of AgNPs are also expected due to changes in the surrounding dielectric medium upon adsorption, but such shifts are often much lower (~5–10 nm) [29]. Surface plasmons are charge density oscillations confined to the surface of the metal NP. When NPs form cluster, the plasmons undergo hybridization due to interparticle interactions [30]. When the interparticle distance between two NPs is within the range of Coulomb interaction, the charge density oscillations from two individual NPs hybridize to form renormalized plasmon energies. We attribute the red shifts of AuNPs and AgNPs upon interactions with GUVs to a coupled effect of NP aggregation and the surface attachment of lipid molecules.

The results from DFM imaging reveal the morphological difference between GUV+AuNPs and GUV+AgNPs (Figures 6 and 7). AuNPs tend to form a continuous golden-brown crust on the membrane surface whereas AgNPs are bejeweled on the membrane as isolated particles or clusters. This difference of NPs in membrane attachment or integration as observed in spatial imaging is also reflected in the significant difference between the spectral linewidth from GUV+AuNPs and GUV+AgNPs. We learned from spectral analysis of all samples in Figure 5 that AuNPs tend to decrease the FWHM of GUVs whereas AgNPs tend to increase the FWHM of GUVs. Overall, GUV+AgNPs have much greater errors than GUV+AuNPs for both peak wavelengths and FWHM. Among the many causes for optical spectral broadening, we find that the inhomogeneous broadening is the most likely cause for the broadening of GUV+AgNPs. The morphological observation does confirm that GUV+AgNPs adopt much varied configurations than GUV+AuNPs do. The surfaces, grain boundaries, and stoichiometry variations are more pronounced in GUV+AgNPs than in GUV+AuNPs. Therefore, the emitting particles in GUV+AgNPs in much diverse local environments would emit at much different frequencies than particles from GUV+AuNPs do, causing a broadening. On the other hand, because of the uniformity of GUV-AuNP interactions, the spectral linewidths are even smaller than those from GUVs alone.

Based on the differences in morphology and spectral responses between GUV+AuNPs and GUV+AgNPs, we hypothesize that there exists a difference in the mechanism of interactions between AuNPs/AgNPs and GUVs. AuNPs are more likely integrated in the lipid bilayer. If AuNPs were simply adsorbed on the lipid bilayer (or co-localization physically), the resulting spectral profile should have been the overlap of those of AuNPs and GUVs alone. Since the average peaks of GUVs or AuNPs alone were ~580 nm and ~590 nm respectively, the expected spectral peak of the overlap should be between 580 and 590 nm. However, the average spectral peak of GUV+AuNPs is ~615 nm, longer than either alone. This may likely be the result of dampened surface plasmonic effect on AuNPs when they are trapped in between the lipid bilayer. When it comes to AgNPs, they mostly get adsorbed on the lipid bilayer as individual NPs or NP clusters. The spectral peaks of the GUV+AuNPs have large variations shifting between 540 nm and 635 nm with respect to ~550 nm for AgNPs alone. The interactions between AgNPs and GUVs are more varied than between AuNPs and GUVs. Integration and adsorption are equally likely for AgNP-GUV interaction, while integration is more prevalent for AuNP-GUV interaction.

The attachment of AuNPs and AgNPs on GUV surfaces is an inevitable first step between NP-membrane interaction because of the electrostatic force between the negatively charged citrate layer around the NPs (Figure 8a) and the cationic amine and anionic phosphate groups in the outer layer of GUVs (Figure 8b,c). The electrostatic interactions are not only essential for NP adhesion, but also critical for disruption of membranes [31]. Despite that Au and Ag are in the same group on the periodic table, Au has a higher effective nuclear charge relative to Ag. For this reason, citrate anions in the particle solvation sphere should bind more tightly to AuNPs than to AgNPs. Therefore, we hypothesize that the ability of cations to disrupt the outer layer of lipid membranes occurs at a much greater extent for AuNPs than for AgNPs. Numerous experimental and simulation studies have reported the disruption of lipid bilayers by AuNPs [27,31–34]. In the mean while when the AuNPs disrupt the outer leaflet of the lipid bilayer, the citrate anions are likely stripped off and left in the aqueous solution as the hydrophobic force between the acyl chains of DMPC and AuNPs outweighs the solvation force. We hypothesize that AuNPs trapped in the hydrophobic bilayer attempt to relieve its excess of surface energy by recruiting more AuNPs from solution (initially assisted by the disruptive electrostatic force) and seeding the growth of crust in the bilayer [32]. Once AuNPs are integrated in the lipid bilayer, the hydrophobic interaction between AuNPs and lipid chains as well as the Van der Waals (VDW) between AuNPs maintain the stability and equilibrium of the formation. No crust formation was observed for AgNPs, because they are not disruptive enough to initiate the entrapment in the bilayer.

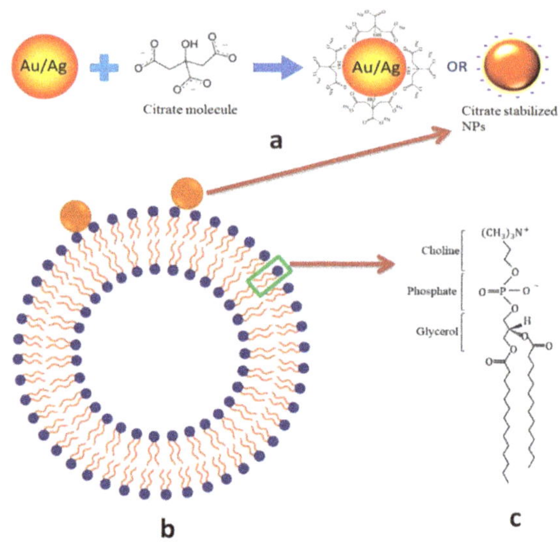

Figure 8. (**a**) Schematic of citrate stabilized NPs. (**b**) Schematic of a giant unilamellar vesicle (GUV) attached with a nanoparticle. (**c**) The molecular structure of DMPC.

GUVs of composition at 20 mol% CHOL:DMPC were identified to exhibit distinct spectral properties as compared to the others. This concentration coincides with the fact that typical nucleated mammalian cell contains between 10% and 20% of cholesterol out of total lipids [35]. Cholesterol plays an important role when it comes to controlling the size and area fraction of lipid phase domains in membranes. In the absence of cholesterol, phospholipid bilayers exist in either a highly ordered gel phase or a liquid-disordered (l_d) phase depending on the temperature. When cholesterol is in presence, an intermediate phase called liquid-ordered (l_o) is formed. Liquid-ordered lipid domains provide anchorage platforms for membrane proteins to form lipid rafts, which are floating microdomains on cell membrane for protein trafficking and cell signal transduction [17]. In previous phase studies of lipid mixture using fluorescence recovery after photobleaching (FRAP), it was found that the percolation threshold concentration of cholesterol is 20–25 mol% in lipid mixture [36]. Percolation threshold is the point where rafts become connected and fluid domains disconnected, when 45–50% of the total membrane is converted to the l_o phase. Beyond 20–25 mol%, cholesterol causes the size of the lipid rafts to decrease [37]. This critical concentration of cholesterol at 20 mol% in a lipid mixture is well coincided in our reflectance spectral analysis of GUVs (Figure 5). At 20 mol% CHOL:DMPC, the peak wavelength of GUVs has a redshift of ~20 nm as compared to other composition and the linewidth is ~25 nm broader than the rest. The 20 mol% CHOL in DMPC seems to be an ideal and economical concentration for GUVs to achieve the best integrity and the least permeability because the liquid-ordered domains have the largest fraction in total membrane area as compared to the other CHOL concentrations. This consistence with other studies also suggests that the hyperspectral analysis technique can be used to characterize the phase and integrity of lipid membranes.

4. Materials and Methods

4.1. Colloidal AuNPs and AgNPs

The AuNPs (10 nm, 0.06 mg/mL, Sigma-Aldrich #752584, sourced from CytoDiagnostics, Inc. Burlington, ON, Canada) used in this work were prepared by reduction of chloroauric acid (HAuCl4) and citrate-stabilized in 0.1 mM PBS [38]. The AgNPs (10 nm, 0.02 mg/mL, Sigma-Aldrich #730785) used in this work were prepared by the reduction of silver nitrate and citrate-stabilized in aqueous buffer [39]. The size of 10 nm was chosen in this study because our earlier work showed that 10-nm AuNPs were most effective in inducing the phase and shape changes in lipid vesicles [27]. The TEM images of AuNPs and AgNPs (Figures S1a and S2) used in this study and more details on size distribution (Figure S1b), concentration and citrate stabilization (Figure S3) can be found in the Supplementary Material.

4.2. Electroformation of GUVs at Various Molar Concentration of Cholesterol vs. DMPC

Electroformation of GUVs is a lab preparation technique for reproducible and controllable production of giant liposomes [40]. It involves the application of an external electric field onto lipid films soaked in hydrating solvent to induce swelling and subsequent vesicle formation. In an improved approach, the alternating (AC) instead of direct (DC) electric field was applied to introduce constant changes in both direction and magnitude of the field intensity, thus agitate lipid molecules to self-assemble into unilamellar bilayer packing and to bud into spherical structures [41]. The AC electroformation of GUVs has enabled various studies on tuning lipid compositions, domain formation and membrane mechanical properties.

The Vesicle Prep Pro apparatus (Nanion Technologies, Munich, Germany, Figure S4) was used for the electroformation of GUVs. A stock solution of phospholipid DMPC (1,2-dimyristoyl-sn-glycero-3-phosphocholine, Sigma-Aldrich) in chloroform ($CHCl_3$) at 6 mg/mL was first prepared. In the case of GUVs at various molar concentration of CHOL to DMPC, a stock solution of CHOL/$CHCl_3$ at 10 mg/mL was mixed into the stock solution of DMPC/$CHCl_3$ at 10, 20, 30, 40 molar percent (mol%) of CHOL:DMPC. Approximately 20 µL of stock solution was dropped on the conducting side of an ITO-coated slide followed by the vacuum evaporation of solvent. A greased O-ring (diameter 28 mm) was then placed around the dried film and filled with 500 µL sorbitol (210 mM). A second ITO slide with conductive side facing down was placed on top of the O-ring to sandwich the soaked film. The ITO slide set was thereafter fit in the electrode chamber of the Vesicle Prep Pro apparatus (Figure S5). Then an alternating voltage of 5 V (p-p) at 5 Hz was applied to the slide chamber at 36 °C. After two hours of running the AC voltage, GUVs are formed and harvested in a vial for future use. More information on the GUV fabrication including the picture of the apparatus and the illustration ITO chamber can be found in Supplementary Material.

4.3. Hyperspectral Dark-Field Microscopy Imaging

20 µL of GUVs of varied composition (DMPC alone and DMPC doped with 10, 20, 30 or 40 mol% of CHOL) were incubated with or without 5 µL of AuNPs or AgNPs (10 nm in size) for two hours prior to the microscopy inspection. To prepare for a sample slide, a drop of 0.5 µL of incubated mixture was streaked on the slide and a cover slip was carefully flapped on the sample to minimize bubble formation. The sample slide was then mounted on the dark-field microscope (CytoViva, Inc., Auburn, AL, USA) for both spatial and hyperspectral imaging. The hyperspectral images were collected and analyzed with ENVI (ENvironment for Visualizing Images, Version 4.8, Harris Geospatial, Boulder, CO, USA), a software application originally designed for the process and analysis of geospatial imagery.

4.4. Analysis of Peak Wavelength and FWHM

The peak wavelength and FWHM (full width at half maximum) were determined from spectral profiles collected from a wide range of samples including AuNPs alone, AgNPs alone, GUVs alone, GUV+AuNPs, and GUV+AgNPs. GUVs tested were of varied composition ranging from DMPC alone to 10, 20, 30, 40 mol% of CHOL:DMPC. For each sample analysis, approximately 15 different regions were selected from several hyperspectral DFM images. In each region of interest, approximately 10 points were selected to produce the average spectral profile. The peak wavelength and FWHM were then determined from the exported spectral data by finding the wavelength corresponding to the maximum intensity and spectral width at half of the maximum intensity. The determination of peak wavelength and

FWHM was performed for all 15 regions of interest for each sample. Then the data of peak wavelength and FWHM were averaged to produce the mean and the standard deviation for each sample.

5. Conclusions

Herein, we present the imaging and spectral results obtained from hyperspectral dark-field microscopy to show the interactions between AuNPs/AgNPs and GUVs. We found that citrate stabilized colloidal AuNPs or AgNPs of 10 nm interact with GUVs very differently. AuNPs tend to integrate in between the lipid bilayer and form a uniform golden-brown crust on vesicles, whereas AgNPs are bejeweled on the vesicle surface as isolated particles or clusters with much varied configurations. The greater ability of AuNPs in disrupting lipid membrane than AgNPs is hypothesized to underlie the different ways AuNPs and AgNPs interact with GUVs. The membrane disruptive ability of AuNPs allows them to entrap in between bilayers and aggregate with other AuNPs to form a golden-brown crust. The permeability of GUVs in general decreases with the increasing molar concentration of CHOL as expected. Among various compositions of GUVs, 20 mol% CHOL:DMPC was found to be an ideal concentration for GUVs to achieve the best integrity and the least permeability, consistent with the finding of other phase studies of lipid mixture that the liquid-ordered domains have the largest area fraction of the entire membrane at 20 mol% of cholesterol. Thereby, hyperspectral analysis is suggested to be a possible technique for phase and integrity characterization of lipid membranes. Upon penetration inside the vesicles at low CHOL concentration, AuNPs were more likely coated with lipid molecules than AgNPs. These results and findings have helped better understand the mechanisms of AuNP/AgNP and membrane interactions, which are fundamental and critical for their future applications in cancer nanotechnology.

Supplementary Materials: Supplementary materials can be found at http://www.mdpi.com/1422-0067/19/4/1014/s1.

Acknowledgments: Anupama Bhat acknowledges the support from the Delaware INBRE program, with a grant from the National Institute of General Medical SciencesNIGMS (P20 GM103446) from the National Institutes of Health and the State of Delaware. Kewei Huan acknowledges the support from the China Scholarship Council (CSC). Tiana Cooks acknowledges the support from the National Science Foundation (NSF-CREST #1242067) and the National Aeronautics and Space Administration (NASA EPSCoR Seed Grant #NNX15AK34A). Hacene Boukari and Qi Lu acknowledge all the grants mentioned above that make this study possible. We also acknowledge the OSCAR Imaging Facility at Delaware State University, which is part of the Delaware Core Facilities Network, for the access of instruments. In particular, Qi Lu acknowledges the insights offered by Robin Helburn from St. Francis College on the physicochemical forces involved in NP-membrane interactions.

Author Contributions: Anupama Bhat and Qi Lu conceived and designed the experiments; Anupama Bhat and Tiana Cooks performed the experiments; Kewei Huan, Tiana Cooks and Qi Lu analyzed the data; Hacene Boukari contributed reagents/materials/analysis tools; Qi Lu and Anupama Bhat wrote the paper.

Conflicts of Interest: The authors declare no conflict of interest.

Abbreviations

AuNPs	Gold nanoparticles
AgNPs	Silver nanoparticles
CHOL	Cholesterol
DFM	Dark-field microscopy
DMPC	Dimyristoyl phosphatidylcholine
FWHM	Full width at half maximum
GUVs	Giant unilamellar vesicles
HSI	Hyperspectral imaging
SWIR	Short wave infrared
VNIR	Visible near infrared

References

1. Connor, E.; Mwamuka, J.; Gole, A.; Murphy, C.; Wyatt, M. Gold Nanoparticles Are Taken Up by Human Cells but Do Not Cause Acute Cytotoxicity. *Small* **2005**, *1*, 325–327. [CrossRef] [PubMed]
2. Bhattacharya, R.; Mukherjee, P. Biological properties of "naked" metal nanoparticles. *Adv. Drug Deliv. Rev.* **2008**, *60*, 1289–1306. [CrossRef] [PubMed]
3. Dykman, L.; Khlebtsov, N. Gold nanoparticles in biomedical applications: Recent advances and perspectives. *Chem. Soc. Rev.* **2012**, *41*, 2256–2282. [CrossRef] [PubMed]
4. Tiwari, P.M.; Vig, K.; Dennis, V.A.; Singh, S.R. Functionalized gold nanoparticles and their biomedical applications. *Nanomaterials* **2011**, *1*, 31–63. [CrossRef] [PubMed]
5. Papasani, M.R.; Wang, G.; Hill, R.A. Gold nanoparticles: The importance of physiological principles to devise strategies for targeted drug delivery. *Nanomed. Nanotechnol. Biol. Med.* **2012**, *8*, 804–814. [CrossRef] [PubMed]
6. Mieszawska, A.J.; Mulder, W.J.M.; Fayad, Z.A.; Cormode, D.P. Multifunctional Gold Nanoparticles for Diagnosis and Therapy of Disease. *Mol. Pharm.* **2013**, *10*, 831–847. [CrossRef] [PubMed]
7. Chuang, Y.-C.; Li, J.-C.; Chen, S.-H.; Liu, T.-Y.; Kuo, C.-H.; Huang, W.-T.; Lin, C.-S. An optical biosensing platform for proteinase activity using gold nanoparticles. *Biomaterials* **2010**, *31*, 6087–6095. [CrossRef] [PubMed]
8. Wang, S.; Chen, K.-J.; Wu, T.-H.; Wang, H.; Lin, W.-Y.; Ohashi, M.; Chiou, P.-Y.; Tseng, H.-R. Photothermal Effects of Supramolecularly Assembled Gold Nanoparticles for the Targeted Treatment of Cancer Cells. *Angew. Chem. Int. Ed.* **2010**, *49*, 3777–3781. [CrossRef] [PubMed]
9. Lee, K.; Lee, H.; Bae, K.H.; Park, T.G. Heparin immobilized gold nanoparticles for targeted detection and apoptotic death of metastatic cancer cells. *Biomaterials* **2010**, *31*, 6530–6536. [CrossRef] [PubMed]
10. Abdelgawad, A.M.; Hudson, S.M.; Rojas, O.J. Antimicrobial wound dressing nanofiber mats from multicomponent (chitosan/silver-NPs/polyvinyl alcohol) systems. *Carbohydr. Polym.* **2014**, *100*, 166–178. [CrossRef] [PubMed]
11. Arvizo, R.R.; Bhattacharyya, S.; Kudgus, R.A.; Giri, K.; Bhattacharya, R.; Mukherjee, P. Intrinsic therapeutic applications of noble metal nanoparticles: Past, present and future. *Chem. Soc. Rev.* **2012**, *41*, 2943–2970. [CrossRef] [PubMed]
12. Jain, S.; Hirst, D.G.; O'Sullivan, J.M. Gold nanoparticles as novel agents for cancer therapy. *Br. J. Radiol.* **2012**, *85*, 101–113. [CrossRef] [PubMed]
13. Jeyaraj, M.; Sathishkumar, G.; Sivanandhan, G.; MubarakAli, D.; Rajesh, M.; Arun, R.; Kapildev, G.; Manickavasagam, M.; Thajuddin, N.; Premkumar, K.; et al. Biogenic silver nanoparticles for cancer treatment: An experimental report. *Colloids Surf. B* **2013**, *106*, 86–92. [CrossRef] [PubMed]
14. Mu, Q.; Jiang, G.; Chen, L.; Zhou, H.; Fourches, D.; Tropsha, A.; Yan, B. Chemical Basis of Interactions Between Engineered Nanoparticles and Biological Systems. *Chem. Rev.* **2014**, *114*, 7740–7781. [CrossRef] [PubMed]
15. Fenz, S.F.; Sengupta, K. Giant vesicles as cell models. *Integr. Biol.* **2012**, *4*, 982–995. [CrossRef] [PubMed]
16. Simons, K.; Ikonen, E. Functional rafts in cell membranes. *Nature* **1997**, *387*, 569–572. [CrossRef] [PubMed]
17. Pike, L.J. Lipid rafts: Bringing order to chaos. *J. Lipid Res.* **2003**, *44*, 655–667. [CrossRef] [PubMed]
18. Resh, M.D. Fatty acylation of proteins: New insights into membrane targeting of myristoylated and palmitoylated proteins. *Biochim. Biophys. Acta* **1999**, *1451*, 1–16. [CrossRef]
19. Hanahan, D.; Weinberg, R. Hallmarks of Cancer: The Next Generation. *Cell* **2011**, *144*, 646–674. [CrossRef] [PubMed]
20. Aruffo, A.; Stamenkovic, I.; Melnick, M.; Underhill, C.B.; Seed, B. CD44 is the principal cell surface receptor for hyaluronate. *Cell* **1990**, *61*, 1303–1313. [CrossRef]
21. Thomas, L.; Byers, H.R.; Vink, J.; Stamenkovic, I. CD44H regulates tumor cell migration on hyaluronate-coated substrate. *J. Cell Biol.* **1992**, *118*, 971–977. [CrossRef] [PubMed]
22. Günthert, U.; Hofmann, M.; Rudy, W.; Reber, S.; Zöller, M.; Hauβmann, I.; Matzku, S.; Wenzel, A.; Ponta, H.; Herrlich, P. A new variant of glycoprotein CD44 confers metastatic potential to rat carcinoma cells. *Cell* **1991**, *65*, 13–24. [CrossRef]

23. Murai, T.; Maruyama, Y.; Mio, K.; Nishiyama, H.; Suga, M.; Sato, C. Low Cholesterol Triggers Membrane Microdomain-dependent CD44 Shedding and Suppresses Tumor Cell Migration. *J. Biol. Chem.* **2011**, *286*, 1999–2007. [CrossRef] [PubMed]
24. Hu, M.; Novo, C.; Funston, A.; Wang, H.; Staleva, H.; Zou, S.; Mulvaney, P.; Xia, Y.; Hartland, G.V. Dark-field microscopy studies of single metal nanoparticles: Understanding the factors that influence the linewidth of the localized surface plasmon resonance. *J. Mater. Chem.* **2008**, *18*, 1949–1960. [CrossRef] [PubMed]
25. Roth, G.A.; Tahiliani, S.; Neu-Baker, N.M.; Brenner, S.A. Hyperspectral microscopy as an analytical tool for nanomaterials. *Wiley Interdiscip. Rev. Nanomed. Nanobiotechnol.* **2015**, *7*, 565–579. [CrossRef] [PubMed]
26. Siddiqi, A.M.; Li, H.; Faruque, F.; Williams, W.; Lai, K.; Hughson, M.; Bigler, S.; Beach, J.; Johnson, W. Use of hyperspectral imaging to distinguish normal, precancerous, and cancerous cells. *Cancer Cytopathol.* **2008**, *114*, 13–21. [CrossRef] [PubMed]
27. Bhat, A.; Edwards, L.W.; Fu, X.; Badman, D.L.; Huo, S.; Jin, A.J.; Lu, Q. Effects of gold nanoparticles on lipid packing and membrane pore formation. *Appl. Phys. Lett.* **2016**, *109*, 263106. [CrossRef] [PubMed]
28. Chen, R.; Choudhary, P.; Schurr, R.N.; Bhattacharya, P.; Brown, J.M.; Ke, P.C. Interaction of lipid vesicle with silver nanoparticle-serum albumin protein corona. *Appl. Phys. Lett.* **2012**, *100*, 013703. [CrossRef] [PubMed]
29. Podila, R.; Chen, R.; Ke, P.C.; Brown, J.; Rao, A. Effects of surface functional groups on the formation of nanoparticle-protein corona. *Appl. Phys. Lett.* **2012**, *101*, 263701. [CrossRef] [PubMed]
30. Nordlander, P.; Oubre, C.; Prodan, E.; Li, K.; Stockman, M.I. Plasmon Hybridization in Nanoparticle Dimers. *Nano Lett.* **2004**, *4*, 899–903. [CrossRef]
31. Xiao, X.; Montaño, G.A.; Edwards, T.L.; Allen, A.; Achyuthan, K.E.; Polsky, R.; Wheeler, D.R.; Brozik, S.M. Surface Charge Dependent Nanoparticle Disruption and Deposition of Lipid Bilayer Assemblies. *Langmuir* **2012**, *28*, 17396–17403. [CrossRef] [PubMed]
32. Montis, C.; Maiolo, D.; Alessandri, I.; Bergese, P.; Berti, D. Interaction of nanoparticles with lipid membranes: a multiscale perspective. *Nanoscale* **2014**, *6*, 6452–6457. [CrossRef] [PubMed]
33. Heikkilä, E.; Martinez-Seara, H.; Gurtovenko, A.A.; Javanainen, M.; Häkkinen, H.; Vattulainen, I.; Akola, J. Cationic Au nanoparticle binding with plasma membrane-like lipid bilayers: Potential mechanism for spontaneous permeation to cells revealed by atomistic simulations. *J. Phys. Chem. C* **2014**, *118*, 11131–11141. [CrossRef]
34. Lin, J.; Zhang, H.; Chen, Z.; Zheng, Y. Penetration of lipid membranes by gold nanoparticles: Insights into cellular uptake, cytotoxicity, and their relationship. *ACS Nano* **2010**, *4*, 5421–5429. [CrossRef] [PubMed]
35. Vance, J.E. Phospholipid synthesis and transport in mammalian cells. *Traffic* **2015**, *16*, 1–18. [CrossRef] [PubMed]
36. Crane, J.M.; Tamm, L.K. Role of cholesterol in the formation and nature of lipid rafts in planar and spherical model membranes. *Biophys. J.* **2004**, *86*, 2965–2979. [CrossRef]
37. Almeida, P.F.; Vaz, W.L.; Thompson, T.E. Percolation and diffusion in three-component lipid bilayers: Effect of cholesterol on an equimolar mixture of two phosphatidylcholines. *Biophys. J.* **1993**, *64*, 399–412. [CrossRef]
38. Toma, H.E.; Zamarion, V.M.; Toma, S.H.; Araki, K. The coordination chemistry at gold nanoparticles. *J. Braz. Chem. Soc.* **2010**, *21*, 1158–1176. [CrossRef]
39. Iravani, S.; Korbekandi, H.; Mirmohammadi, S.; Zolfaghari, B. Synthesis of silver nanoparticles: Chemical, physical and biological methods. *Res. Pharm. Sci.* **2014**, *9*, 385–406. [PubMed]
40. Mikelj, M.; Praper, T.; Demic, R.; Hodnik, V.; Turk, T.; Anderluh, G. Electroformation of giant unilamellar vesicles from erythrocyte membranes under low-salt conditions. *Anal. Biochem.* **2013**, *435*, 174–180. [CrossRef] [PubMed]
41. Angelova, M.I.; Soléau, S.; Méléard, P.; Faucon, F.; Bothorel, P.H.F. Preparation of giant vesicles by external AC electric fields. Kinetics and applications. In *Trends in Colloid and Interface Science VI*; Helm, C., Lusche, M., Muhwald, H., Eds.; Steinkopff: Darmstadt, Germany, 1992; Volume 89, pp. 127–131. ISBN 978-3-7985-0913-9.

International Journal of
Molecular Sciences

MDPI

Article

Challenges and Contradictions of Metal Nano-Particle Applications for Radio-Sensitivity Enhancement in Cancer Therapy

Eva Pagáčová [1], Lenka Štefančíková [1,2], Franz Schmidt-Kaler [3], Georg Hildenbrand [3,4], Tomáš Vičar [5], Daniel Depeš [1], Jin-Ho Lee [3], Felix Bestvater [6], Sandrine Lacombe [2], Erika Porcel [2], Stéphane Roux [7], Frederik Wenz [4], Olga Kopečná [1], Iva Falková [1], Michael Hausmann [3,*] and Martin Falk [1,*]

1 Czech Academy of Sciences, Institute of Biophysics, v.v.i., Kralovopolska 135, 612 65 Brno, Czech Republic; pagacova@ibp.cz (E.P.); StefancikovaL@seznam.cz (L.S.); depesd26@gmail.com (D.D.); kopecna@ibp.cz (O.K.); ivafalk@ibp.cz (I.F.)
2 Institute des Sciences Moléculaires d'Orsay (ISMO), Université Paris Saclay, Université Paris Sud, CNRS, 91405 Orsay Cedex, France; sandrine.lacombe@u-psud.fr (S.L.); erika.porcel@u-psud.fr (E.P.)
3 Kirchhoff-Institute for Physics, University of Heidelberg, Im Neuenheimer Feld 227, 69120 Heidelberg, Germany; franzschmidtkaler@web.de (F.S.-K.); hilden@kip.uni-heidelberg.de (G.H.); jin-ho.lee@kip.uni-heidelberg.de (J.-H.L.)
4 Department of Radiation Oncology, Universitätsmedizin Mannheim, Medical Faculty Mannheim, Heidelberg University, 68159 Mannheim, Germany; Frederik.Wenz@medma.uni-heidelberg.de
5 Brno University of Technology, Department of Biomedical Engineering, Technická 3082/12, 61600 Brno, Czech Republic; tomasvicar@gmail.com
6 German Cancer Research Center (DKFZ), Im Neuenheimer Feld 280, 69120 Heidelberg, Germany; f.bestvater@dkfz.de
7 Institute UTINAM, UMR CNRS 6213-Université de Bourgogne Franche-Comté, 25020 Besançon Cedex, France; stephane.roux@univ-fcomte.fr
* Correspondence: hausmann@kip.uni-heidelberg.de (M.H.); falk@ibp.cz (M.F.); Tel.: +49-6221-549824 (M.H.); +420-541517116 (M.F.)

Received: 17 January 2019; Accepted: 24 January 2019; Published: 30 January 2019

Abstract: From the very beginnings of radiotherapy, a crucial question persists with how to target the radiation effectiveness into the tumor while preserving surrounding tissues as undamaged as possible. One promising approach is to selectively pre-sensitize tumor cells by metallic nanoparticles. However, though the "physics" behind nanoparticle-mediated radio-interaction has been well elaborated, practical applications in medicine remain challenging and often disappointing because of limited knowledge on biological mechanisms leading to cell damage enhancement and eventually cell death. In the present study, we analyzed the influence of different nanoparticle materials (platinum (Pt), and gold (Au)), cancer cell types (HeLa, U87, and SKBr3), and doses (up to 4 Gy) of low-Linear Energy Transfer (LET) ionizing radiation (γ- and X-rays) on the extent, complexity and reparability of radiation-induced γH2AX + 53BP1 foci, the markers of double stand breaks (DSBs). Firstly, we sensitively compared the focus presence in nuclei during a long period of time post-irradiation (24 h) in spatially (three-dimensionally, 3D) fixed cells incubated and non-incubated with Pt nanoparticles by means of high-resolution immunofluorescence confocal microscopy. The data were compared with our preliminary results obtained for Au nanoparticles and recently published results for gadolinium (Gd) nanoparticles of approximately the same size (2–3 nm). Next, we introduced a novel super-resolution approach—single molecule localization microscopy (SMLM)—to study the internal structure of the repair foci. In these experiments, 10 nm Au nanoparticles were used that could be also visualized by SMLM. Altogether, the data show that different nanoparticles may or may not enhance radiation damage to DNA, so multi-parameter effects have to be considered to better interpret the radiosensitization. Based on these findings, we discussed on conclusions and contradictions related to the effectiveness and presumptive mechanisms of the cell radiosensitization by nanoparticles. We also demonstrate that SMLM offers new perspectives to study internal structures of repair foci with the goal to better evaluate potential differences in DNA damage patterns.

Keywords: metal nanoparticles; cancer radiotherapy; tumor cell radiosensitization; DNA damage; DNA repair; DNA double strand breaks (DSBs); super-resolution microscopy; single-molecule localization microscopy (SMLM); DNA repair foci; damage to lysosomes

---

1. Introduction

More than a half of all cancer patients are currently treated with radiotherapy [1] that, together with chemotherapy, still represents the most efficient curative approach for many cancer types. The therapeutic window of radiotherapy (and chemotherapy) [2] is based on different capacities of normal and cancer cells to repair DNA damage. Because of defects in cell cycle checkpoints and/or repair pathways [3], cancer cells more or less suffer from genomic instability and are more susceptible than normal cells to DNA-damaging agents. Some tumors are highly radioresistant though, making them difficult to eradicate while preserving the surrounding normal tissues undestroyed [4–6]. A crucial part of cancer treatment development therefore concerns a question of how to deliver the radiation effectiveness into the tumor while preserving the normal surrounding tissues as much as possible. This issue becomes of fundamental importance for radioresistant tumors and/or tumors located in close proximity of vital organs or structures. An illustrative example could be the most aggressive and radioresistant tumor [7] starting in the brain—glioblastoma—leading us to select U87 glioblastoma cells as a model in the present study. HeLa cells, an often used model in bio-medical research, were included in the present study as a different cancer cell type for their lower radioresistance and different origin. SkBr3 cells [8] were involved as a model for breast cancer with Her2/neu up-regulation, on which the radiation effects are studied in combination with antibody and/or chemo-treatment [9].

Several promising strategies are continuously being developed to improve radiotherapy. For instance, spatial dose fractionation, time dose fractionation, micro/mini-beam irradiation, heavy-ion irradiation [10–17], and application of normal cell radio-protectants [18,19] and/or tumor cell radiosensitizers [20] could already be used in practice and eventually combined. One of the radiosensitizing approaches proposed is to selectively potentiate radiation toxicity for tumor cells by metal nanoparticles [21–28]. Due to their high electron content and photoelectric absorption cross-section, metal (high atomic number = high-Z material) nanoparticles emit showers of secondary electrons upon irradiation [29,30]. Launched electrons then generate clouds of high ionization densities, capable of enhancing radiation-induced cell damage and death rates [31].

The cell nucleus and DNA located therein are sensitive to many stressors [32–36] and can be highly damaged with relatively low doses of ionizing radiation [37,38]. As deleterious effects of ionizing radiation on (cancer) cells are mostly mediated through fragmentation of nuclear chromatin by inserting double strand breaks (DSBs) into the DNA molecule [39], nanoparticle radiosensitizing effects have primarily been ascribed to an increased number and/or complexity of DSBs generated by radiation in presence of nanoparticles [23]. Clustered (complex) DSBs can only be repaired with difficulty [16,40,41] and were recognized as the main factor responsible for the superior radiobiological efficiency (RBE) of densely ionizing radiations. Hence, according to this hypothesis, at a given absorbed dose and irradiation parameters, nanoparticles boost cell killing by locally amplifying the dose [42] and, in turn, DNA damage. Indeed, increased numbers relative to untreated samples of single-stranded breaks (SSBs) and DSBs were measured in DNA irradiated in the solution with various metal nanoparticles [43]. Since nanoparticles are preferentially internalized and accumulated by cancer cells, even passively due to mechanisms collectively known as the so-called Enhanced Permeability and Retention (EPR) effect, these enhancement effects of radiotherapy could be selectively targeted to tumors [44–50]. Moreover, some nanoparticles exert dual multiple benefits in cancer treatment at the same time—they can be used as contrast agents in theranostics [51] and/or vehicles for delivery of various chemotherapeutics or biological treatment compounds to the tumor. Nanoparticles can be also functionalized (surface material modification, attached antibodies, size, shape, etc.) to better identify and infiltrate the tumor [23,48]. Moreover, nanoparticles can be used as imaging tags especially where photo-bleaching has to be avoided [52,53].

Aforementioned physical predictions on the mechanism of nanoparticle-mediated radiosensitization were confirmed experimentally [51]. As already noticed, isolated DNA showed increased fragmentation after being irradiated in presence of various nanoparticles [43]. In other experiments, nanoparticles also increased cell dying when being added to cell cultures prior to irradiation [21,43,54]. Nevertheless, it is in fact not so easy to explain the nanoparticle-mediated cell radiosensitization, despite seemingly ideal correspondence between the theoretical predications and experimental results. The Achilles' heel of the current "mainstream" hypothesis followed from in situ/in vivo experiments showing that nanoparticles, even those of very small dimensions (e.g., of 2–3 nm in diameter, as used in this work), penetrate the cells but not the cell nucleus [21,23,53,55] unless they are specifically modified for this purpose [42]. Nanoparticles of different materials and sizes, entering the cells by pinocytosis (reviewed in [56]), thus remain retained inside the cytoplasm, where they accumulate especially in endoplasmic vesicles (endosomes) and lysosomes [21,55,57]. Under some circumstances, nanoparticles may co-localize preferentially with the endoplasmic reticulum (ER) [23,58] and Golgi apparatus (reviewed in [56]). Interestingly, mitochondria, the only cytoplasmic organelles in human cells that contain their own DNA, do not represent a primary target for nanoparticles, though some nanoparticles can also be targeted to these structures (reviewed in [56]). These findings put into play a plethora of various cellular processes potentially participating in nanoparticle-mediated tumor cell radiosensitization. It is therefore possible that different nanoparticles do not share a common mode of action, both in terms of the type of cell damage and its underlying mechanism (reviewed in [59]).

In the present work, we analyzed for different metal nanoparticles, whether their extra-nuclear [21–23,55] presence in cells can, by itself or upon cell irradiation, enhance damage of the nuclear DNA. In addition, we followed in detail nanoparticle effects on the kinetics and efficiency of DNA repair in cells exposed to low-Linear Energy Transfer (LET) ionizing radiation (γ- and X-rays). We used high-resolution immunofluorescence confocal microscopy (ICM) and single molecule localization microscopy (SMLM) [38,60] to quantify γH2AX/53BP1 DSB repair foci formation [61] and disassembly during a long period of time post-irradiation (PI) in cells exposed to different doses of γ/X-rays after being or being not incubated with platinum nanoparticles (Pt-NPs). On the basis of these data, preliminary data for gold nanoparticles (Au-NPs), and our earlier data for gadolinium nanoparticles (Gd-NPs) [21], we discussed here on what is known about metal nanoparticle effects on cells and potential mechanisms of nanoparticle-mediated radiosensitization.

In general, the aim of the following study was to verify whether cell radiosensitization by metal nanoparticles is correlated with escalation of DNA damage and/or affection of DNA damage repair capacity. The results should contribute to a better understanding of the mechanism by which various nanoparticles (different materials and sizes) radiosensitize cells with future attempt to rationally design therapeutically more efficient nanoparticles.

2. Results

2.1. Experimental Conditions and Approaches

We explored how platinum (Pt) and gold (Au) nanoparticles influence DNA DSB induction and repair in three different cancer cell types, U87 glioblastoma cells, HeLa cervix cancer cells and SkBr3 breast cancer cells, exposed to γ-(^{137}Cs) or X-radiation. U87 glioblastoma cells were selected for their high resistance to radiotherapy. HeLa cells, showing relatively lower resistance to radiation, were then involved into the study to explore how tumor cell types of different radiosensitivities and origins respond to nanoparticle uptake and nanoparticle uptake followed by irradiation. The SkBr3 model is known to be more radioresistant in comparison to HeLa cells and was especially taken for super-resolution localization microscopy. The mechanism and kinetics of nanoparticle internalization were in detail evaluated in our previous studies with Gd-NPs (3 ± 1 nm) [21,55] and Au-NPs (10 nm) [53] of a comparable size to Pt-NPs (2.6 nm) and Au-NPs (2.4 nm) used in the present study. We showed that these ultrafine Gd-NPs as well as the larger Au-NPs efficiently penetrate into the cell cytoplasm but remain restricted from the cell nucleus. Even short (2 h) incubation with nanoparticles was proved to be sufficient to ensure their internalization and cell radiosensitization upon irradiation with γ/X-rays [53]. Therefore, we used the compatible conditions also here, though lower concentration and longer incubation period (0.5 mM/6 h) were preferred to ensure sufficient cell accumulation but minimize the potential cytotoxicity.

Modern, top-tech microscopy approaches were used in the present study to analyze DNA damage and repair with high precision. ICM allowed for quantification of DSBs during a long PI time period [16] (Figure 1). When co-localized γH2AX and 53BP1 repair foci are used as DSB markers, the sensitivity of the method is clearly superior over other modern methods, including fluorescence COMET assay (single cell gel electrophoresis) on single cells [20]. Newly developed SMLM, used here for super-resolution ultra-structural analyses of repair foci [61], offers even better resolution (up to 10–20 nm) and sensitivity than ICM. Nevertheless, since higher numbers of cells can be currently analyzed by ICM, we took advantage of this method to determine the extent of DSB induction and DSB repair kinetics in statistically relevant numbers of spatially (three-dimensionally = 3D) fixed cells, still with a very high credibility and fidelity of analysis. To further increase the credibility of our study, we scored the repair foci both manually and automatically. This also allowed us to compare the positives and negatives of both approaches and to determine the influence of γH2AX/53BP1 focus scoring method on the results.

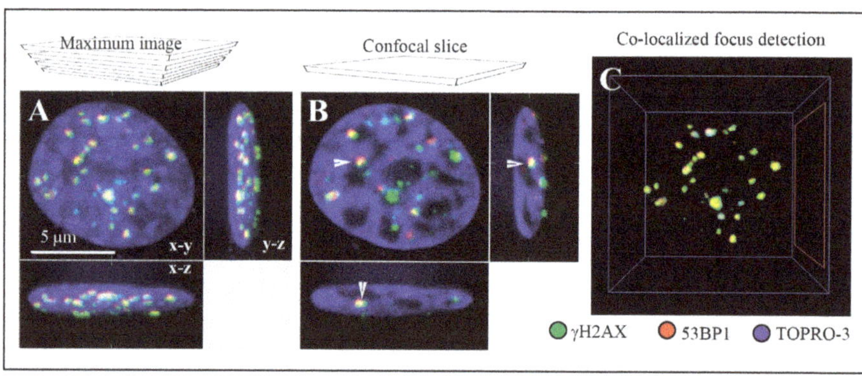

Figure 1. The ability of immunofluorescence confocal microscopy to quantify DSBs (Double Strand Breaks) in cells incubated with nanoparticles or incubated with nanoparticles and consecutively irradiated. DSBs were quantified by the means of immunofluorescence detection of co-localized γH2AX (green) and 53BP1 (red) repair foci, the DSB markers. The nucleus of an illustrative U87 cell exposed to 2 Gy of γ-rays and spatially (three-dimensionally = 3D) fixed at 2 h post-irradiation (PI) is shown as: (**A**) a maximum intensity projection of 40 confocal slices (0.3 μm thick; "maximum image") or (**B**) a single confocal slice (0.3 μm thick) intersecting the indicated (white arrow) γH2AX/53BP1 focus. Images are displayed in all three (in the x-y, x-z and y-z) planes, and chromatin is counterstained with TO-PRO-3 (artificially blue). (**C**) An example of computational detection of co-localized (yellow) γH2AX (green) and 53BP1 (red) repair foci in 3D space (Aquarium Software).

2.2. Pt-NP and Au-NP Short-Term Genotoxicity—the Effect on Nuclear DNA in Non-Irradiated Cells

Firstly, we analyzed potential negative influence of 2.6 nm Pt-NPs and 2.4 nm Au-NPs on the nuclear DNA of U87 and HeLa cells before irradiation. Cells were cultured with Pt-NPs or Au-NPs in the concentration of 0.5 mM for 6 h and potential induction of γH2AX/53BP1 (DSB) foci was studied as an indicator of nanoparticle-mediated genotoxicity. The repair foci have been present in both U87 and HeLa cell types already prior to incubation with nanoparticles and γH2AX foci mostly co-localized with 53BP1 protein. This observation points to a permanent existence of DSBs in U87 and HeLa cells, which is in accordance with their tumorous nature associated with genomic instability (Figure 2). U87 cells carried higher numbers of the foci than HeLa cells, with the mean values of 3.47 and 2.03, respectively. Figures 3–5 (0 min post irradiation (PI) in all graphs) show that the average/median numbers of the foci per nucleus were almost identical (manual analysis, Figure 3) or increased slightly (automatic analysis, Figures 4 and 5) after incubation of cells with Pt-NPs. The mean numbers of co-localized γH2AX/53BP1 foci per nucleus, provided by the automatic analyses, were 4.34 for U87 (Figure 3) and 3.88 HeLa cells (Figure 5). Such a differences, statistically significant though (U87: p = 0.010; HeLa: p = 0.003), are not supportive of biologically more relevant genotoxicity of the nanoparticles studied (2.6 nm Pt-NPs, and 2.4 nm Au-NPs; Figure 6), at least in terms of increased DNA fragmentation, consequently leading to genome rearrangements. Nevertheless, our studies limited to DSB induction cannot exclude a "milder" effect of nanoparticles on the DNA molecule, manifested for instance as oxidative base modifications. This kind of DNA damage may appear due to nanoparticle-mediated production of reactive oxygen species (ROS), which was frequently reported in the literature as the main cause of nanoparticle cytotoxicity. Moreover, especially in the context of what will follow, a negative potential of cytoplasmically localized nanoparticles may be preferentially or even exclusively targeted to the cytoplasmic structures. To summarize, our observations did not reveal more prominent genotoxicity of 2.6 nm platinum nanoparticles after short-term (6 h) incubation with U87 and HeLa cells, but more experiments are needed to comprehend potential cytotoxic effects of these nanoparticles in a more comprehensive way. Preliminary results seem to confirm this conclusion also for 2.4 nm Au-NPs.

2.3. DSB Induction and Repair in U87 Cancer Cells Treated or Not-Treated with Metal Nanoparticles Prior to Irradiation

After excluding the possibility that the studied 2.6 nm Pt-NPs and 2.4 nm Au-NPs markedly increase γH2AX/53BP1 focus (DSB) formation even by themselves, i.e., already in non-irradiated cells, we analyzed whether these nanoparticles can enhance DSB induction or affect DSB repair capacity of U87 and HeLa cells upon irradiation. The situation was compared for two γ-ray doses, 2 Gy and 4 Gy. We decided for a 2 Gy dose since this exposure is frequently used in clinical practice as a single fraction dose delivered to patients during a fractionated therapy. The higher dose of 4 Gy was applied in order to generate larger numbers of DSBs and explore differences between samples with better sensitivity (since the differences in DSB numbers per nucleus may be only small for low doses and therefore distinguishable from natural variability only with difficulty).

Figure 2 compares the γH2AX/53BP1 focus (DSB) formation and repair kinetics for U87 cells treated or not-treated with 2.6 nm Pt-NPs prior to irradiation with 4 Gy of γ-rays. Representative cell nuclei of both cell populations are displayed for different periods of time PI up to 48 h PI. Independently of the nanoparticle treatment, it is evident from Figure 2 that γH2AX foci are only incompletely formed in U87 cells early after irradiation (5–30 min PI) and also their co-localization with 53BP1 repair protein is very low. Correspondingly, the background signals (i.e., the proportions of γH2AX and especially 53BP1 molecules outside foci) are often high. A similar "picture" has also been reported for U87 cells exposed to heavy ions [62,63]. With ongoing time after irradiation, γH2AX and 53BP1 foci grow both in number and size and their mutual co-localization increases too. For both cell types (U87, and HeLa) and radiation doses (4 Gy, and 2 Gy), the number of co-localized γH2AX and 53BP1 foci reached the maximum between 30 min and 1 h PI. Later on, the number of foci started to decrease, while the size of foci gradually increased and the extent of co-localization between γH2AX and 53BP1 remained very high. Importantly (as quantified later), we did not observe any visual difference between nanoparticle-treated cells and their untreated counterparts with regard to the extent of DSB induction and repair kinetics.

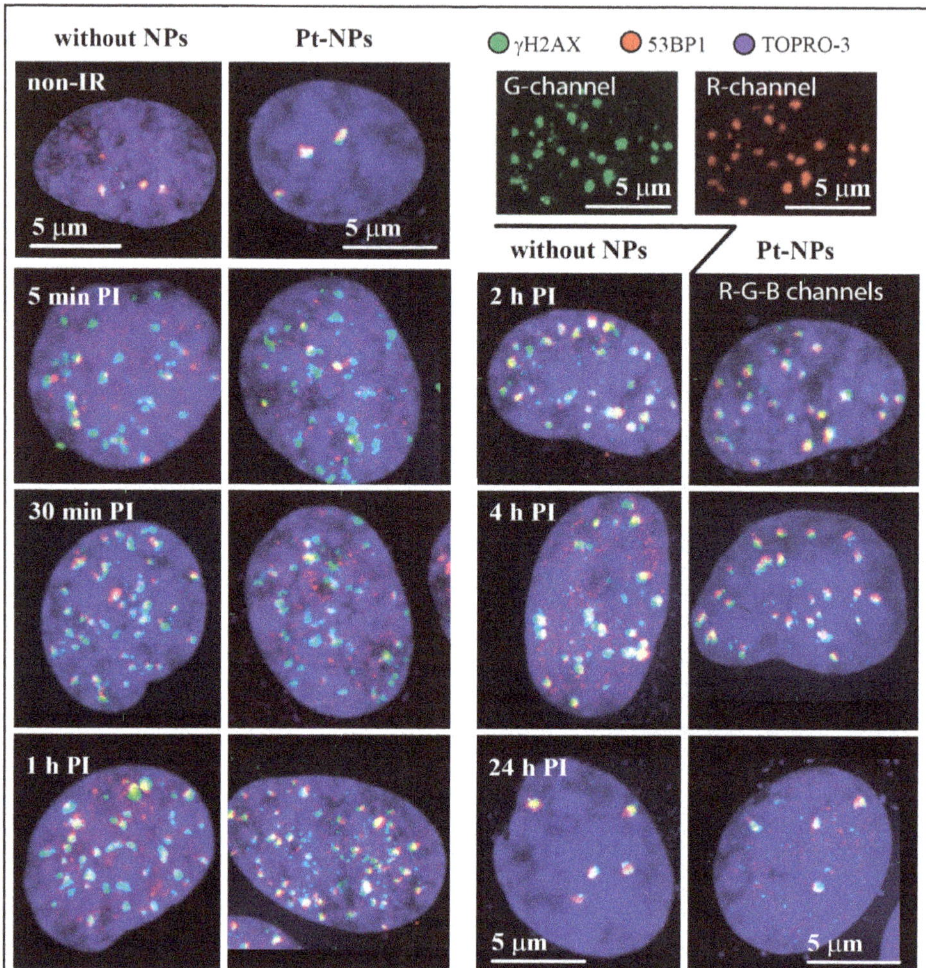

Figure 2. γH2AX/53BP1 foci (DSB) formation and repair kinetics in U87 cells incubated or not incubated with 2.6 nm platinum nanoparticles (Pt-NPs; 0.5 mM for 6 h) and consequently irradiated with 4 Gy of γ-rays. Maximum images (see Figure 1) are displayed for representative nuclei of cells that were spatially (3D) fixed in the indicated periods of time PI. For the nucleus fixed at 2 h PI, γH2AX foci (inserted G-channel panel) and 53BP1 foci (inserted R-channel panel) are also shown separately to demonstrate their mutual co-localization. γH2AX (green), 53BP1 (red), and chromatin counterstained with TO-PRO-3 (artificially blue). None-IR figures correspond to non-irradiated cells.

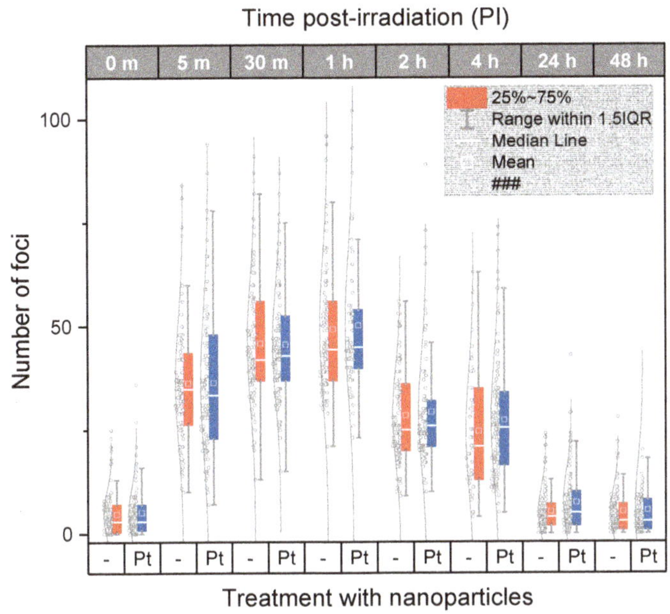

Figure 3. Manual analysis of the extent of γH2AX+53BP1 focus (DSB) induction and repair kinetics in U87 glioblastoma cells irradiated with 4 Gy of γ-rays compared with cells treated (0.5 mM for 6 h) and not treated prior to irradiation with 2.6 nm platinum nanoparticles (Pt-NPs). The average and median numbers of co-localized γH2AX + 53BP1 repair foci (i.e., DSBs) per nucleus are shown for different periods of time PI, together with the focus number distributions in each cell population. The boxes include 50% of the values (25th to 75th percentile) centered on the median (the horizontal line through the box). The mean values are represented by the squares within the boxes. The outliers were identified according to the 1.5*IQR method (IQR = interquartile range). Pt—samples treated with platinum nanoparticles, m—the period of time after irradiation in minutes, 0 m—non-irradiated samples.

The quantitative results obtained for different ways of analysis (i.e., manual and automated) and two radiation doses (4 Gy and 2 Gy) are summarized in Figures 3–5. Figure 3 compares the average/median numbers of γH2AX/53BP1 foci per nucleus together with the focus number distributions as gained by manual analysis for U87 cells exposed to 4 Gy of γ-rays in presence and absence of Pt-NPs, respectively. Except for two late time points PI (4 h and 24 h PI), all statistical characteristics (means, medians, and distributions) are almost identical for nanoparticle-treated and untreated cells.

The automated image analysis (Figure 4a) of the same cells that were previously evaluated manually provided much lower numbers of γH2AX/53BP1 foci compared to that in the manual analysis, especially at the early periods of time PI (up to 1 h PI). The maximum numbers of foci per nucleus were detected at 1 h PI in all samples, irrespective of the nanoparticle treatment and the way of analysis. During this period of time, about 50 foci per nucleus were counted manually while this value decreased to about 35 with the automated analysis. Taking into account previous reports showing that 1 Gy of γ-rays generates ~9–35 γH2AX foci per nucleus on average, depending on the cell type, the results of the manual analysis (mean = 12.5 foci/nucleus/Gy) can be considered as more realistic in terms of absolute numbers. A lower sensitivity of automatic analysis follows from the fact that computational parameters of focus scoring were set very strictly, just to detect only well-developed foci with an extensive overlap between γH2AX and 53BP1. The reason for this setting was to eliminate potential uncertainty with identification of small and/or immature foci since these foci could not be often easily separated from the background signal. Consistently, more prominent differences between the manual and automated analysis appeared at the shorter time points PI (up to 1 h PI), i.e., during the period of time when the representation of immature foci was high, especially in U87 cells. Under such conditions, automatic software analysis is still extremely difficult and manual analysis promises more precise results, especially in terms of counting the absolute focus numbers. On the other hand, computational analysis ensures detection of only precisely specified foci and therefore high reproducibility and objectivity of results, independently of the observer experience.

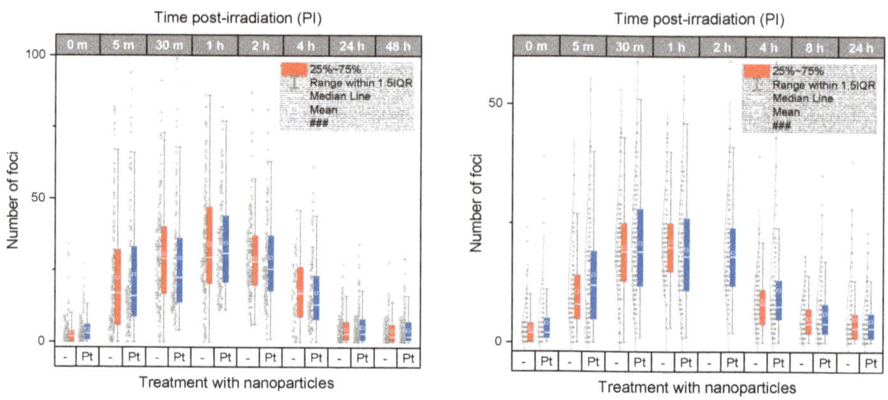

(**a**) (**b**)

Figure 4. Software analysis of the extent of γH2AX+53BP1 focus (DSB) induction and repair kinetics in U87 glioblastoma cells irradiated with 4 Gy (**a**) or 2 Gy (**b**) of γ-rays compared with cells treated (0.5 mM for 6 h) or not treated prior to irradiation with 2.6 nm platinum nanoparticles (Pt-NPs). The average and median numbers of co-localized γH2AX + 53BP1 repair foci (i.e., DSBs) per nucleus are shown for different periods of time PI, together with the focus number distributions in each cell population. The boxes include 50% of the values (25th to 75th percentile) centered on the median (the horizontal line through the box). The mean values are represented by the squares within the boxes. The outliers were identified according to the 1.5*IQR method (IQR = interquartile range). Pt—samples treated with platinum nanoparticles, m—the period of time after irradiation in minutes, 0 m—non-irradiated samples.

Another motivation to restrict the automated analysis selectively on well-developed foci followed from the question whether nanoparticles in irradiated cells may differently influence generation or repair of small and large γH2AX foci (the smaller foci were scored as DSBs by the manual analysis but not automated analysis). Except as described, both approaches provided very similar results despite of the different characters of manual and automated focus counting. Importantly, as for the manual analysis, the average numbers, medians, and distributions of γH2AX/53BP1 foci varied only inappreciably between U87 cells irradiated (4 Gy) with Pt-NPs present or absent. Very similar results for nanoparticle-treated and untreated samples were found also at 4 h PI and 24 h PI (Figure 4a), making the differences obtained for these time points by the manual analysis rather a deviation from otherwise tightly "overlapping" γH2AX/53BP1 focus profiles in time PI than a biologically relevant result.

For the lower radiation dose of 2 Gy of γ-rays (equivalent to a common single daily dose in fractionated radiotherapy), the same results as for the higher dose of 4 Gy were acquired (Figure 4b). Again, very similar numbers of γH2AX+53BP1 foci per nucleus were counted in irradiated U87 cells, irrespective of their incubation with Pt-NPs. Slightly higher mean numbers of γH2AX foci per nucleus were recognized in nanoparticle-treated cells compared to those in untreated ones only at 8 h and 24 h PI; however, comparable medians of the compared samples do not support existence of significant differences between Pt-NP-containing cells and controls even at these periods of time.

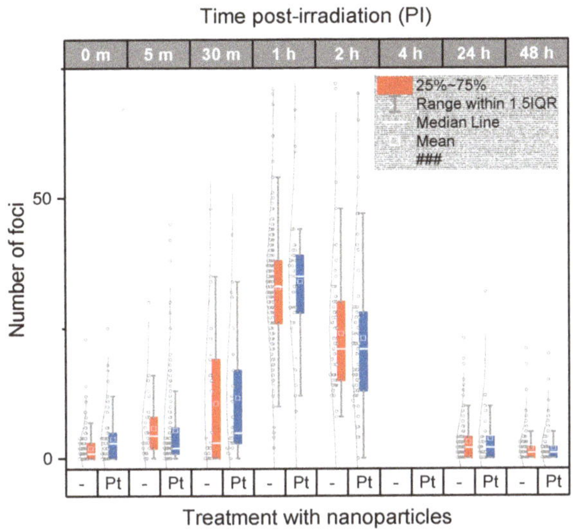

Figure 5. Automated analysis of the extent of γH2AX + 53BP1 focus (DSB) induction and repair kinetics compared for HeLa cells irradiated with 4 Gy of γ-rays in presence (0.5 mM for 6 h) or absence of 2.6 nm Pt-NPs. The average and median numbers of co-localized γH2AX+53BP1 foci (i.e., DSBs) per nucleus are shown for different periods of time PI, together with the focus number distributions in each cell population. The boxes include 50% of the values (25th to 75th percentile) centered on the median (the horizontal line through the box). The mean values are represented by the squares within the boxes. The outliers were identified according to the 1.5*IQR method (IQR = interquartile range). Pt—samples treated with platinum nanoparticles, m—the period of time after irradiation in minutes, 0 m—non-irradiated samples.

Figure 6. An illustrative Single Molecule Localization Microscopy (SMLM) image of an SkBr3 cell after uptake of 10 nm Au-NPs in the cytosol. The Au-NPs show a fluorescent blinking after laser illumination at 594 nm. Each point thus represents a single Au nanoparticle. Whereas the cytosol seems to be full of nanoparticles, the nucleus is empty. The points of low intensity seemingly covering the nucleus in the image either are the background or belong to out-of-focus image planes above or below the nucleus. Scale bar 1 μm.

2.4. DSB Induction and Repair in HeLa Cancer Cells Treated or not-Treated with Metal Nanoparticles Prior to Irradiation

In the next step, we performed the same experiments as described in the previous chapter for U87 cells also with HeLa cervix carcinoma cells that differ from U87 cell by their origin and relatively lower radioresistance. Involvement of the two cell types into the study is important since the same nanoparticles may behave unequally in dependence of specific cell characteristics. Results for HeLa cells irradiated with 4 Gy of γ-rays in presence or absence of 2.6 nm Pt-NPs are compared in Figure 5. Though some differences in the extent of γH2AX/53BP1 foci formation and kinetics of their disappearance appeared between U87 and HeLa cells, 2.6 nm Pt-NPs added to HeLa cells cultures prior to irradiation (0.5 mM, 6 h-incubation) had no effect on DNA damage and repair, confirming thus our findings for U87 cells. Cell-type-specific extent of γH2AX/53BP1 foci induction and repair capacity might be attributed to different radiosensitivities of U87 and HeLa cells. In the present study, however, different levels of radioresistance and other characteristics of U87 and HeLa cells did not influence the processes initiated by nanoparticles in both non-irradiated and irradiated cells. Similarly, as described above for non-irradiated U87 cells incubated with 2.6 nm PT-NPs, addition of Pt-NPs by itself slightly increased γH2AX/53BP1 focus numbers per nucleus also in HeLa cells (i.e., without irradiation). This can be considered as a sign of potential genotoxicity of Pt-NPs, but biological relevance of this finding does not seem to be high.

2.5. SkBr3 Cancer Cells Treated or Not-Treated with Gold Nanoparticles Prior to Irradiation—Studying γH2AX Arrangement and Focus Formation by Single Molecule Localization Microscopy

In the next step, we studied internal molecule arrangements and focus formation of γH2AX repair foci at the nanoscale by using SMLM [61]. These experiments can provide important new insights into the character of DSB damage generated by ionizing radiation in cells incubated or not incubated with metal nanoparticles. For these data, we assumed that the antibody tags against the H2AX phosphorylation sites represent the spatial topology of the foci. In the first approach, we therefore measured distance frequencies between labelling points and verified dose–efficiency curves on the point numbers in comparison to our recent approach [61].

SkBr3 cells were irradiated with 6 MeV X-rays at doses of 0, 0.5, 1, 2, or 4 Gy. For each dose, a specimen with and without 10 nm Au-NPs was irradiated. In order to ensure a maximum uptake and incorporation of these larger Au-NPs, incubation of cells was hold for 16 h prior to irradiation (Figure 6). Forty-five minutes after irradiation, the specimens were fixed and subjected to SMLM followed by software analysis of the H2AX labelling tags and their mutual distances. In Figure 7, typical next-neighbor density images are shown. In contrast to raw SMLM images showing just the positioning of fluorochromes with high precision (10–20 nm), these images encode the density of next neighbors in a 1000 nm environment by intensity. At a first glimpse, it seems that the cells with incorporated Au-NPs form more intensive foci, i.e., foci with more point signals than the cells that were irradiated with the same dose but without Au-NPs. In the case of the non-irradiated control, a random distribution may be supported by the visual impression, which contrasts with signal clustering in all irradiated cells.

A more quantitative analysis based on Ripley's K- and L-values [38] revealed a non-random distance distribution in all irradiated cell samples as it is shown for a case after 500 mGy radiation exposure without Au-NPs (Figure 8). This result indicates that, in all cases, clustering of γH2AX labelling tags can be expected. Therefore, we further studied the distance frequencies in order to find out whether the general γH2AX pattern is differing for the radiation doses and/or nanoparticle treatment conditions (Figure 9). In all irradiated cells, the average distance between γH2AX points was between 20 and 25 nm. Importantly, this γH2AX pattern did not change in specimens treated with Au-NPs.

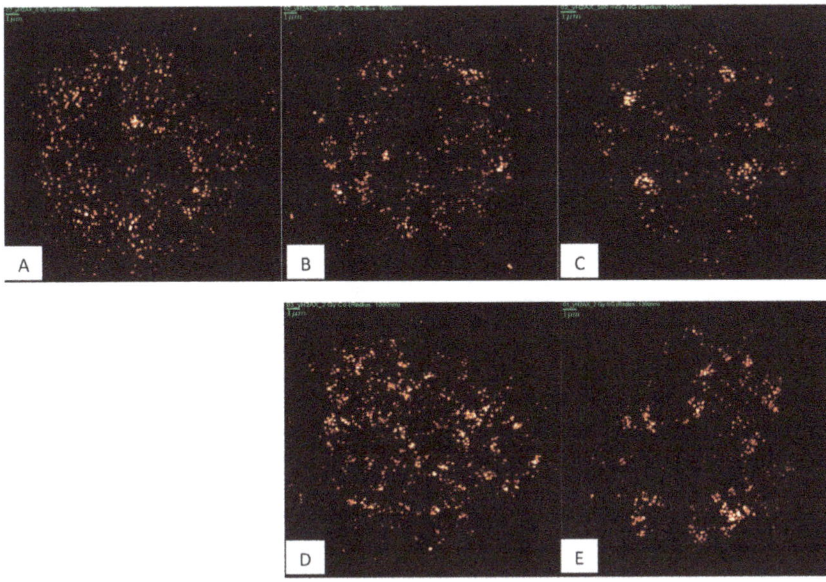

Figure 7. Illustrative SMLM next-neighbor density images comparing γH2AX labelling tag numbers and distributions in SkBr3 cells after irradiation (**B**,**D**) without particle incubation and after irradiation preceded by uptake of 10 nm Au-NPs into the cytosol (**C**,**E**). The intensity of the points represents the number of next neighbors in a 1000 nm radius environment. The control without any treatment (no NP-incubation, and no irradiation) is shown in (**A**), indicating some repair activity also in untreated cells. Images of irradiated cell were taken at 45 min PI. Scale bar: 1 μm.

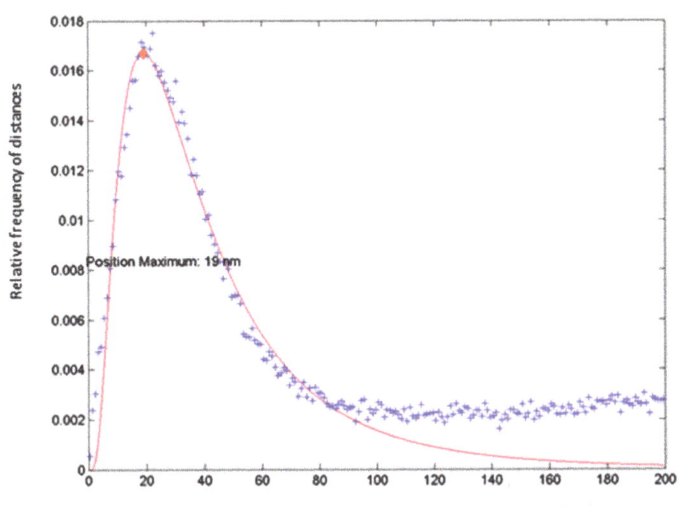

Figure 8. Example of a distance frequency distribution obtained for γH2AX labelling tags in an SkBr3 cell nucleus of the control (no NPs) specimen exposed to 0.5 Gy of X-rays. In all cases independent of the treatment, compatible distributions were obtained, indicating a characteristic non-random distance distribution. (blue crosses: number of measured distances; red diamond: peak maximum; red curve: fit curve)

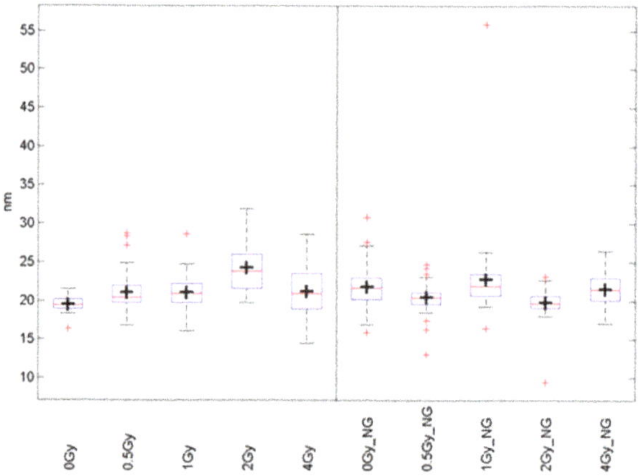

Figure 9. Boxplots of distance frequencies between γH2AX labelling tags in SkBr3 cell nuclei of irradiated specimens (left) and Au-NP-incorporated and irradiated specimens (NG, right). (black cross: mean value; red line: median value; red cross: outliner; blue box: first quartile; dashed line: standard deviation)

The numbers of γH2AX labelling tags can be used to determine the dependence of DNA damage extent on radiation dose and presence of nanoparticles [61]. Hence, we constructed preliminary dose–efficiency curves for X-ray doses up to 4 Gy and compared the numbers of γH2AX signal points in cells incubated or not incubated with Au-NPs (Figure 10). A slight linear increase in the number of γH2AX points was registered up to 2 Gy. In this dose interval, the curves were comparable for cells with and without Au-NP incorporation. Interestingly, a steep increase of the curve appeared between 2 Gy and 4 Gy after Au-NP incorporation, which was not observed in the control. In addition, the SMLM data indicate that the dose enhancement effects, as indicated by γH2AX signals, may be small, especially in dose ranges up to 2 Gy, which supports the data obtained above by ICM and foci counting. However, further experiments with other cell lines seem to be necessary for making the final conclusions.

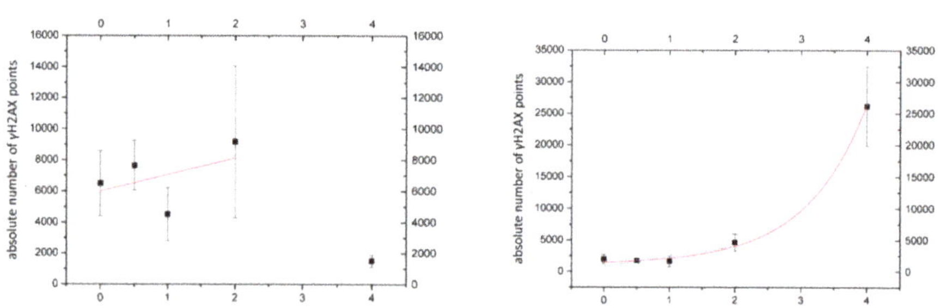

Figure 10. Dose–efficiency curves (number of γH2AX labelling tags vs. dose counted by SMLM). For the irradiated SkBr3 cell nuclei without Au-NP incorporation (left graph, a linear increase (red fit curve) can be observed at doses between 0 and 2 Gy. This was compatible to the blue linear fitting curve for irradiated SkBr3 cell nuclei with Au-NP incorporation (right graph. For the higher dose values, an exponential growth (red fit curve) or quadratic increase could be fitted to the values. (black square: mean value; error bar: standard deviation)

In any case, we show here that the microscopic tools for nano-architecture analysis are available and adaptable to the challenges of NP-modified radiation treatment. The techniques of nano-probing and localization microscopy can be further improved by topological analyses of other repair foci (e.g.,

53BP1 or Mre11) or analyses of chromatin conformation changes that may be induced by additional NP treatment.

2.6. Compared Effects of Pt, Au and Gd Nanoparticles—Preliminary Results

Finally, despite a preliminary character of the Au-NP data, we attempted here to compare DNA effects for three types of ultrafine (2–3 nm) metal nanoparticles composed of platinum, gold, and gadolinium, respectively. The size and incubation parameters were kept as similar as possible for all experiments to isolate only the effect of the nanoparticle material. The values for gadolinium(III) containing nanoparticles presented in Figure 11 were taken from our previous study performed with the same cells (U87) and under comparable experimental conditions [21]. As it is evident from Figure 11, the differences in DNA damage and repair between U87 cells exposed to 4 Gy of γ-rays after being or being not incubated with nanoparticles are quite small for all nanoparticles—platinum, gold, and gadolinium—studied. This means that 2.6 nm Pt, 2.4 nm Au and 2.0 nm Gd nanoparticles of given composition neither intensify DSB induction by ionizing radiation nor affect consequent repair of these lesions.

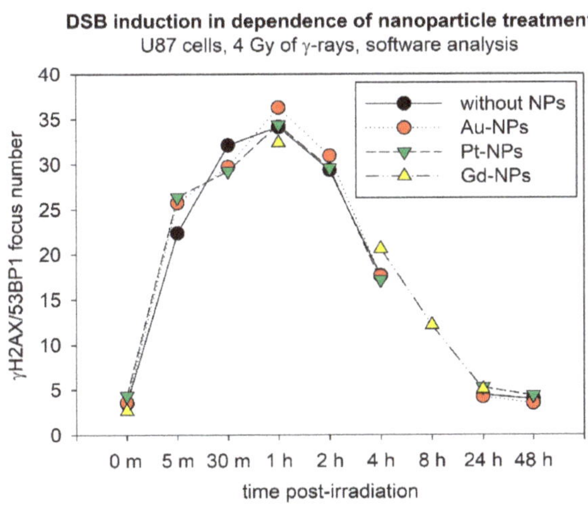

Figure 11. Comparison of γH2AX/53BP1 focus (DSB) formation and repair in U87 cells irradiated with 4 Gy of γ-rays in absence or presence of 2.6 nm Pt-NPs, 2.4 nm Au-NPs or 2.0 nm Gd-NPs. The results of an automated software analysis are shown as mean numbers of foci per nucleus measured at the indicated periods of time PI. Black circles—without NPs, green triangles—Pt-NPs (0.5 mM, 6 h-incubation), and red circles—Au-NPs (0.5 mM, 6 h-incubation; preliminary results). The data are also compared to our earlier results [21] for Gd-NPs (1 mM for 1 h, ^{60}Co-irradiation, 4 Gy) (yellow triangles). X-axis: m = minutes, h = hours; 0 min = non-irradiated samples.

Nevertheless, some indications can be recognized in our summarized data, eventually pointing to a delay in DSB repair, though the overall repair capacity of U87 cells has remained uninfluenced. Such a delay could be theoretically explained by a higher complexity of DSBs generated in presence of NPs. Therefore, as a rough estimation of DSB complexity, we quantified by Immune Fluorescence Microscopy) IFM the γH2AX focus areas for U87 cells irradiated (4 Gy) in presence or absence of Pt-NP nanoparticles. The results are presented in Figure 12. The curves for nanoparticle-treated cells and untreated controls seem to diverge starting with 4 h PI, indicating increased volumes of γH2AX foci in cells incubated with 2.6 nm Pt-NPs. Though these data are rather preliminary and experiments are needed for more nanoparticle types, well compatible results came also from SMLM nano-analyses, showing, compared to irradiated but untreated cells, more intensive γH2AX foci composed of more γH2AX signals in cells irradiated in presence of 10 nm Au-NPs. However, it remains difficult to explain why the complexity but not the extent of DSB damage increased in presence of NPs. Alternatively, cytoplasmically located NPs may enhance radiation damage to the cytoplasm. Consequent suboptimal condition of cells may indirectly decrease DSB repair. This could be supported by the observation that potential indications of a slower repair in nanoparticle-treated cells appeared only in later periods of time PI. However, it should be emphasized that, as a whole, our results rather support the no-difference scenario for nanoparticle-treated and untreated irradiated cells.

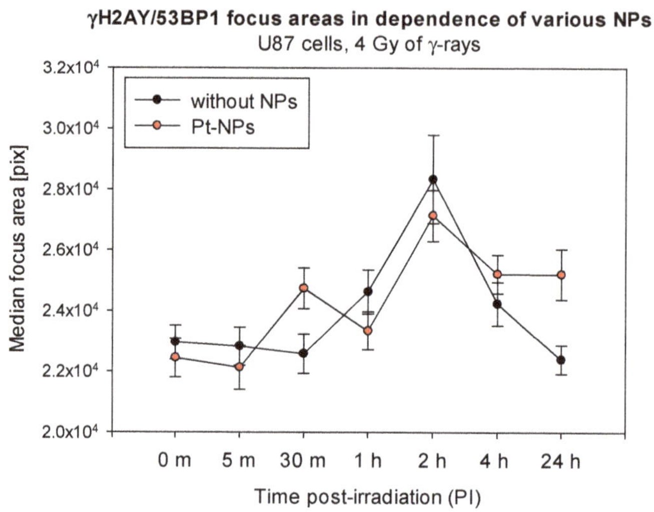

Figure 12. Comparison of γH2AX/53BP1 focus areas at different periods of time PI compared for U87 cells irradiated with 4 Gy of γ-rays in absence or presence of Pt-NPs. The results of an automated software analysis are shown as mean numbers of pixels per focus. Black circles—without NPs, and red circles—Pt-NPs (0.5 mM, 6 h-incubation; preliminary results). Error bars = standard error, m = minutes, pix = pixels, 0 m = non-irradiated samples.

3. Discussion

The radiosensitizing effect of metal nanoparticles on tumor cells has been widely reported in the literature [22]. However, the mechanism or even multiple mechanisms of this potential radiotherapy improvement remains unknown. From the potential of physics [31], nanoparticles would have multiple benefits in cancer diagnosis and radiation treatment. They have been used as contrast agents [51] and locally for tumor damaging [24,64]. Functionalized nanoparticles can be used as vehicles for bio-molecules and drugs to infiltrate a tumor [23,48]. According to the cell-killing mechanism of ionizing radiation, which is based on DNA fragmentation through DSB induction, and the capability of metal nanoparticles to locally amplify the absorbed radiation dose at the microscale, a hypothesis on nanoparticle-mediated cell radiosensitization has been proposed and increased cell dying confirmed by colony-forming assays [42]. It has been well documented that irradiated nanoparticles, preferentially sequestered by tumor cells due to the so-called EPR effect and other effects, emit showers of secondary electrons that consequently increase water radiolysis around the sites of nanoparticle accumulations and damage important biomolecules, mainly the nuclear DNA. The Achilles' heel of this otherwise very logical idea poses in a well-proved fact that while DNA is located in the cell nucleus, the nucleus is inaccessible even for nanoparticles of ultrafine dimensions (~2.5–10 nm) as used in the present study [53]. At the same time, the action radius of most secondary electrons kicked-off from cytoplasmically located nanoparticles is quite short [31]. Since some amounts of nanoparticles become concentrated around the cell nucleus or are specifically directed to the endoplasmic vesicles and reticulum, some secondary electrons may surely reach and damage the chromatin [23]. However, to what extent these rather rare acts of damage could contribute (increase) to cell killing remains a subject of debates. Moreover, the research on this topic is largely complicated by tremendous variability in the nanoparticle design (material, composition, size, shape, surface functionalization, etc.), cell-type-specific behavior and experimental conditions (type of radiation, radiation doses, nanoparticle concentrations and incubation times, etc.). Based on this situation, the aim of this article was to show by improved techniques of light microscopy whether cell radiosensitization by metal nanoparticles is correlated with an escalation of DNA damage expressed by the number of repair foci and/or affection of DNA damage repair capacity expressed by the maintenance of repair foci.

In the present work, we analyzed effects on DNA DSB induction and repair exerted by ultrafine nanoparticles composed of three different materials—2.6 nm Pt-NPs, 2.4 nm Au-NPs, 10 nm Au-NPs and 2.0 Gd-NPs—with an emphasis on result precision. All experiments were performed under the comparable conditions and in three different cancer cell lines (U87, HeLa and SkBr3) exposed to different doses (up to 4 Gy) of γ-rays or X-rays to reduce a potential bias of specific experimental conditions. U87 cells show very high radioresistance, which makes them an ideal candidate for a potential nanoparticle-enhanced radiotherapy. HeLa cells, on the other hand, are more radiosensitive and SkBr3 cells lie between HeLa cells and U87 cells. To monitor DSB induction and repair in a more comprehensive way, we quantified DSB numbers per nucleus at several time points PI, up to 24 h PI. This allowed us to compare the samples, not only the initial extent of DSB induction but also the kinetics and final efficiency of DSB repair. In addition, we were able to eliminate false differences between samples that could possibly appear if only one or two periods of time were followed. We used currently the most sensitive and accurate approach for DSB quantification—ICM of γH2AX and 53 BP1 repair foci in spatially (3D) fixed cells. γH2AX foci and 53 BP1 foci were evaluated in parallel and only co-localized foci of both DSB markers were considered as DSBs to further improve the quality and relevance of results. For the same reason, and to study small and larger γH2AX/53BP1 foci separately, we scored the foci both manually and automatically, by using novel software that is based on machine learning and has been purposefully developed and calibrated in our laboratory for the present analyses.

Taken all the ICM results together, we cannot confirm a significant effect of any nanoparticle studied (Pt, Au, and Gd) on the introduced number or repair efficiency of DSBs in irradiated cells. This conclusion holds true for both cell types (U87, and HeLa), radiation doses (4 Gy, and 2 Gy), and means of analysis used in the present study. Moreover, nano-scale SMLM studies on SkBr3 breast cancer cells with incorporated 10 nm Au-NPs also indicated that the spatial organization of γH2AX labelling tags seems not to be influenced by the presence of NPs in cells irradiated with different doses (0.5–4 Gy) of X-rays. The exception from this conclusion could be a slight delay of DSB repair in cells treated with 2.6 nm Pt and 2.4 nm Au nanoparticles in later (>4 h PI) time points PI. This difference in repair kinetics might be related to a larger size/higher intensity of γH2AX foci in nanoparticle-treated cells as observed for 2.6 nm Pt-NPs by ICM and for 10 nm Au-NPs by SMLM. However, it should be noted that the reported differences between nanoparticle-treated and untreated irradiated cells were only minor and non-systematic. We can therefore reasonably conclude that while nanoparticle-mediated radiosensitization has often been related to escalated DNA damage, the results presented here for ultrafine Pt and Au nanoparticles and also our earlier data for Gd nanoparticles [21] do not support this idea as a general mechanism responsible for the radiosensitizing phenomenon. Only for 10 nm Au-NPs and doses of 2 Gy or higher, some increase of γH2AX focus number was observed by SMLM, especially after nanoparticle modification for specific targeting to the ER [23], whereby this has only been observed for one cell line.

Our data suggest that there are at least some nanoparticles that increase cell killing upon irradiation [21,42] while they have none or a negligible effect on nuclear DNA break regions highlighted by H2AX phosphorylation sites. This confirms our intuition on the action mode of the radiosensitizing nanoparticles we developed. In other experiments (unpublished), we observed a great increase in the life span of animals bearing tumor (9L cell gliosarcoma in brain) or the inhibition of tumor growth (A375sc melanoma in flank) when the animals were treated by radiotherapy after intravenous (9L gliosarcoma) or intratumoral (melanoma) injection whereas the majority of the nanoparticles in the tumor were suspected to be outside the cells. Moreover, we also observed in preliminary experiments that the number of γH2AX is almost the same when irradiation is performed in presence or in absence of the radiosensitizing nanoparticles (unpublished results). Hence, these nanoparticles seem to sensitize cells to radiation through cytoplasmic effects that are independent of DNA damage and/or repair. While our findings do not exclude the possibility that some types of nanoparticles support radiation cell killing through the "classic" DNA damage-based mechanism, they open the door to exiting research of new mechanisms that could be dominant under some circumstances, as for instance chromatin topology-related effects and re-arrangements of compaction forms. Furthermore, accumulation of nanoparticles in endosomes and lysosomes as revealed in our earlier reports [21,55] could result in damage of these structures with important consequences. While lysosomes were originally thought only as cellular dustbins, recent studies involve lysosomes in important cell signaling pathways, eventually initiating apoptosis (see [32] and citations therein). In addition, even simple disruption of a larger amount of lysosomes due to their membrane damage by locally amplified radiation effects, mediated by intra-lysosomal nanoparticle accumulations, may result in massive leakage of lytic enzymes from these "suicide bags" [65,66] and extensive cytoplasmic damage. This can also initiate cell death. Indeed, the destabilization of lysosomes via lysosomal membrane permeabilization (LMP), leading to release of their aggressive content into the cytoplasm, is currently intensively studied as a potentially efficient way of therapeutic cell death triggering [56].

Cytoplasmically located nanoparticles may also influence organelles or structures which they do not co-localize with. For instance, increased production of ROS has been frequently reported in the literature as a main cause of nanoparticles' cytotoxicity. Therefore, ROS generated by nanoparticles in extensive amounts upon irradiation may damage organelles located in close proximity to nanoparticle location sites, for instance mitochondria. Among other cytoplasmic targets, mitochondria are especially attractive since they are critical for cell survival (energy metabolism) and represent the only extracellular structures having their own DNA. Therefore, nanoparticle-mediated fragmentation of mitochondrial DNA may represent an elegant modification of the "classic" DNA damage-based hypothesis on cell radiosensitization by nanoparticles, returning this idea into the game. It should also be noted that ROS are effective signaling molecules with a strong potential to directly influence biochemical cellular pathways.

The Endoplasmic Reticulum (ER) may represent another target for nanoparticle effects. While the efficient functioning of the ER is essential for most cellular activities and survival, it may be under some modifications also invaded by nanoparticles [23]. Moreover, ER plays an important role in the response to oxidative stress-induced damage and is quite sensitive to ROS [67]. Hence, irradiated nanoparticles may exert cytotoxic effects on cells by modulating ER stress [67]. For instance, Ag-NPs resulted in cytotoxicity and cell death by apoptotic, which was associated with (secondary) DNA fragmentation [67]. This observation not only explains how nanoparticles may initiate cell death through disturbing functions of ER, but also stresses the importance of time when interpreting the nanoparticle-mediated DNA effects. In this light, it is possible that in some studies the nanoparticle-mediated effects on DNA can rather reflect this secondary apoptotic DNA fragmentation than primary enhancement of DSB induction by radiation. The mechanism, how ER stress can lead to apoptosis, has been described by [68]. A disruption of ER function leads to accumulation and aggregation of unfolded proteins accompanied with stress signaling. The stress signals are detected by transmembrane receptors, which in turn initiate the unfolded protein response (UPR) trying to restore normal ER functions. However, if the stress persists too long, apoptotic cell death ensues [68].

Altogether, we show that the radiosensitizing effect of at least some metal nanoparticles may rely on cytoplasmic processes rather than DNA damaging events. Based on the available literature, we also outline the way of how damage of the most relevant cytoplasmic structures may initiate cell death. Though we did not observe different responses to nanoparticles or irradiation in presence of nanoparticles for the two studied cell types (U87, and HeLa), we emphasize the necessity to analyze in detail each particular combination of nanoparticles and the cell type planned to be therapeutically targeted. This imperative follows from extensive controversies that are still present in the literature on the nanoparticle-mediated irradiation effects. For instance, Au-NPs induced apoptosis in MCF-7 and N87 cancer cell lines by disrupting lysosomes and mitochondria, but this effect did not appear in normal Chinese hamster ovary (CHO) and 293T cell lines. This observation further supports our conclusion that nanoparticle-mediated cell killing enhancement may be located in the cytoplasm, but more importantly gives a perspective of selective nanoparticle toxicity for tumor cells [64,69]. Interestingly, from the opposite point of view, some radio-protective chemicals (amifostine) protect normal cells from radiation effects but delay DSB repair in tumor cells [20].

The final question remains whether it is in principle a good or bad massage finding that nanoparticles damage the cells without affecting DNA. On the one hand, it could be beneficial since nanoparticles located outside the tumor will not increase the risk of genome damage and secondary malignancies induction in normal tissues surrounding the tumor. On the other hand, the radiosensitizing mechanism operating through DNA damage could be more efficient. A solution of this dilemma could be based on selective targeting of nanoparticles to specific genome sequences, like oncogenes, using appropriately designed oligo-nucleotides as being available for radio-emitters [70]. With techniques of COMBO-FISH [71,72] and PNA probe combinations [73], NPs may be transferred to cell nuclei and specifically addressed to given chromatin targets. This could be achieved by adding a nuclear localization signal (NLS) peptide motif and a specific PNA oligonucleotide probe to the surface of nanoparticles [74]. Using such sophisticated approaches of specific targeting of genome aberrations like multiple gene copies would open new aspects in tumor therapy.

4. Materials and Methods

4.1. Cells and Cell Culturing

Three cancer cell types were studied in the present study: highly radioresistant U87 glioblastoma cancer cells, radioresistant SkBr3 breast cancer cells and relatively less radioresistant HeLa cervix cancer cells. U87 and HeLa cells were obtained from ATCC (American Type Culture Collection, Manassas, VA, USA). SkBr3 was commercially available and used for several SMLM studies in our laboratory. U87 and HeLa cells were grown in Dulbecco's modified essential medium (Thermo Fisher Scientific, Waltham, MA, USA) supplemented with 10% heat-inactivated fetal calf serum (Thermo Fisher Scientific, Waltham, MA, USA), 100 U/mL penicillin (PAA), 100 µg/mL streptomycin (PAA), and 1% NEAA (Thermo Fisher Scientific, Waltham, MA, USA). Cell cultures were kept in T-25 cell flasks at 37 °C in a humidified atmosphere with 5% CO_2.

For the experiments with SMLM, SkBr3 cells were prepared as described in detail elsewhere [72]. SkBr3 cells were grown in McCoy's 5A cell medium, containing 10% fetal bovine serum (FBS) and 1% penicillin/streptomycin. Cells were cultivated and maintained at 37 °C in a humidified atmosphere at 95% air/5% CO_2. Then, the cells were trypsinized and transferred to coverslips, put into six-well plates and further cultivated (about three passages, i.e., about 38 h) until 80% confluence.

4.2. Nanoparticles and Incubation of Cells with Nanoparticles

Platinum nanoparticles (Pt-PEG-17, referred to as Pt-NPs) were prepared as explained in the recently submitted French patent (FR 1900008). Briefly: Pt-NPs were synthetized by γ-ray water radiolysis of Pt containing salt and embedded with polyethylene glycol (PEG) to increase their biocompatibility. Pt-NPs were mainly spherical with an average platinum core diameter of 2.6 nm. Preliminary results were obtained also for gold nanoparticles (Au-NPs) which are composed of a Au core of 2.4 nm encapsulated by the dithiolated polyaminocarboxylate (DTDTPA) shell. For SkBr3

Au-NPs incorporation, 8 µL of 10 nm-sized gold particles (Aurion, Wageningen, The Netherlands) were added to the medium in each well 16 h prior to irradiation in order to obtain a maximum uptake in the cell cytoplasm via diffusion (Figure 6) [53,75]. In other experiments 2.6 nm Pt-NPs or 2.4 nm Au-NPs were added to the medium 6 h before irradiation at 0.5 mM concentration.

4.3. Cell Irradiation

Cells were irradiated in 6-well culture plates containing a culture medium and 2.0×10^4–2.0×10^5 cells per well. Consecutively, cells were exposed to 2 or 4 Gy of γ-rays (1 Gy/min), delivered by a ^{137}Cs irradiator at room temperature (RT). During irradiation, the samples were kept in thermo-isolating boxes to prevent sample infection and temperature changes, and then immediately returned to the incubator (37 °C, 5% CO_2) until taken for the experiment. For SMLM experiments, 10 nm Au-NPs were incubated 16 h before irradiation. Then, the cells were simultaneously exposed with and without Au-NPs using a 6 MeV Linac radiation source (Artiste, Siemens, Erlangen, Germany). The exposure doses of 0.5, 1, 2 and 4 Gy were obtained by changing the irradiation time at the same dose rate.

4.4. Immunodetection of γH2AX/53BP1 Foci and Double Strand Break Quantification

DNA DSBs were quantified in spatially (three-dimensionally = 3D) fixed cells by the means of high-resolution ICM detection of co-localized γH2AX and 53BP1 repair foci as described earlier [16,20]. Briefly, cells were fixed with 4% paraformaldehyde (10 min, at room temperature RT) prior to irradiation (0 min PI, non-irradiated controls) and in several time points PI covering a long (48 h) PI period (5 min, 30 min, 1 h, 2 h, 4 h, 8 h, 24 h and 48 h PI). Cells were permeabilized in 0.2% Triton X-100/PBS (15 min, RT) and immunoassayed with mouse antiphospho-H2AX (serine 139) (Merck, Darmstadt, Germany, cat. no.: 05-636) and rabbit anti-53BP1 (Cell Signaling Technology, Danvers, MA, USA, cat. no.: 4937) primary antibodies to simultaneously detect the γH2AX and 53BP1. Antiphospho-H2AX antibody was visualized with the secondary FITC-conjugated donkey anti-mouse antibody and anti-53BP1 antibody with Cy3-conjugated donkey anti-rabbit antibody (both Jackson Laboratory, West Grove, PA, USA, cat. no.: 715-095-150 and 711-165-152). Chromatin was counterstained with 1 µM TO-PRO-3 (Molecular Probes, Eugene, OR, USA) prepared in 2× saline sodium citrate (SSC). After brief washing in 2× SSC, Vectashield medium (Vector Laboratories, Burlington, Ontario, Canada) was used for sample mounting.

4.5. Fixation and Immunostaining of γH2AX for Single Molecule Localization Microscopy

45 min after irradiation, the cells were fixed in order to obtain an early response of the biological system to damage. The cells were washed in 1× Phosphate-Buffered Saline (PBS) with $MgCl_2$ (0.901 mM)/$CaCl_2$ (0.493 mM) and fixed in 3.7% formaldehyde (in 1× PBS + Mg/Ca; freshly prepared from paraformaldehyde) for 20 min at RT. After washing twice with 1× PBS + Mg/Ca, the cells were stored in 3.7% formaldehyde (in 1× PBS + Mg/Ca) at 4 °C. After 4-weeks storage, the formaldehyde was replaced by 1× PBS (+ 0.1% sodium azide). After removing the sodium azide from the coverslips, the cell membranes were permeabilized by 0.2% Triton-X100 three times for 5 min. After washing three times in 1× PBS (+ Mg/Ca), the cells were blocked in 2% bovine serum albumin (BSA) for half an hour and incubated in 100 µL of the primary antibody solution (mouse anti-phospho-histone H2A.X (Ser139) antibody; Merck Chemicals, Darmstadt, Germany; dilution: 1:500) at 37 °C for 18 h in a humidified chamber. Thereafter the coverslips with the cells were washed three times for 5 min with 1× PBS (+ Mg/Ca) to remove the remaining, unbound primary antibodies. Afterwards the secondary AlexaFluor 647 goat anti-mouse antibody was incubated in a humidified chamber at 37 °C for 30 min. Then, the cells were fixed in 2% formaldehyde (in 1× PBS (+ Mg/Ca) at 37 °C for 10 min. Finally, the cells were counterstained with 4′19,6-diamidin-2-phenylindol (DAPI; Sigma Aldrich, now Merck, Darmstadt, Germany) for 5 min in darkness and were, after washing twice with 1× PBS (+ Mg/Ca) for 5 min each, embedded in 20 µL ProLong Gold embedding medium (ThermoFisher Scientific, Waltham,

MA, USA, ProLong Gold Antifade Mountant, P36930). The specimen was sealed and stored at 4 °C in complete darkness until SMLM application.

4.6. Confocal Microscopy

An automated high-resolution confocal fluorescence microscopic system Leica DM RXA [76–78], equipped with a CSU10a Nipkow disc (Yokogawa, Tokyo, Japan), an oil immersion Plan Fluotar objective (100×/NA1.3), a CoolSnap HQ CCD camera (Photometrix, Tucson, AZ, USA), and an Ar/Kr laser (Innova 70C Spectrum, Coherent, Santa Clara, CA, USA), was used for image acquisition [79,80]. About 40 individual confocal slices with 0.3 µm z-step increments across the nuclei were captured for each cell. Obtained images were analyzed using Acquiarium software [80] which enabled the three-dimensional reconstruction of images and inspection of individual γH2AX and 53BP1 foci in 3D space. Co-localized γH2AX/53BP1 foci were considered as DSBs to increase the precision of DSB detection, especially in the early-stage PI (with a higher background of signals) and also the probability that only unrepaired DSBs are still evaluated in later and very late periods of time PI.

4.7. Single Molecule Localization Microscopy

As described in detail elsewhere [61,81], the localization microscope used was equipped with four lasers to excite different fluorophores. The wavelength and the intensity of illumination were chosen by an acousto–optical tunable filter (AOTF). For our experiments, the 642 nm laser with 140 mW output power was used to stimulate the dye molecules to blink. A 100x/NA 1.46 oil immersion objective was used. The fluorescence of the specimen was separated from the illumination light by two quadband interference filters and was recorded by an EMCCD camera (Andor iXon Ultra 897, Belfast, UK). The EMCCD camera was operated at a gain of 100 and a series of 2000 up to 6000 image frames was recorded for each cell nucleus. Prior to the SMLM measurement, a widefield image was taken in the DAPI channel and the γH2AX channel with 10% laser intensity. Thereafter, the γH2AX image stack was recorded at 70% illumination intensity. Cells were chosen to have consistent size and form, a distinctive edge, a good staining signal-to-background ratio and a certain minimal distance to the next cell. The acquired data stacks were evaluated as described in detail in [61]. γH2AX labelling molecules were counted and distances between each point were determined.

Data displayed in box graphs (Figure 9) show the frequency distributions of distances of γH2AX-labelling molecules. The boxes include 50% of the values (25th to 75th percentile) centered on the median (the horizontal line through the box). The mean values are represented by the squares within the boxes. The vertical lines begin at the 5th percentile and end at the 95th percentile.

4.8. Data Analysis and Statistical Evaluation after Confocal Microscopy

The SigmaPlot 14.0 (Systat Software Inc., San Jose, CA 95131 USA) and Origin 2018b (OriginLab Corporation, Northampton, MA 01060, USA) were used for data analysis and processing. The Mann–Whitney rank sum test was employed to compare γH2AX/53BP1 focus (DSB) numbers in untreated and nanoparticle-treated cells at all the particular periods of time PI. The results were considered as statistically significant at $p < 0.05$. The foci numbers were quantified both manually and automatically. In manual analysis, around 100 nuclei in each single experiment were blind-inspected (no information about the sample treatment) by eye by an experienced evaluator. For computational analysis, between 100 and 250 nuclei were scored. Because there is not a suitable tool fulfilling our demands on automatic γH2AX/53BP1 foci counting with our specific data, a custom program for fast and accurate foci counting, calibrated to our data, has been developed. The program works in a semi-automatic manner, where it allows for a visual inspection with the possibility to make quick manual adjustments and corrections, if necessary. The algorithm is composed of 3 steps—nucleus segmentation, foci segmentation and final foci classification, in order to eliminate false detections. Convolution Neural Network (with SegNet topology) was trained for robust nuclei segmentation, followed by splitting of touching nuclei with watershed transform applied on the distance transform of the segmented binary image. Inside a bounding box of each nucleus, the foci are segmented with a maximally stable extremal region detector, which is fast and invariant to image intensity values. The detector is set to high recall in order to obtain all possible foci for the classifier. Classification of true foci is done with Support Vector Machine classifiers on some extracted features (e.g., foci mean intensity and foci size). The program allows user to adjust the classifier bias value (to set classifier sensitivity), because the properties of foci are very heterogeneous between different samples and measurements. Besides the count of foci, it also allows exporting some other features for following analysis (cell size, foci sizes, foci intensities, etc.) (the full description of the software will be published separately). In Figures 3–5, the data are displayed as box graphs also showing the distributions of DSBs foci per nucleus. The boxes include 50% of the values (25th to 75th percentile) centered on the median (the horizontal line through the box). The mean values are represented by the squares within the boxes. The vertical lines begin at the 5th percentile and end at the 95th percentile.

5. Conclusions

In the present study, we demonstrate that ultrafine (2–10 nm) platinum and gold nanoparticles do not escalate DNA damage or compromise DSB repair in irradiated tumor cells of different types. This confirms our recent findings for 2.0 nm gadolinium nanoparticles [21]. However, 10 nm Au-NPs may potentially influence the character of DNA damage at the nanoscale, as it was discovered by using SMLM [61,72,81]. Some indications in this sense have been obtained also by ICM for 2 nm Pt-NPs. While these findings are difficult to be interpreted in terms of biological relevance, contradictions still persist in the literature on the enhancement of nuclear DNA damage in cells irradiated in presence of metal nanoparticles. At the current stage of knowledge, it is reasonable to conclude that different mechanisms, involving an enhancement of DNA damage on the one side and cytoplasmic effects on the other side, participate in radiosensitization exerted by metal nanoparticles. More mechanisms probably contribute to the final radiosensitizing effect, involvement of which depends on the nanoparticle characteristics (material, size, composition, and surface functionalization), cell type and experimental conditions. Therefore, many questions on nanoparticle-mediated radiosensitization remain open, emphasizing the importance of more systematic future research. Methodologically, we demonstrate current possibilities and usefulness of the newly developed super-resolution microscopy technique (SMLM) that together with appropriate nano-probing technologies has a potential to shift our studies on DNA damage and repair to nanoscale dimensions. Mutual comparison of micro- and nano-scale results may provide a clue on many important processes taking part in cells and their molecular mechanisms.

Author Contributions: Research conceptualization, M.F., M.H., G.H., L.S. and S.L; methodology development, M.F., M.H., S.L., L.S. and S.R.; experiment performance, E.P. (Eva Pagáčová), L.S., O.K., I.F., F.S.-K., J.-H.L. and S.R.; software development, T.V.; validation, D.D., E.P. (Eva Pagáčová), J.-H.L., G.H., M.H. and M.F.; formal analysis, D.D, E.P. (Eva Pagáčová), I.F., M.H., M.F.; investigation, E.P. (Eva Pagáčová), S.L., M.H. and M.F.; resources, F.B., F.W., M.H. and M.F.; data curation, M.F., D.D., E.P. (Eva Pagáčová), T.V.; writing of the original draft preparation, F.S.-K., M.F. and M.H.; writing of review and editing, G.H., S.L., M.F. and M.H.; visualization, E.P. (Eva Pagáčová), F.S.-K., D.D., I.F., O.K.; supervision, E.P. (Erika Porcel), S.L., M.H. and M.F.; project administration, M.F. and M.H.; funding acquisition, M.H. and M.F.

Funding: The work was supported by the Ministry of Health of the Czech Republic (AZV grant no. 16-29835A), the Czech Science Foundation (project 16-12454S), the Heidelberg University Mobility Grant for International Research Cooperation within the excellence initiative II of the Deutsche Forschungsgemeinschaft (DFG) to M.H., and from the grants from the Czech Republic to the Joint Institute for Nuclear Research, Dubna (Projects of the Czech Plenipotentiary and the 3 + 3 Projects). The financial support by Deutsche Forschungsgemeinschaft and Ruprecht-Karls-Universität Heidelberg within the funding program Open Access Publishing is gratefully acknowledged. The research leading to these results has also received funding from the People Programme (Marie Curie Actions) of the European Union's Seventh Framework Programme (FP7/2007–2013) under REA Grant Agreement No [624370].

Acknowledgments: The authors thank Emanuel Maus (Kirchhoff-Institute for Physics) for providing an SMLM image. The authors acknowledge Daniela Salado (Institute des Sciences Moléculaires d'Orsay, Université Paris Sud 11, Orsay Cedex, France) and Hynd Remita (Laboratoire de Chimie Physique, Orsay, France) for synthesizing platinum and 2.4 nm gold nanoparticles. Furthermore, we thank Jin-Hau Ewwer, Institute of Research Rating and Enhancement (IRRE), Altenburschla, Germany, and Paul I. M. Prinz Zippl, University of Vienna, Austria, for always finding the right way of haziness in constructive discussions.

Conflicts of Interest: The authors declare no conflicts of interest. The funders had no role in the design of the study; in the collection, analyses, or interpretation of data; in the writing of the manuscript, and in the decision to publish the results.

Abbreviations

DSB	double strand break
SMLM	single molecule localization microscopy
ICM	immunofluorescence confocal microscopy
Pt-NPs	platinum nanoparticles
Au-NPs	gold nanoparticles
PI	post-irradiation
EPR	enhanced permeability and retention

References

1. Atun, R.; Jaffray, D.A.; Barton, M.B.; Bray, F.; Baumann, M.; Vikram, B.; Hanna, T.P.; Knaul, F.M.; Lievens, Y.; Lui, T.Y.M.; et al. Expanding global access to radiotherapy. *Lancet Oncol.* **2015**, *16*, 1153–1186. [CrossRef]
2. Wenz, F.; Tiefenbacher, U.; Willeke, F.; Weber, K.-J. Auf der Suche nach der *Therapeutischen breite* in der Radioonkologie. *Oncol. Res. Treat.* **2001**, *24*, 51–55. [CrossRef] [PubMed]
3. Löffler, H.; Bochtler, T.; Fritz, B.; Tews, B.; Ho, A.D.; Lukas, J.; Bartek, J.; Krämer, A. DNA Damage-Induced Accumulation of Centrosomal Chk1 Contributes to its Checkpoint Function. *Cell Cycle* **2007**, *6*, 2541–2548. [CrossRef] [PubMed]
4. Tomita, K.; Kuwahara, Y.; Takashi, Y.; Igarashi, K.; Nagasawa, T.; Nabika, H.; Kurimasa, A.; Fukumoto, M.; Nishitani, Y.; Sato, T. Clinically relevant radioresistant cells exhibit resistance to H_2O_2 by decreasing internal H_2O_2 and lipid peroxidation. *Tumour Biol.* **2018**, *40*. [CrossRef]
5. Lam, W.W.; Oakden, W.; Murray, L.; Klein, J.; Iorio, C.; Screaton, R.A.; Koletar, M.M.; Chu, W.; Liu, S.K.; Stanisz, G.J. Differentiation of Normal and Radioresistant Prostate Cancer Xenografts Using Magnetization Transfer-Prepared MRI. *Sci. Rep.* **2018**, *8*, 10447. [CrossRef]
6. Gu, H.; Huang, T.; Shen, Y.; Liu, Y.; Zhou, F.; Jin, Y.; Sattar, H.; Wei, Y. Reactive Oxygen Species-Mediated Tumor Microenvironment Transformation: The Mechanism of Radioresistant Gastric Cancer. *Oxid. Med. Cell. Longev.* **2018**, *2018*, 5801209. [CrossRef]
7. Zhou, H.; Miki, R.; Eeva, M.; Fike, F.M.; Seligson, D.; Yang, L.; Yoshimura, A.; Teitell, M.A.; Jamieson, C.A.M.; Cacalano, N.A. Reciprocal Regulation of SOCS1 and SOCS3 Enhances Resistance to Ionizing Radiation in Glioblastoma Multiforme. *Clin. Cancer Res.* **2007**, *13*, 2344–2353. [CrossRef]
8. Engel, L.W.; Young, N.A. Human breast carcinoma cells in continuous culture: A review. *Cancer Res.* **1978**, *38*, 4327–4339.
9. Lacroix, M.; Leclercq, G. Relevance of breast cancer cell lines as models for breast tumours: An update. *Breast Cancer Res. Treat.* **2004**, *83*, 249–289. [CrossRef]
10. Durante, M.; Orecchia, R.; Loeffler, J.S. Charged-particle therapy in cancer: Clinical uses and future perspectives. *Nat. Rev. Clin. Oncol.* **2017**, *14*, 483–495. [CrossRef]
11. González, W.; Prezado, Y. Spatial fractionation of the dose in heavy ions therapy: An optimization study. *Med. Phys.* **2018**, *45*, 2620–2627. [CrossRef] [PubMed]
12. Jánváry, L.Z.; Ferenczi, Ö.; Takácsi-Nagy, Z.; Bajcsay, A.; Polgár, C. Application of CyberKnife stereotactic radiosurgery in the treatment of head and neck cancer. *Magy. Onkol.* **2018**, *62*, 180–185. [PubMed]
13. Zhang, H.; Wan, C.; Huang, J.; Yang, C.; Qin, Y.; Lu, Y.; Ma, J.; Wu, B.; Xu, S.; Wu, G.; et al. In Vitro Radiobiological Advantages of Hypofractionation Compared with Conventional Fractionation: Early-Passage NSCLC Cells are Less Aggressive after Hypofractionation. *Radiat. Res.* **2018**. [CrossRef] [PubMed]

14. Sammer, M.; Greubel, C.; Girst, S.; Dollinger, G. Optimization of beam arrangements in proton minibeam radiotherapy by cell survival simulations. *Med. Phys.* **2017**, *44*, 6096–6104. [CrossRef] [PubMed]
15. Prezado, Y.; Renier, M.; Bravin, A. A new method of creating minibeam patterns for synchrotron radiation therapy: A feasibility study. *J. Synchrotron Radiat.* **2009**, *16*, 582–586. [CrossRef] [PubMed]
16. Jezkova, L.; Zadneprianetc, M.; Kulikova, E.; Smirnova, E.; Bulanova, T.; Depes, D.; Falkova, I.; Boreyko, A.; Krasavin, E.; Davidkova, M.; et al. Particles with similar LET values generate DNA breaks of different complexity and reparability: A high-resolution microscopy analysis of γH2AX/53BP1 foci. *Nanoscale* **2018**, *10*, 1162–1179. [CrossRef]
17. Girst, S.; Greubel, C.; Reindl, J.; Siebenwirth, C.; Zlobinskaya, O.; Walsh, D.W.M.; Ilicic, K.; Aichler, M.; Walch, A.; Wilkens, J.J.; et al. Proton Minibeam Radiation Therapy Reduces Side Effects in an In Vivo Mouse Ear Model. *Int. J. Radiat. Oncol.* **2016**, *95*, 234–241. [CrossRef]
18. Hofer, M.; Hoferová, Z.; Depeš, D.; Falk, M. Combining Pharmacological Countermeasures to Attenuate the Acute Radiation Syndrome-A Concise Review. *Molecules* **2017**, *22*, 834. [CrossRef]
19. Hofer, M.; Hoferová, Z.; Falk, M. Pharmacological Modulation of Radiation Damage. Does It Exist a Chance for Other Substances than Hematopoietic Growth Factors and Cytokines? *Int. J. Mol. Sci.* **2017**, *18*, 1385. [CrossRef]
20. Hofer, M.; Falk, M.; Komůrková, D.; Falková, I.; Bačíková, A.; Klejdus, B.; Pagáčová, E.; Štefančíková, L.; Weiterová, L.; Angelis, K.J.; et al. Two New Faces of Amifostine: Protector from DNA Damage in Normal Cells and Inhibitor of DNA Repair in Cancer Cells. *J. Med. Chem.* **2016**, *59*, 3003–3017. [CrossRef]
21. Štefančíková, L.; Lacombe, S.; Salado, D.; Porcel, E.; Pagáčová, E.; Tillement, O.; Lux, F.; Depeš, D.; Kozubek, S.; Falk, M. Effect of gadolinium-based nanoparticles on nuclear DNA damage and repair in glioblastoma tumor cells. *J. Nanobiotechnol.* **2016**, *14*, 63. [CrossRef] [PubMed]
22. Ngwa, W.; Boateng, F.; Kumar, R.; Irvine, D.J.; Formenti, S.; Ngoma, T.; Herskind, C.; Veldwijk, M.R.; Hildenbrand, G.L.; Hausmann, M.; et al. Smart Radiation Therapy Biomaterials. *Int. J. Radiat. Oncol. Biol. Phys.* **2017**, *97*, 624–637. [CrossRef] [PubMed]
23. Hildenbrand, G.; Metzler, P.; Pilarczyk, G.; Bobu, V.; Kriz, W.; Hosser, H.; Fleckenstein, J.; Krufczik, M.; Bestvater, F.; Wenz, F.; et al. Dose enhancement effects of gold nanoparticles specifically targeting RNA in breast cancer cells. *PLoS ONE* **2018**, *13*, e0190183. [CrossRef] [PubMed]
24. Ngwa, W.; Kumar, R.; Sridhar, S.; Korideck, H.; Zygmanski, P.; Cormack, R.A.; Berbeco, R.; Makrigiorgos, G.M. Targeted radiotherapy with gold nanoparticles: Current status and future perspectives. *Nanomedicine* **2014**, *9*, 1063–1082. [CrossRef] [PubMed]
25. Lux, F.; Tran, V.L.; Thomas, E.; Dufort, S.; Rossetti, F.; Martini, M.; Truillet, C.; Doussineau, T.; Bort, G.; Denat, F.; et al. AGuIX® from bench to bedside-Transfer of an ultrasmall theranostic gadolinium-based nanoparticle to clinical medicine. *Br. J. Radiol.* **2018**. [CrossRef]
26. Kuncic, Z.; Lacombe, S. Nanoparticle radio-enhancement: Principles, progress and application to cancer treatment. *Phys. Med. Biol.* **2018**, *63*, 02TR01. [CrossRef] [PubMed]
27. Li, S.; Porcel, E.; Remita, H.; Marco, S.; Réfrégiers, M.; Dutertre, M.; Confalonieri, F.; Lacombe, S. Platinum nanoparticles: An exquisite tool to overcome radioresistance. *Cancer Nanotechnol.* **2017**, *8*, 4. [CrossRef] [PubMed]
28. Sancey, L.; Lux, F.; Kotb, S.; Roux, S.; Dufort, S.; Bianchi, A.; Crémillieux, Y.; Fries, P.; Coll, J.-L.; Rodriguez-Lafrasse, C.; et al. The use of theranostic gadolinium-based nanoprobes to improve radiotherapy efficacy. *Br. J. Radiol.* **2014**, *87*. [CrossRef] [PubMed]
29. Nikjoo, H.; Uehara, S.; Emfietzoglou, D.; Brahme, A. Heavy charged particles in radiation biology and biophysics. *New J. Phys.* **2008**, *10*, 075006. [CrossRef]
30. Hossain, M.; Su, M. Nanoparticle location and material dependent dose enhancement in X-ray radiation therapy. *J. Phys. Chem. C Nanomater. Interfaces* **2012**, *116*, 23047–23052. [CrossRef]
31. Zygmanski, P.; Liu, B.; Tsiamas, P.; Cifter, F.; Petersheim, M.; Hesser, J.; Sajo, E. Dependence of Monte Carlo microdosimetric computations on the simulation geometry of gold nanoparticles. *Phys. Med. Biol.* **2013**, *58*, 7961–7977. [CrossRef] [PubMed]
32. Falk, M.; Hausmann, M.; Lukášová, E.; Biswas, A.; Hildenbrand, G.; Davídková, M.; Krasavin, E.; Kleibl, Z.; Falková, I.; Ježková, L.; et al. Determining Omics spatiotemporal dimensions using exciting new nanoscopy techniques to assess complex cell responses to DNA damage: Part A—Radiomics. *Crit. Rev. Eukaryot. Gene Expr.* **2014**, *24*, 205–223. [CrossRef] [PubMed]

33. Falk, M.; Hausmann, M.; Lukášová, E.; Biswas, A.; Hildenbrand, G.; Davídková, M.; Krasavin, E.; Kleibl, Z.; Falková, I.; Ježková, L.; et al. Determining Omics spatiotemporal dimensions using exciting new nanoscopy techniques to assess complex cell responses to DNA damage: Part B—Structuromics. *Crit. Rev. Eukaryot. Gene Expr.* **2014**, *24*, 225–247. [CrossRef] [PubMed]
34. Rittich, B.; Spanová, A.; Falk, M.; Benes, M.J.; Hrubý, M. Cleavage of double stranded plasmid DNA by lanthanide complexes. *J. Chromatogr. B Analyt. Technol. Biomed. Life Sci.* **2004**, *800*, 169–173. [CrossRef] [PubMed]
35. Kratochvílová, I.; Golan, M.; Pomeisl, K.; Richter, J.; Sedláková, S.; Šebera, J.; Mičová, J.; Falk, M.; Falková, I.; Řeha, D.; et al. Theoretical and experimental study of the antifreeze protein AFP752, trehalose and dimethyl sulfoxide cryoprotection mechanism: Correlation with cryopreserved cell viability. *RSC Adv.* **2017**, *7*, 352–360. [CrossRef] [PubMed]
36. Freneau, A.; Dos Santos, M.; Voisin, P.; Tang, N.; Bueno Vizcarra, M.; Villagrasa, C.; Roy, L.; Vaurijoux, A.; Gruel, G. Relation between DNA double-strand breaks and energy spectra of secondary electrons produced by different X-ray energies. *Int. J. Radiat. Biol.* **2018**, 1–10. [CrossRef] [PubMed]
37. Falk, M.; Lukasova, E.; Kozubek, S. Higher-order chromatin structure in DSB induction, repair and misrepair. *Mutat. Res.* **2010**, *704*, 88–100. [CrossRef]
38. Hausmann, M.; Ilić, N.; Pilarczyk, G.; Lee, J.-H.; Logeswaran, A.; Borroni, A.; Krufczik, M.; Theda, F.; Waltrich, N.; Bestvater, F.; et al. Challenges for Super-Resolution Localization Microscopy and Biomolecular Fluorescent Nano-Probing in Cancer Research. *Int. J. Mol. Sci.* **2017**, *18*, 2066. [CrossRef]
39. Schipler, A.; Iliakis, G. DNA double-strand-break complexity levels and their possible contributions to the probability for error-prone processing and repair pathway choice. *Nucleic Acids Res.* **2013**, *41*, 7589–7605. [CrossRef]
40. Mladenov, E.; Magin, S.; Soni, A.; Iliakis, G. DNA double-strand-break repair in higher eukaryotes and its role in genomic instability and cancer: Cell cycle and proliferation-dependent regulation. *Semin. Cancer Biol.* **2016**, *37–38*, 51–64. [CrossRef]
41. Mladenov, E.; Magin, S.; Soni, A.; Iliakis, G. DNA double-strand break repair as determinant of cellular radiosensitivity to killing and target in radiation therapy. *Front. Oncol.* **2013**, *3*, 113. [CrossRef] [PubMed]
42. Burger, N.; Biswas, A.; Barzan, D.; Kirchner, A.; Hosser, H.; Hausmann, M.; Hildenbrand, G.; Herskind, C.; Wenz, F.; Veldwijk, M.R. A method for the efficient cellular uptake and retention of small modified gold nanoparticles for the radiosensitization of cells. *Nanomed. Nanotechnol. Biol. Med.* **2014**, *10*, 1365–1373. [CrossRef] [PubMed]
43. Porcel, E.; Liehn, S.; Remita, H.; Usami, N.; Kobayashi, K.; Furusawa, Y.; Le Sech, C.; Lacombe, S. Platinum nanoparticles: A promising material for future cancer therapy? *Nanotechnology* **2010**, *21*, 85103. [CrossRef] [PubMed]
44. Maeda, H. Tumor-selective delivery of macromolecular drugs via the EPR effect: Background and future prospects. *Bioconjug. Chem.* **2010**, *21*, 797–802. [CrossRef]
45. Maeda, H.; Matsumura, Y. EPR effect based drug design and clinical outlook for enhanced cancer chemotherapy. *Adv. Drug Deliv. Rev.* **2011**, *63*, 129–130. [CrossRef]
46. Fang, J.; Nakamura, H.; Maeda, H. The EPR effect: Unique features of tumor blood vessels for drug delivery, factors involved, and limitations and augmentation of the effect. *Adv. Drug Deliv. Rev.* **2011**, *63*, 136–151. [CrossRef]
47. Prabhakar, U.; Maeda, H.; Jain, R.K.; Sevick-Muraca, E.M.; Zamboni, W.; Farokhzad, O.C.; Barry, S.T.; Gabizon, A.; Grodzinski, P.; Blakey, D.C. Challenges and key considerations of the enhanced permeability and retention effect for nanomedicine drug delivery in oncology. *Cancer Res.* **2013**, *73*, 2412–2417. [CrossRef] [PubMed]
48. Bertrand, N.; Wu, J.; Xu, X.; Kamaly, N.; Farokhzad, O.C. Cancer nanotechnology: The impact of passive and active targeting in the era of modern cancer biology. *Adv. Drug Deliv. Rev.* **2014**, *66*, 2–25. [CrossRef]
49. Chithrani, D.B. Nanoparticles for improved therapeutics and imaging in cancer therapy. *Recent Pat. Nanotechnol.* **2010**, *4*, 171–180. [CrossRef]
50. Chithrani, D.B.; Jelveh, S.; Jalali, F.; van Prooijen, M.; Allen, C.; Bristow, R.G.; Hill, R.P.; Jaffray, D.A. Gold nanoparticles as radiation sensitizers in cancer therapy. *Radiat. Res.* **2010**, *173*, 719–728. [CrossRef]
51. Hainfeld, J.F.; Smilowitz, H.M.; O'Connor, M.J.; Dilmanian, F.A.; Slatkin, D.N. Gold nanoparticle imaging and radiotherapy of brain tumors in mice. *Nanomedicine* **2013**, *8*, 1601–1609. [CrossRef]

52. He, H.; Xie, C.; Ren, J. Nonbleaching fluorescence of gold nanoparticles and its applications in cancer cell imaging. *Anal. Chem.* **2008**, *80*, 5951–5957. [CrossRef] [PubMed]
53. Moser, F.; Hildenbrand, G.; Müller, P.; Al Saroori, A.; Biswas, A.; Bach, M.; Wenz, F.; Cremer, C.; Burger, N.; Veldwijk, M.R.; et al. Cellular Uptake of Gold Nanoparticles and Their Behavior as Labels for Localization Microscopy. *Biophys. J.* **2016**, *110*, 947–953. [CrossRef]
54. Lacombe, S.; Porcel, E.; Scifoni, E. Particle therapy and nanomedicine: State of art and research perspectives. *Cancer Nanotechnol.* **2017**, *8*, 9. [CrossRef]
55. Stefančíková, L.; Porcel, E.; Eustache, P.; Li, S.; Salado, D.; Marco, S.; Guerquin-Kern, J.-L.; Réfrégiers, M.; Tillement, O.; Lux, F.; et al. Cell localisation of gadolinium-based nanoparticles and related radiosensitising efficacy in glioblastoma cells. *Cancer Nanotechnol.* **2014**, *5*, 6. [CrossRef] [PubMed]
56. Yameen, B.; Choi, W.I.; Vilos, C.; Swami, A.; Shi, J.; Farokhzad, O.C. Insight into nanoparticle cellular uptake and intracellular targeting. *J. Control. Release* **2014**, *190*, 485–499. [CrossRef] [PubMed]
57. Fernando, L.P.; Kandel, P.K.; Yu, J.; McNeill, J.; Ackroyd, P.C.; Christensen, K.A. Mechanism of cellular uptake of highly fluorescent conjugated polymer nanoparticles. *Biomacromolecules* **2010**, *11*, 2675–2682. [CrossRef] [PubMed]
58. Cartiera, M.S.; Johnson, K.M.; Rajendran, V.; Caplan, M.J.; Saltzman, W.M. The uptake and intracellular fate of PLGA nanoparticles in epithelial cells. *Biomaterials* **2009**, *30*, 2790–2798. [CrossRef]
59. Fröhlich, E. Cellular targets and mechanisms in the cytotoxic action of non-biodegradable engineered nanoparticles. *Curr. Drug Metab.* **2013**, *14*, 976–988. [CrossRef]
60. Lemmer, P.; Gunkel, M.; Baddeley, D.; Kaufmann, R.; Urich, A.; Weiland, Y.; Reymann, J.; Müller, P.; Hausmann, M.; Cremer, C. SPDM: Light microscopy with single-molecule resolution at the nanoscale. *Appl. Phys. B* **2008**, *93*, 1–12. [CrossRef]
61. Hausmann, M.; Wagner, E.; Lee, J.-H.; Schrock, G.; Schaufler, W.; Krufczik, M.; Papenfuß, F.; Port, M.; Bestvater, F.; Scherthan, H. Super-resolution localization microscopy of radiation-induced histone H2AX-phosphorylation in relation to H3K9-trimethylation in HeLa cells. *Nanoscale* **2018**, *10*, 4320–4331. [CrossRef] [PubMed]
62. Depes, D.; Lee, J.-H.; Bobkova, E.; Jezkova, L.; Falkova, I.; Bestvater, F.; Pagacova, E.; Kopecna, O.; Zadneprianetc, M.; Bacikova, A.; et al. Single-molecule localization microscopy as a promising tool for γH2AX/53BP1 foci exploration. *Eur. Phys. J. D* **2018**, *72*. [CrossRef]
63. Reindl, J.; Girst, S.; Walsh, D.W.M.; Greubel, C.; Schwarz, B.; Siebenwirth, C.; Drexler, G.A.; Friedl, A.A.; Dollinger, G. Chromatin organization revealed by nanostructure of irradiation induced γH2AX, 53BP1 and Rad51 foci. *Sci. Rep.* **2017**, *7*, 40616. [CrossRef] [PubMed]
64. Sun, H.; Jia, J.; Jiang, C.; Zhai, S. Gold Nanoparticle-Induced Cell Death and Potential Applications in Nanomedicine. *Int. J. Mol. Sci.* **2018**, *19*, 754. [CrossRef] [PubMed]
65. Duve, C. Lysosomes revisited. *Eur. J. Biochem.* **1983**, *137*, 391–397. [CrossRef] [PubMed]
66. Wattiaux, R.; Coninck, S.W.-D.; Jadot, M.; Hamer, I.; Bielande, V.; Beauloye, V. Lysosomes as Suicide Bags. In *Endocytosis*; Courtoy, P.J., Ed.; Springer: Berlin/Heidelberg, Germany, 1992; pp. 433–437, ISBN 978-3-642-84297-9.
67. Zhang, R.; Piao, M.J.; Kim, K.C.; Kim, A.D.; Choi, J.-Y.; Choi, J.; Hyun, J.W. Endoplasmic reticulum stress signaling is involved in silver nanoparticles-induced apoptosis. *Int. J. Biochem. Cell Biol.* **2012**, *44*, 224–232. [CrossRef] [PubMed]
68. Szegezdi, E.; Logue, S.E.; Gorman, A.M.; Samali, A. Mediators of endoplasmic reticulum stress-induced apoptosis. *EMBO Rep.* **2006**, *7*, 880–885. [CrossRef]
69. Zhang, F.; Zhu, X.; Gong, J.; Sun, Y.; Chen, D.; Wang, J.; Wang, Y.; Guo, M.; Li, W. Lysosome—Mitochondria-mediated apoptosis specifically evoked in cancer cells induced by gold nanorods. *Nanomedicine* **2016**, *11*, 1993–2006. [CrossRef]
70. Dahmen, V.; Kriehuber, R. Cytotoxic effects and specific gene expression alterations induced by I-125-labeled triplex-forming oligonucleotides. *Int. J. Radiat. Biol.* **2012**, *88*, 972–979. [CrossRef]
71. Hausmann, M.; Winkler, R.; Hildenbrand, G.; Finsterle, J.; Weisel, A.; Rapp, A.; Schmitt, E.; Janz, S.; Cremer, C. COMBO-FISH: Specific labeling of nondenatured chromatin targets by computer-selected DNA oligonucleotide probe combinations. *BioTechniques* **2003**, *35*, 564–577. [CrossRef]

72. Krufczik, M.; Sievers, A.; Hausmann, A.; Lee, J.-H.; Hildenbrand, G.; Schaufler, W.; Hausmann, M. Combining Low Temperature Fluorescence DNA-Hybridization, Immunostaining, and Super-Resolution Localization Microscopy for Nano-Structure Analysis of ALU Elements and Their Influence on Chromatin Structure. *Int. J. Mol. Sci.* **2017**, *18*, 1005. [CrossRef]
73. Müller, P.; Schmitt, E.; Jacob, A.; Hoheisel, J.; Kaufmann, R.; Cremer, C.; Hausmann, M. COMBO-FISH enables high precision localization microscopy as a prerequisite for nanostructure analysis of genome loci. *Int. J. Mol. Sci.* **2010**, *11*, 4094–4105. [CrossRef] [PubMed]
74. Zanta, M.A.; Belguise-Valladier, P.; Behr, J.P. Gene delivery: A single nuclear localization signal peptide is sufficient to carry DNA to the cell nucleus. *Proc. Natl. Acad. Sci. USA* **1999**, *96*, 91–96. [CrossRef] [PubMed]
75. Chithrani, B.D.; Ghazani, A.A.; Chan, W.C.W. Determining the size and shape dependence of gold nanoparticle uptake into mammalian cells. *Nano Lett.* **2006**, *6*, 662–668. [CrossRef] [PubMed]
76. Falk, M.; Lukášová, E.; Štefančíková, L.; Baranová, E.; Falková, I.; Ježková, L.; Davídková, M.; Bačíková, A.; Vachelová, J.; Michaelidesová, A.; et al. Heterochromatinization associated with cell differentiation as a model to study DNA double strand break induction and repair in the context of higher-order chromatin structure. *Appl. Radiat. Isot.* **2014**, *83*, 177–185. [CrossRef] [PubMed]
77. Ježková, L.; Falk, M.; Falková, I.; Davídková, M.; Bačíková, A.; Štefančíková, L.; Vachelová, J.; Michaelidesová, A.; Lukášová, E.; Boreyko, A.; et al. Function of chromatin structure and dynamics in DNA damage, repair and misrepair: γ-rays and protons in action. *Appl. Radiat. Isot.* **2014**, *83*, 128–136. [CrossRef] [PubMed]
78. Sevcik, J.; Falk, M.; Macurek, L.; Kleiblova, P.; Lhota, F.; Hojny, J.; Stefancikova, L.; Janatova, M.; Bartek, J.; Stribrna, J.; et al. Expression of human BRCA1Δ17-19 alternative splicing variant with a truncated BRCT domain in MCF-7 cells results in impaired assembly of DNA repair complexes and aberrant DNA damage response. *Cell. Signal.* **2013**, *25*, 1186–1193. [CrossRef]
79. Kozubek, M.; Kozubek, S.; Lukásová, E.; Bártová, E.; Skalníková, M.; Matula, P.; Matula, P.; Jirsová, P.; Cafourková, A.; Koutná, I. Combined confocal and wide-field high-resolution cytometry of fluorescent in situ hybridization-stained cells. *Cytometry* **2001**, *45*, 1–12. [CrossRef]
80. Matula, P.; Maška, M.; Daněk, O.; Matula, P.; Kozubek, M. Acquiarium: Free Software for the Acquisition and Analysis of 3D Images of Cells in Fluorescence Microscopy. In Proceedings of the IEEE International Symposium on Biomedical Imaging, Boston, MA, USA, 28 June–1 July 2009; pp. 1138–1141, ISBN 978-1-4244-3932-4.
81. Eryilmaz, M.; Schmitt, E.; Krufczik, M.; Theda, F.; Lee, J.-H.; Cremer, C.; Bestvater, F.; Schaufler, W.; Hausmann, M.; Hildenbrand, G. Localization Microscopy Analyses of MRE11 Clusters in 3D-Conserved Cell Nuclei of Different Cell Lines. *Cancers* **2018**, *10*, 25. [CrossRef]

International Journal of
Molecular Sciences

MDPI

Article

Lipid Nanoparticles Decorated with TNF-Related Aptosis-Inducing Ligand (TRAIL) Are More Cytotoxic than Soluble Recombinant TRAIL in Sarcoma

Ana Gallego-Lleyda [1,2], Diego De Miguel [1,3], Alberto Anel [1,2] and Luis Martinez-Lostao [2,4,5,6,*]

1 Departamento de Bioquímica, Biología Molecular y Celular, Universidad de Zaragoza, 50009 Zaragoza, Spain; annaiss89@hotmail.com (A.G.-L.); diego_demiguel@hotmail.com (D.D.M.); anel@unizar.es (A.A.)
2 Instituto de Investigación Sanitaria de Aragón (ISS), 50009 Zaragoza, Spain
3 Cell Death, Cancer and Inflammation, University College of London, London WC1E 6BT, UK
4 Servicio de Inmunología, Hospital Clínico Universitario Lozano Blesa, 50009 Zaragoza, Spain
5 Departamento de Microbiología, Medicina Preventiva y Sąlud Pública, Universidad de Zaragoza, 50009 Zaragoza, Spain
6 Instituto de Nanociencia de Aragón, 50009 Zaragoza, Spain
* Correspondence: lumartin@unizar.es; Tel.: +34-976-76-88-59

Received: 15 April 2018; Accepted: 11 May 2018; Published: 13 May 2018

Abstract: Sarcomas are rare and heterogeneous cancers classically associated with a poor outcome. Sarcomas are 1% of the cancer but recent estimations indicate that sarcomas account for 2% of the estimated cancer-related deaths. Traditional treatment with surgery, radiotherapy, and chemotherapy has improved the outcome for some types of sarcomas. However, novel therapeutic strategies to treat sarcomas are necessary. TNF-related apoptosis-inducing ligand (TRAIL) is a death ligand initially described as capable of inducing apoptosis on tumor cell while sparing normal cells. Only few clinical trials have used TRAIL-based treatments in sarcoma, but they show only low or moderate efficacy of TRAIL. Consequently, novel TRAIL formulations with an improved TRAIL bioactivity are necessary. Our group has developed a novel TRAIL formulation based on tethering this death ligand on a lipid nanoparticle surface (LUV-TRAIL) resembling the physiological secretion of TRAIL as a trasmembrane protein inserted into the membrane of exosomes. We have already demonstrated that LUV-TRAIL shows an improved cytotoxic activity when compared to soluble recombinant TRAIL both in hematological malignancies and epithelial-derived cancers. In the present study, we have tested LUV-TRAIL in several human sarcoma tumor cell lines with different sensitivity to soluble recombinant TRAIL, finding that LUV-TRAIL was more efficient than soluble recombinant TRAIL. Moreover, combined treatment of LUV-TRAIL with distinct drugs proved to be especially effective, sensitizing even more resistant cell lines to TRAIL.

Keywords: sarcoma; TRAIL; flavopiridol; immunotherapy; lipid nanoparticles

---

1. Introduction

Sarcomas include a heterogeneous group of complex cancers which derive from mesenchymal transformed cells. Sarcomas are rare, making up for about 1% of all cancers [1]. It is well known that prognosis of sarcomas is poor, with a 5-year survival rate of less than 15% when metastasis happens [2]. Over the last years, a great effort for improving treatment outcome has been made, but the complexity of sarcomas makes that advancements in developing new therapies are extremely slow [3]. Therefore, there is an urgent need to develop and test novel therapeutic strategies in order to improve prognosis of sarcoma patients [4]. Among the novel anti-tumor agents used, apoptosis ligand 2/TNF-related apoptosis-inducing ligand (Apo2L/TRAIL) has been tested both in pre-clinical studies and clinical trials in some types of sarcoma [5].

Apoptosis ligand 2/TNF-related apoptosis-inducing ligand (Apo2L/TRAIL) is a TNF family member described as capable of inducing apoptosis in a wide variety of transformed cells while sparing normal cells [6,7]. This fact made TRAIL to be considered as a promising anti-tumor agent. In fact, TRAIL-based therapeutic strategies were developed and used in several phase I/II clinical trials on a wide variety of human cancers [8–10]. However, although TRAIL-based therapies were proven safe, about 50% of human cancers were intrinsically resistant to TRAIL since the clinical studies using TRAIL showed limited therapeutic activity of this death ligand in different types of cancer [9,11,12] including sarcoma [5]. Therefore, novel formulations of TRAIL with improved bioactivity and combination of TRAIL with sensitizing agents could be plausible strategies to overcome TRAIL resistance [13–16].

Our group generated artificial lipid nanoparticles containing membrane-bound TRAIL (LUV-TRAIL), resembling the natural TRAIL-containing exosomes physiologically released by activated human T cells upon re-stimulation [17,18]. We demonstrated that LUV-TRAIL are more effective than soluble recombinant TRAIL (sTRAIL), inducing apoptosis both in vitro and in vivo in hematological malignancies [19–21] as well as in epithelial-derived cancers [22–24].

In the present work, we have tested LUV-TRAIL in several human sarcoma cell lines of distinct origin with different sensitivity to sTRAIL. LUV-TRAIL were more efficient inducing apoptosis in all sarcoma cells tested than sTRAIL. However, although LUV-TRAIL showed a greater pro-apoptotic potential compared to sTRAIL, some sarcoma cell lines still remained fairly resistant to LUV-TRAIL. In this line, a wide range of sensitizing strategies have been described to overcome TRAIL-resistance in tumor cells. Here, we have tested the therapeutic combination of LUV-TRAIL with several drugs previously described as sensitizing agents [23–30], in order to improve the pro-apoptotic ability of LUV-TRAIL. Our results showed that, among all the combinations tested, flavopiridol induced the strongest sensitizing effect in all sarcoma cell lines tested, by inducing a down-regulation of the anti-apoptotic protein FLIP.

In summary, LUV-TRAIL showed an improved cytotoxicity against sarcoma cells, overcoming the intrinsic resistance of these cells to sTRAIL. This was further potentiated when they were combined with sensitizing agents such as flavopiridol (FVP), opening the door to future clinical applications as anti-tumor therapy in sarcoma.

2. Results

2.1. LUV-TRAIL Showed an Enhanced In Vitro Cytotoxic Activity Compared to Soluble Recombinant TRAIL in Human Sarcoma Cells

First, dose-response assays were carried out to compare the in vitro bioactivity of sTRAIL and LUV-TRAIL, and cell viability was analyzed by the MTT assay (Figure 1a). Whereas A673 cells showed a great sensitivity both to sTRAIL and LUV-TRAIL, HT-1080 and RD cell lines showed a moderate sensitivity to sTRAIL However, LUV-TRAIL was capable of inducing a significant decrease of cell viability both in HT-1080 and RD cells in comparison with sTRAIL. To assess whether the decrease in cell viability observed after treatment with both forms of TRAIL (sTRAIL and LUV-TRAIL) was due to the onset of cell death, annexin-V staining was performed (Figure 1b). In all sarcoma cell lines tested, LUV-TRAIL induced a remarkable increase of cell death at higher dose used (1000 ng/mL) when compared to sTRAIL. Importantly, the cell death observed in all sarcoma cell lines was specifically attributable to TRAIL-receptor activation by TRAIL, as cell death was completely inhibited when cells were pre-incubated with the TRAIL neutralizing antibody RIK2 before treatment with sTRAIL and LUV-TRAIL (Figure 1b). In this line, LUVs alone (without TRAIL anchoring on their surface), did not exert any cytotoxic effect in any sarcoma cell lines tested (see black bars on control points in Figure 1b).

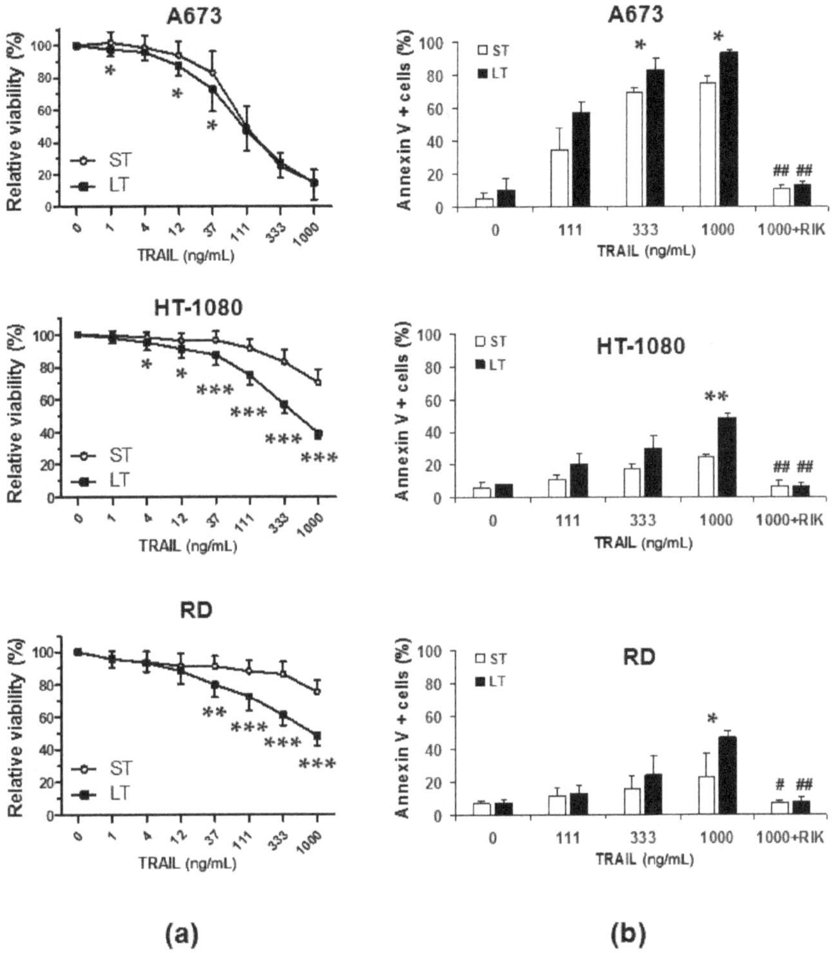

Figure 1. (**a**) Analysis of cell viability after treatment with LUV-TRAIL on human sarcoma cell lines. Dose-response assays using the indicated doses of sTRAIL (ST) and LUV-TRAIL (LT) were performed on A673 cells, HT-1080 cells, and RD cells. Cells were treated with ST or LT for 24 h. Then, cell viability was measured by the MTT assay method. Graphs show the mean ± SD of at least three independent experiments. * $p < 0.05$, ** $p < 0.01$, *** $p < 0.001$; (**b**) Cytotoxicity assays on human sarcoma cell lines. Cells were treated with indicated doses of sTRAIL (ST) or LUV-TRAIL (LT) for 24 h and annexin V positive cells were quantified by flow cytometry. When cells were treated with 1000 ng/mL, they were previously pre-incubated in presence or absence of the anti-TRAIL blocking mAb, RIK2 (500 ng/mL). Graphics show the percentage of annexin-V positive cells analyzed expressed as the mean ± SD of at least three experiments. * $p < 0.05$. (ST versus LT). # $p < 0.05$, ## $p < 0.01$ (ST versus ST + RIK2 and, LT versus LT + RIK2). TRAIL, TNF-related apoptosis-inducing ligand; LUV-TRAIL, TRAIL on a lipid nanoparticle surface; sTRAIL, soluble recombinant TRAIL.

2.2. LUV-TRAIL Activated the Caspase Cascade More Efficiently than sTRAIL in Human Sarcoma Cells

Next, the implication of caspases in the cytotoxicity induced by LUV-TRAIL in sarcoma cells was assessed. For that purpose, sarcoma cells were incubated with sTRAIL or LUV-TRAIL and activation of the main caspases involved in the extrinsic apoptotic pathway was analyzed by Western blot. Activation of both caspase-8 and caspase-3 was clearly increased when sarcoma cells were treated with LUV-TRAIL compared to sTRAIL, as evidenced by the disappearance of the pro-forms of both caspases (Figure 2a). Moreover, cleavage of the specific caspase-3 substrate, PARP-1, and the specific caspase-8 substrate, Bid, correlated with the activation of both caspases -3 and -8, respectively, indicating a fully functional activation of the extrinsic apoptotic pathway upon LUV-TRAIL treatment. When time course assays were performed (Figure 2b), caspase activation was faster in A673 cells when they were treated with LUV-TRAIL, although, as seen previously, both formulations of TRAIL present similar cytotoxicity at 24 h. In HT-1080 cells, similar kinetics was observed at shorter times when they were treated both with sTRAIL and LUV-TRAIL. However, as shown in Figure 2a, caspase activation was greater when HT-1080 cells were treated with LUV-TRAIL in comparison with sTRAIL after 24 h of treatment. These data reflect that LUV-TRAIL required longer time of incubation to induce a greater caspase activation and, hence, a greater cytotoxicity than sTRAIL in HT-1080 cells. In case of RD cells, although no obvious differences could be observed in caspase activation after treatment with sTRAIL or LUV-TRAIL, Bid and PARP-1 degradation was faster when cells were treated with LUV-TRAIL. Finally, to fully assess and characterize the role of caspases in LUV-TRAIL induced cell death, cell death-inhibition assays were performed using the general caspase inhibitor z-VAD-fmk (Figure 2c). As expected, caspase inhibition fully abrogated cell death induced not only by sTRAIL but also by LUV-TRAIL. Moreover, when cells were pre-incubated with the specific caspase-8 inhibitor IETD-fmk, cell death induced by LUV-TRAIL was also fully abrogated, proving that cell death was fully dependent on the activation of the canonical extrinsic apoptotic pathway, ruling out any other form of cell death that could be triggered by TRAIL, such as necroptosis.

2.3. Combination of LUV-TRAIL with Anti-Cancer Agents Improved its Cytotoxicity against Human Sarcoma Cells

Although LUV-TRAIL showed a greater cytotoxic effect than sTRAIL and was able to overcome sTRAIL resistance in HT-1080 and RD cells, LUV-TRAIL was only capable of inducing a decrease of cell viability of around 50% in both sarcoma cell lines at the 1000 ng/mL dose. Therefore, we decided to combine LUV-TRAIL with several anti-cancer agents such as flavopiridol (FVP), doxorubicin (DOX), trabectedin (TRAB), and bortezomib (BORT) in order to improve LUV-TRAIL-induced cytotoxicity. First of all, dose-response assays were carried out to analyze the cytotoxic effect of all anti-cancer agents used alone (Figure 3). Aiming for a sensitizing effect rather than an additive effect, sub-toxic doses of the different drugs were selected by choosing the maximum concentrations not inducing a decrease of cell viability greater than 20% (marker doses for each anti-cancer drug in Figure 3).

After that, dose-response assays were carried out by pre-incubating sarcoma cells with the selected doses of each anti-cancer drug for 2 h before treating with either sTRAIL or LUV-TRAIL.

In the case of A673 cells, although both sTRAIL and LUV-TRAIL induced a great decrease of cell viability at the highest dose used (1000 ng/mL), all different anti-cancer drugs were able to sensitize to LUV-TRAIL at intermediate doses (Figure 4a).

In HT-1080 cells, although FVP greatly sensitized to both sTRAIL and LUV-TRAIL, the combination with LUV-TRAIL was the most cytotoxic (Figure 4b). Combination of TRAIL with DOX sensitized to both sTRAIL and LUV-TRAIL to the same extent. Similarly, BORT equally sensitized the cells to both sTRAIL and LUV-TRAIL, whereas TRAB did not show any sensitizing effect whatsoever (Figure 4b).

Finally, in RD cells, combination of both sTRAIL and LUV-TRAIL with FVP or TRAB induced similar cytotoxocity levels, whereas DOX and BORT did not sensitize to neither sTRAIL nor LUV-TRAIL (Figure 4c).

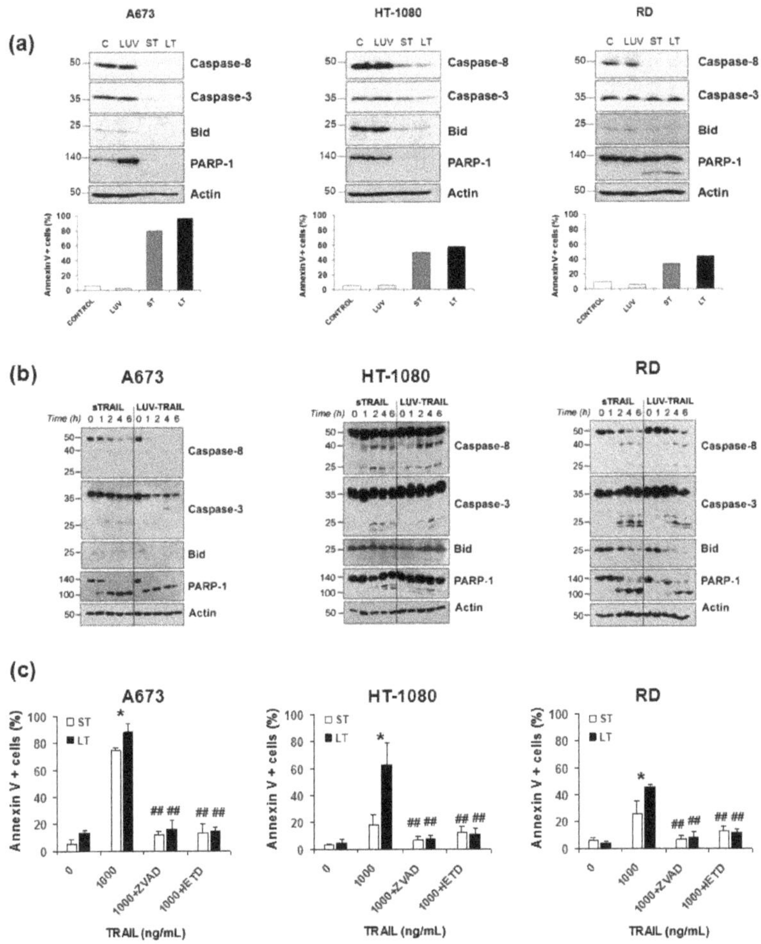

Figure 2. (**a**) Analysis of caspase activation in human sarcoma cells. Cells were untreated (Control, designed as C), or treated with LUVs without TRAIL (LUV), sTRAIL (ST), and LUV-TRAIL (LT) at 1000 ng/mL for 24 h. After that, cells were lysed, and lysates were subjected to SDS-PAGE and to Western blot analysis. Levels of caspase-8, caspase-3, Bid, and PARP-1 were analyzed using specific antibodies. Level of actin levels was used as a control for equal protein loading. Cell death was measured in parallel by flow cytometry after annexin-V staining (bottom graphs); (**b**) Analysis of time-course caspase activation in human sarcoma cells. Cells were treated with sTRAIL or LUV-TRAIL at 1000 ng/mL at the indicated times. After that, cells were lysed, and lysates were subjected to SDS-PAGE and to Western blot analysis. Levels of caspase-8, caspase-3, Bid, and PARP-1 were analyzed using specific antibodies. Level of actin levels was used as a control for equal protein loading. Caspase activation was evidenced by the disappearance of the pro-forms shown in the Western blot; (**c**) Analysis of cell death inhibition by caspase inhibition. Human sarcoma cells were treated with 1000 ng/mL of both sTRAIL (ST) and LUV-TRAIL (LT) for 24 h previously incubated in presence or absence of the pan-caspase inhibitor z-VAD-fmk (30 μM) and of the specific caspase-8 inhibitor IETD-fmk (30 μM). Graphics show the mean ± SD of the cell death of treated cells expressed as percentage of at least three experiments. * $p < 0.05$. (ST versus LT). ## $p < 0.01$ (ST versus ST + caspase inhibitors and, LT versus LT + caspase inhibitors).

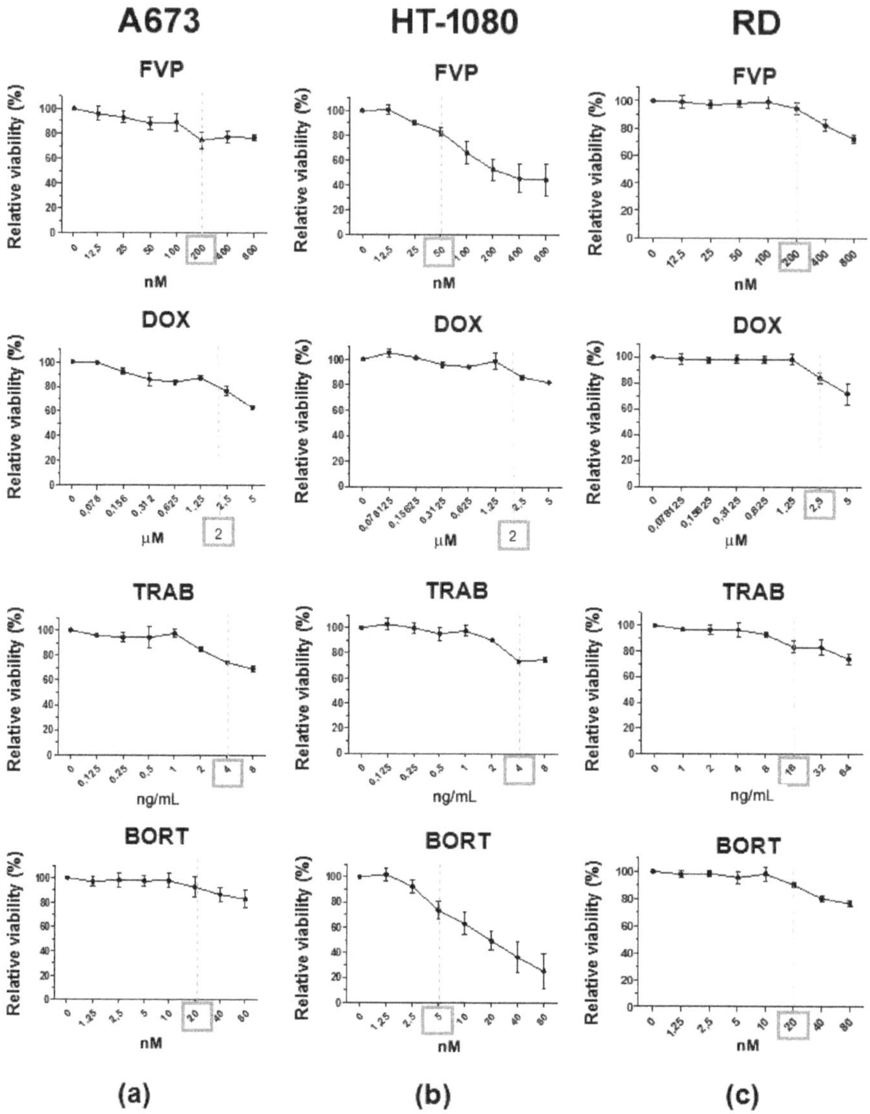

Figure 3. Analysis of the effect of sensitizer agents in cell viability of human sarcoma cells. Dose-response assays using the sensitizer agents: flavopiridol (FVP), doxorubicin (DOX), trabectedin (TRAB), and bortezomib (BORT) were performed on A673 cells (**a**), HT-1080 cells (**b**), and RD cells (**c**). Cell viability was assessed by MTT assay after 24 h. The results were expressed as the mean ± SD of at least three experiments. Marked doses of each drugs were selected for further experiments of sensitization.

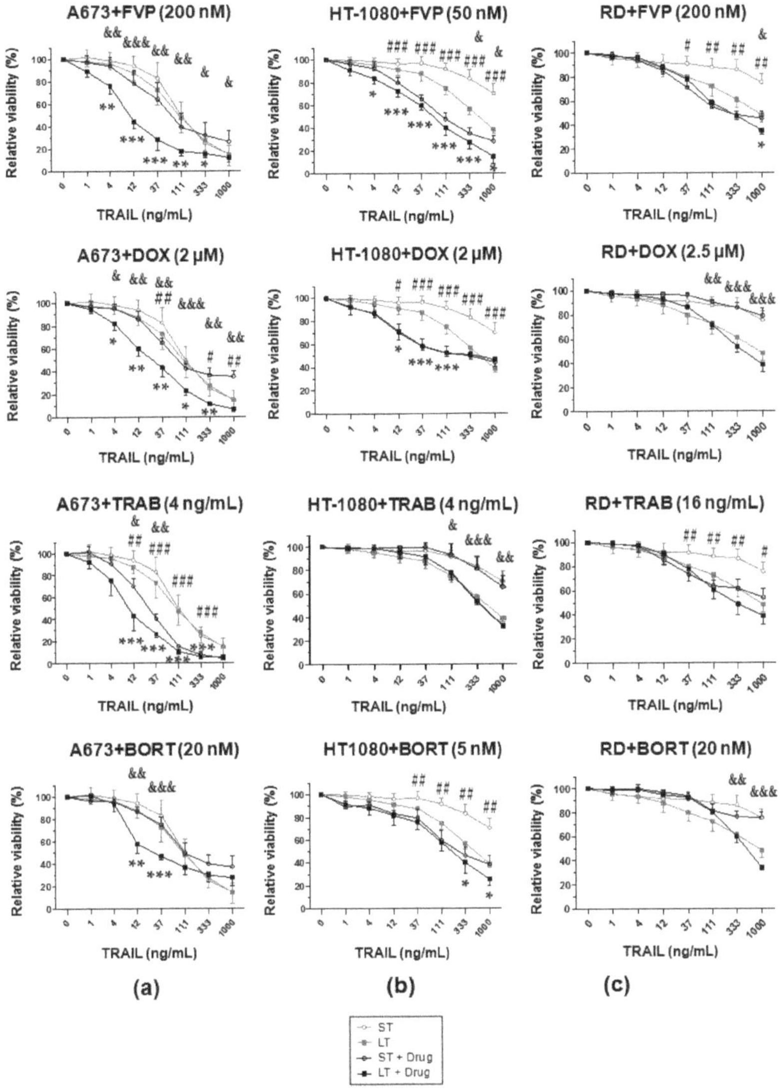

Figure 4. Analysis of cell viability of human sarcoma cell lines after treatment with sTRAIL or LUV-TRAIL in combination with sensitizer agents. Cells were pre-incubated for 2 h with sensitizer drugs: flavopiridol (FVP), doxorubicin (DOX), trabectedin (TRAB), and bortezomib (BORT) at indicated doses. Afterwards, sTRAIL (ST) or LUV-TRAIL (LT) were added at the indicated concentrations and left overnight. The following day, cell viability was measured by the MTT assay. In every graph, the results from treatments with ST and LT in the absence of drugs are superposed in light grey to show the sensitizing effect more clearly. Graphs show the mean ± SD of at least independent experiments. Asterisk signs indicate significance between LT alone with LT in combination with a given drug (* $p < 0.05$, ** $p < 0.005$, *** $p < 0.001$). Pound signs indicate significance between ST alone with ST in combination with a given drug (# $p < 0.05$, ### $p < 0.005$, ### $p < 0.001$). A673 cells (**a**), HT-1080 cells (**b**), and RD cells (**c**). Ampersand signs indicate significance between ST in combination with a given drug with LT in combination with same drug. (& $p < 0.05$, && $p < 0.005$, &&& $p < 0.001$).

2.4. Combination of FVP with LUV-TRAIL Induced Apoptotic Cell Death and Decreased Long-Term Clonogenic Survival of Human Sarcoma Cells

According to the results obtained, combination of LUV-TRAIL with FVP greatly enhanced their cytotoxic ability in all sarcoma cell lines tested. Next, we sought to confirm whether the decrease of cell viability observed in Figure 4 was due to the induction of apoptosis or to a cytostatic effect. First, appearance of typical apoptotic features was assessed by microscopy (Figure 5a). In all cases, cells treated with sTRAIL and LUV-TRAIL alone or in combination with FVP exhibited the typical apoptotic nuclear morphological changes such as blebbing, nuclear fragmentation, and chromatin condensation (Figure 5a). Moreover, these nuclear apoptotic morphological changes were more pronounced when both forms of TRAIL were combined with FVP compared with treatments with TRAIL alone (both sTRAIL and LUV-TRAIL). Finally, apoptotic morphological changes correlated in all cases with induction of apoptosis carried out in parallel using annexin V staining (Figure 5a, showed as percentage of annexin V positive cells below each picture).

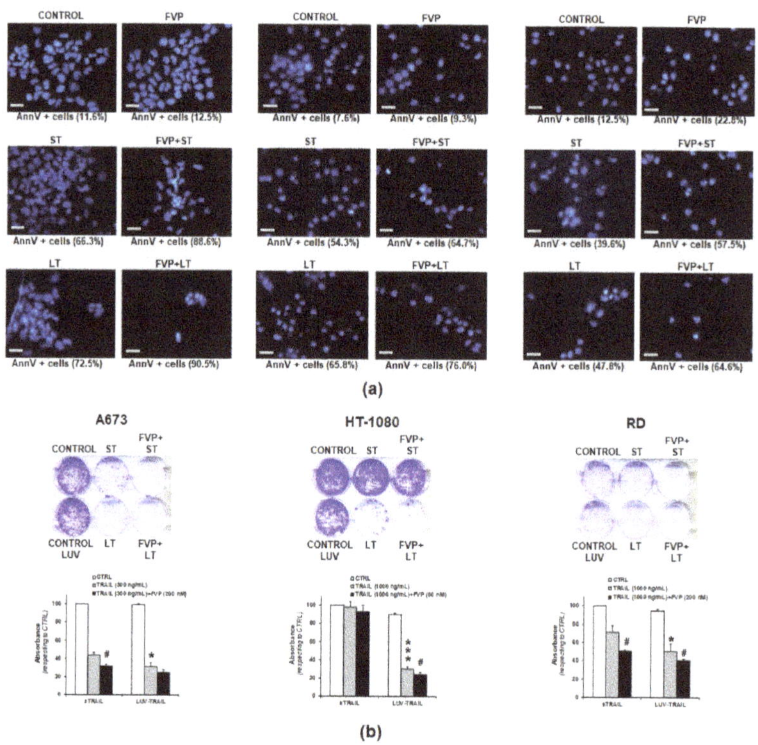

Figure 5. (**a**) Analysis of nuclear morphological changes after combination of LUV-TRAIL with FVP. Human sarcoma cells were pre-incubated for 1 h with flavopiridol (FVP, 200 nM for A673 and RD cells, 50 nM for HT-1080 cells). Afterwards, sTRAIL (ST) or LUV-TRAIL (LT) were added (300 ng/mL for A673 cells and 1000 ng/mL for HT-1080 and RD cells) and left overnight. The following day, nuclear staining was performing by using Hoechst 33342. As control, cells were untreated (CONTROL) or treated only with FVP (FVP) at above indicated doses. Original magnification at 400×. Scale bar = 20 µM.

Cell death was measured in parallel by flow cytometry after annexin-V staining (bottom graphs); (**b**) Analysis of clonogenic survival after combination of LUV-TRAIL with FVP. Clonogenic assay was performed to analyze long-term survival in human sarcoma cells after treatment with sTRAIL (ST) or LUV-TRAIL (LT) for 24 h (300 ng/mL for A673 cells and 1000 ng/mL for HT-1080 and RD cells). As indicated, cells were pre-incubated with FVP (FVP, 200 nM for A673 and RD cells, 50 nM for HT-1080 cells) for 1 h. Upper panels show 6-well plates seeded with human sarcoma cells and stained with crystal violet after 11 days. Bottom panels show the quantification of crystal violet absorbance after solubilizing in DMSO and measuring absorbance at 550 nm. Graphic shows the mean ± SD of the absorbance of treated cells expressed as percentage with respect to the untreated cells (control). Asterisk signs indicate significance between LT and LT (* $p < 0.05$, ** $p < 0.005$, *** $p < 0.001$). Pound signs indicate significance between TRAIL (ST or LT) alone with TRAIL in combination with FVP (# $p < 0.05$).

Next, we sought to analyze the long-term cytotoxic effect of the combinations of both forms of TRAIL alone or in combination with FVP. For this purpose, clonogenic assays were carried out. (Figure 5b). Notably, liposomes alone (without TRAIL anchored on their surface) were added as controls to rule out any long-term cytotoxic effect of the lipid nanoparticles (see controls for LUV-TRAIL in Figure 5b). Treatment with LUV-TRAIL for 24 h significantly delayed clonogenic survival of sarcoma cells when compared to sTRAIL, indicating that LUV-TRAIL not only showed a greater cytotoxic short-term effect than sTRAIL but also showed a long-term effect affecting to survival and proliferation of sarcoma cells. Remarkably, when both forms of TRAIL were combined with FVP, the clonogenic suvival of sarcoma cells was enhanced compared to TRAIL alone, being again more pronounced in the case of LUV-TRAIL.

2.5. Combination of FVP with LUV-TRAIL Activated the Caspase Cascade More Efficiently than with sTRAIL in Human Sarcoma Cells

After assessing the greater cytotoxic effect of the combination of LUV-TRAIL with FVP at both short and long term, activation of the main caspases involved in the extrinsic apoptotic pathway (caspase-8 and caspase-3) was analyzed by Western blot. As aforementioned (Figure 2), both sTRAIL and LUV-TRAIL induced the activation of the extrinsic apoptotic pathway, as shown by the clear decrease of the pro-forms of both caspases-8 and -3, as well as of Bid and PARP-1, in the three sarcoma cell lines tested (Figure 6a). Moreover, caspase activation correlated in all cases with induction of apoptosis as indicated by annexin V staining performed in parallel (Figure 6a, lower panels). Pretreatment with FVP greatly increased the disappearance of all proteins, proving that FVP sensitization to both sTRAIL and LUV-TRAIL relies on an enhanced activation of the extrinsic apoptotic pathway. Importantly, FVP did not induce any detectable caspase activation in any sarcoma cell tested (Figure 6a, lower panels).

On the other hand, cell death induced by combination of either sTRAIL or LUV-TRAIL with FVP was fully inhibited by the general caspase inhibitor z-VAD-fmk and the specific caspase inhibitor IETD-fmk. These data demonstrated that cell death induced by the combination of FVP with TRAIL (sTRAIL and LUV-TRAIL) was a caspase-dependent apoptotic process through the activation of the extrinsic apoptotic pathway by caspase-8 (Figure 6b). In this line, cell death induced by the combination of FVP with both formulations of TRAIL was exclusively dependent of TRAIL since blocking TRAIL-signaling with the TRAIL-blocking antibody RIK2 entirely abrogated apoptosis induced by combination of FVP with both sTRAIL and LUV-TRAIL (Figure 6b).

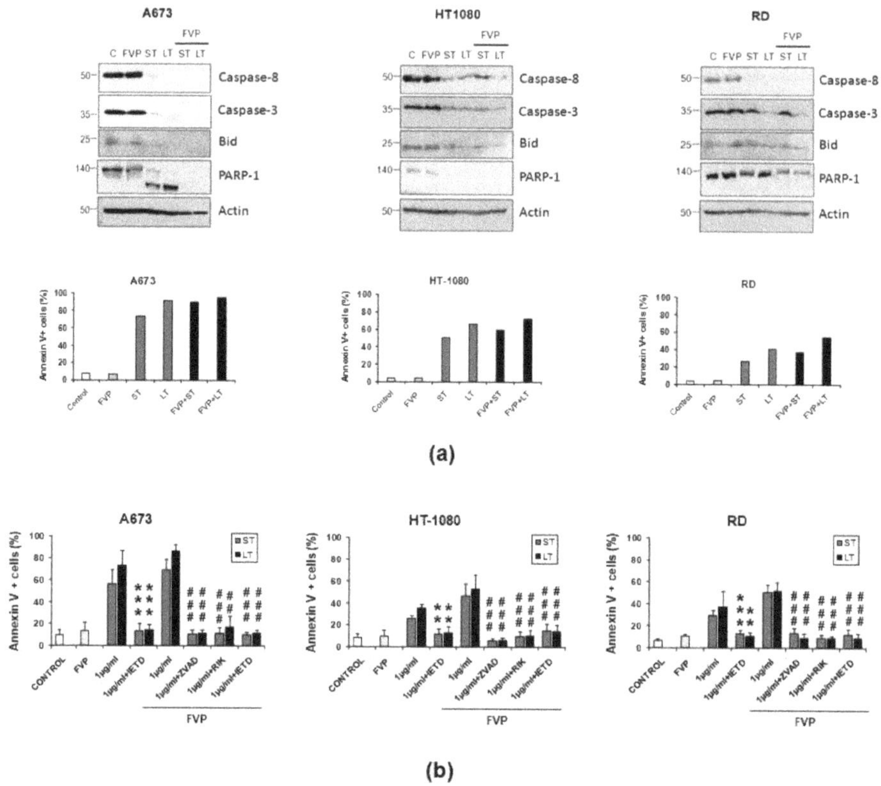

Figure 6. (**a**) Caspase activation induced by combined treatments in human breast tumor cells. Cells were pre-incubated for 1 h with FVP (200 nM for A673 and RD cells, 50 nM for HT-1080 cells) and then treated with indicated doses of sTRAIL (ST) or LUV-TRAIL (LT) overnight. After that, cells were lysed and levels of caspase-8, caspase-3, Bid, and PARP-1 were analyzed by Western blot analysis using specific antibodies. Actin levels were also determined as a control for equal protein loading (upper panels). As control, cells were untreated (C) or treated only with FVP at above indicated doses. An aliquot of untreated and treated human sarcoma cells was collected in parallel and apoptosis was analyzed by annexin V staining using flow cytometry (lower panels); (**b**) Analysis of apoptosis inhibition by caspase inhibition. Human sarcoma cells were treated with indicated doses of sTRAIL (ST) and LUV-TRAIL (LT) previously pre-incubated in presence or absence of FVP (200 nM for A673 and RD cells, 50 nM for HT-1080 cells). Combined treatment was also performed pre-incubating with the TRAIL-blocking antibody RIK2 (500 ng/mL), with the pan-caspase inhibitor z-VAD-fmk (30 μM) and with the specific caspase-8 inhibitor IETD-fmk (30 μM). Graphics show the percentage of annexin-V positive cells analyzed by flow cytometry as the mean ± SD of at least three experiments. ** $p < 0.01$, *** $p < 0.001$ (ST versus ST + RIK and, LT versus LT + RIK). ## $p < 0.01$, ### $p < 0.001$ (ST versus ST + caspase inhibitors and, LT versus LT + caspase inhibitors).

2.6. FVP Diminished Expression of Anti-Apoptotic Proteins in Human Sarcoma Cells

After assessing the enhanced cytotoxic activity showed by combined treatments of both TRAIL formulations with FVP, we analyzed the underlying mechanism of FVP-induced sensitization to TRAIL. Analysis of FVP effect on the surface expression of pro-apoptotic receptors, DR4 and DR5, and decoy receptors, DcR1 and DcR2, showed no significant changes of death receptors (DR) upon treatment with FVP (Figure 7a).

Therefore, we investigated the expression of different proteins involved in the regulation of TRAIL-induced apoptosis in sarcoma cells by Western blot. As previously observed, we confirmed that FVP did not significantly induce apoptosis at the doses used (Figure 7b, bottom panels). Flavopiridol induced a clear decrease of the anti-apoptotic protein FLIP (mainly $cFLIP_S$) in all sarcoma cell lines tested (Figure 7b, upper panels). Moreover, in A673 cells a clear decrease of the anti-apoptotic protein XIAP was observed. Finally, expression of other anti-apoptotic proteins such as Mcl-1 and $Bcl\text{-}X_L$ did not vary after FVP treatment in any sarcoma cell line tested (Figure 7b).

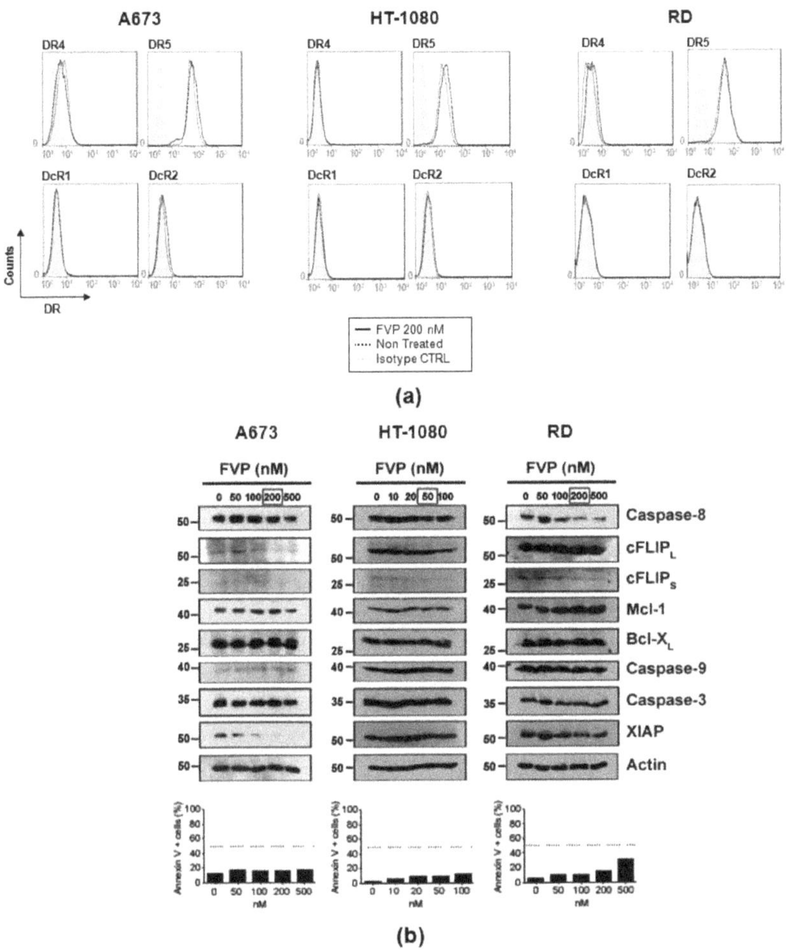

Figure 7. (**a**) Effect of FVP in death receptor (DR) surface expression in human sarcoma cells. Surface expression of DR4, DR5, DcR1, and DcR2 in A673, HT-1080, and RD cells untreated (black dotted line) or treated with 200 nM FVP for 16 h (black line) was analyzed by flow cytometry. Grey dotted line indicates isotype control and grey histogram indicates unlabeled cells. (**b**) Effect of FVP in protein expression in human sarcoma cells. Flavopiridol (FVP) were used at indicated doses was used to treat human sarcoma cell lines for 16 h. After that, cells 10^6 cells were lysed and expression of caspase-8, FLIP, Mcl-1, $Bcl\text{-}X_L$, caspase-9, caspase-3, and XIAP were analyzed by Western blot using specific antibodies. Actin levels were also determined as a control for equal protein loading (upper panels). An aliquot of untreated and treated human sarcoma cells was collected in parallel and labeled with annexin-V. Apoptosis was analyzed by flow cytometry (lower panels).

3. Discussion

Sarcomas are relatively rare malignant tumors of mesenchymal origin and constitute about 1% of all cancers. Current treatment of sarcomas implies a multidisciplinary approach including surgery, chemotherapy, and radiotherapy [31]. This has led to an improvement in prognosis of patients with sarcoma (overall survival about 50%) [32]. However, the implementation of novel therapies could help to improve the survival of patients suffering from sarcoma in the future. In this line, TRAIL has been tested in distinct types of sarcoma both in pre-clinical studies and clinical trials [5]. However, the potential of TRAIL as possible treatment in sarcomas has been explored in all cases using the soluble form of TRAIL [28,33,34].

Our group previously generated a new TRAIL formulation (LUV-TRAIL) based on tethering human recombinant TRAIL on the surface of artificial LUV-type liposomes. The improved bioactivity of LUV-TRAIL has been validated in a broad panel of human cancer cells derived from hematological malignancies [19–21], as well as epithelial cancer cells both in vitro and in vivo [22–24]. Notably, although LUV-TRAIL was more cytotoxic than sTRAIL against cancer cells, it lacked toxicity against normal cells both in vitro [19] and in vivo [23]. The enhanced bioactivity of LUV-TRAIL compared with sTRAIL relied on its capability of forming supra-trimeric populations of high molecular order which were not present in sTRAIL [24]. These supra-trimeric populations of high molecular order present in LUV-TRAIL formulation promoted the clear formation of DR5 oligomers on the target cells. In short, liposome-bound TRAIL induced superior DR5 clustering, enhancing DISC recruitment and, consequently, triggering caspase activation more efficiently than the sTRAIL [20,24].

In the present work, we have extended the study of the anti-tumor potential of LUV-TRAIL to sarcomas. LUV-TRAIL was capable of inducing cell death more efficiently than sTRAIL both in TRAIL-sensitive sarcoma cells (A673) cells and TRAIL-resistant sarcoma cells (HT-1080 and RD). Cell death induced by LUV-TRAIL in sarcoma cells was specifically due to TRAIL since pre-incubation with the neutralizing antibody RIK2 before treatment fully abrogated cell death induced by LUV-TRAIL. In this line, LUVs without TRAIL anchoring on their surface did not show any cytotoxicity in sarcoma cells, confirming that LUV-TRAIL-induced cell death was in fact fully attributable to TRAIL. Moreover, cell death induced by LUV-TRAIL was a caspase-dependent apoptotic process through the activation of the extrinsic apoptotic pathway by caspase-8 as evidenced by the fact that the pan-caspase inhibitor zVAD-fmk and the specific caspase-8 inhibitor IETD-fmk completely inhibited LUV-TRAIL-induced cell death.

It is interesting that in the several studies that have explored the anti-tumor potential of TRAIL in sarcomas, this death ligand has been used in combination with other drugs, indicating that some sarcoma types are partially resistant to the sTRAIL [27,28,33,35–37]. In this line, although LUV-TRAIL showed more pro-apoptotic potential than sTRAIL in sarcoma cell lines, they only induced a moderate cytotoxic effect in sTRAIL-resistant HT-1080 and RD cells. Therefore, we decided to combine LUV-TRAIL with several drugs described to sensitize sarcoma cells to TRAIL-induced apoptosis such as doxorubicin (DOX) [27], trabectedin (TRAB) [28], bortezomib (BORT) [38], and flavopiridol (FVP). Previous studies of our group on a breast cancer model, sensitization experiments also using flavopiridol were performed pre-incubating cancer cells before treatment with TRAIL (sTRAIL or LUV-TRAIL), as well as simultaneous treatment with FVP and both forms of TRAIL [23]. On that model, no differences were observed when cells were previously treated with FVP and then with TRAIL in comparison with simultaneous treatment. In this line, other studies carried out by us using multiple myeloma cells do not show differences between pre-incubation strategy and simultaneous treatment when TRAIL (sTRAIL or LUV-TRAIL) is combined with drugs (data not shown).

Among them, FVP was proven as the most efficient in the three sarcoma cell lines tested. Flavopiridol is a semisynthetic flavone that has showed a potent inhibitory effect on cell proliferation in sarcoma cells [39–41]. Flavopiridol has also been used in clinical trials, and even though was not effective as monotherapy [42], FVP potentiated the anti-tumor activity of other anti-tumor agents when was used in combination [43,44].

Our results demonstrate that FVP greatly sensitized not only to sTRAIL but also, and largely, to LUV-TRAIL. The cytotoxic effect of LUV-TRAIL in combination with FVP was fully attributed to TRAIL since the selected doses of FVP did not induce cell death per se. In this line, pre-incubation with the neutralizing antibody RIK2 fully inhibited cell death induced by the combined treatment. Cell death induced by the combination of FVP with TRAIL showed the typical morphological changes of apoptotic cell death such as blebbing, nuclear fragmentation, and chromatin condensation. Furthermore, this sensitization resulted in an increased activation of both caspase-8 and -3, which was completely inhibited by the pan-caspase inhibitor z-VAD-fmk and also by the caspase-8 inhibitor IETD-fmk. Altogether, these data indicate that combination of FVP with LUV-TRAIL was a caspase-dependent apoptotic process through the activation of the extrinsic apoptotic pathway. Moreover, when long-term effect of LUV-TRAIL was analyzed, this novel TRAIL formulation inhibited clonogenic cell growth largely than sTRAIL. Moreover, combination of FVP with both forms of TRAIL, mainly with LUV-TRAIL, enhanced long-term effect indicating that the combined treatment not only was more effective at short-term but also was prolonged over time as shown in clonogenic assays.

Several mechanisms have been described to explain FVP-induced TRAIL-sensitization, among them up-regulation of DR expression [23] or inactivation of anti-apoptotic proteins such as FLIP, XIAP, Mcl-1, or survivin [25,45,46]. Trying to ascertain the mechanisms involved in the synergy observed between FVP and LUV-TRAIL, we firstly analyzed DR expression upon FVP treatment. We previously demonstrated that DR up-regulation was a decisive sensitizing mechanism of FVP to TRAIL-induced apoptosis in breast cancer cells [23]. However, FVP did not modify either the expression of the pro-apoptotic TRAIL receptors or that of the decoy receptors in sarcoma cell lines, underscoring that sensitization mechanisms are different depending on the cancer type studied. In this line, we also did not observe DR up-regulation in lung cancer cells upon FVP treatment [22]. On the other hand, it has been described that FVP promotes degradation of anti-apoptotic proteins such as cFLIP, Mcl-1, and XIAP [25,45,46]. In agreement with previous studies, we observed that FVP induced a clear decrease of FLIP in all sarcoma cell lines tested. In fact, FVP induced a clear down-regulation of the short isoform of cFLIP ($cFLIP_S$), while no changes were observed for the long isoform of cFLIP ($cFLIP_L$), with the exception of A673 cells. It is noteworthy to point out that $cFLIP_S$ is considered to be solely an anti-apoptotic protein promoting caspase-8 inhibition, while the role of $cFLIP_L$ is not clear, and seems that the long isoform of cFLIP promotes or inhibits apoptosis depending on the relative amounts of both caspase-8 and $cFLIP_L$ [47,48]. Interestingly, FVP also induced a clear decrease of XIAP expression in A463 cells, which also resulted to be the most sensitive cell line to the combined treatment of FVP with both TRAIL formulations. Although we have not performed specific experiments to address the precise contribution of XIAP down-regulation to FVP-mediated sensitization to TRAIL, this could at least partially explain the higher sensitizing effect of FVP in these cells compared to the other cell lines.

Summarizing, this study shows that LUV-TRAIL significantly improves the bioactivity of sTRAIL on sarcoma cells. Furthermore, the combination of LUV-TRAIL with FVP increased even more the cytotoxic potential of LUV-TRAIL, opening the door to new TRAIL-sensitization strategies of which LUV-TRAIL mainly could benefit. In conclusion, the present study validates our novel formulation of TRAIL based on anchoring this death ligand on liposome surface in sarcoma and could be of relevance in a future clinical application of TRAIL in this type of cancer.

4. Materials and Methods

4.1. Preparation of Lipid Nanoparticles Decorated with Soluble Recombinant TRAIL

LUV (Large Unilamellar Vesicles)-type lipid nanoparticles with soluble recombinant TRAIL (sTRAIL) tethered on their surface was performed as previously described [19,49]. Briefly, a mixture of phosphatidylcholine (PC), sphingomyelin (SM), cholesterol (CHOL), and 1,2-dioleoyl-sn-glycero-3-{[*N*-(5-amino-1-carboxypentyl)-iminodiacetic acid]succinyl} (nickel salt) (DOGS-NTA-Ni) (all from Avanti Polar Lipids, Alabaster, AL; USA) in the weight ratio of 55:30:10:5 were firstly dried under a nitrogen and next under vacuum. Lipid mixture with a composition resembling that of natural exosomes was resuspended in KHE buffer (100 mM KCl, 10 mM HEPES, pH 7.0, containing 0.1 m MEDTA). After that, resuspended lipids were freeze-thawed 10 times and extruded 10 times through two polycarbonate membranes with a pore size of 0.2 μm (Whatman, Maidstone, UK) using an extruder (Northern Lipids, Burnaby, BC, Canada). LUV were incubated in KHE buffer for 30 min at 37 °C with soluble recombinant TRAIL (sTRAIL), corresponding to amino acids 95–281 with a 6-histidine tag in its N-terminal extreme cloned into the pET-28c plasmid (Novagen, kindly provided by Dr. Marion MacFarlane) [50]. Then LUV with sTRAIL tethered on their surface (LUV-TRAIL) were ultracentrifugated for 6 h at 100,000 revolutions per minute at 4 °C, supernatant was removed, and finally, the pellet containing LUV-TRAIL was resuspended in an equal volume of sterile KHE buffer.

4.2. Cell Culture and Cytotoxicity Assays

A673 cells (derived from Ewing's sarcoma), HT-1080 cells (derived from fibrosarcoma), and RD cells (derived from rhabdomyosarcoma) were obtained from American Type Culture Colection (ATCC, Manassas, VA, USA). Sarcoma cell lines were routinely cultured in Dulbecco's Modified Eagle Medium (DMEM) medium supplemented with 10% fetal bovine serum (FBS), 2 mM L-glutamine and penicillin/streptomycin (i.e., complete medium) at 37 °C with a 5% CO_2.

4.3. Cell Viability Assays

For cell viability quantification, cells (2.5×10^4 cells/well) were seeded in 96-well plates (100 μL/well) in complete medium and left overnight to be attached to the bottom. Cells were then treated with different concentrations (1–1000 ng/mL) of sTRAIL or LUV-TRAIL for 24 h. Cell viability was evaluated by a modification of 3-[4,5-di-methylthiazol-2-yl]-2,5-diphenyltetrazolium bromide (MTT) method of Mosmann as previously described [51]. Data was expressed as the percentage of cell viability with respect to control cells (untreated cells for sTRAIL, and cells treated with LUVs without TRAIL for LUV-TRAIL).

4.4. Cytotoxicity Assays

Cytotoxicity assays were performed as follows: 2×10^4 cells were seeded in 96-well plates in complete medium and left overnight to be attached to the bottom. After that, cells were treated with different concentrations of sTRAIL or LUV-TRAIL (1–1000 ng/mL) for 24 h. Then, apoptosis was measured by analyzing phosphatidyl-serine exposure on cell surface was analyzed to quantify apoptosis. For that, cells were incubated with 0.5 μg/mL annexin-V-APC, Immunostep, Salamanca, Spain) in annexin-binding buffer (ABB, 140 mM NaCl, 2.5 mM $CaCl_2$, 10 mM HEPES/NaOH, pH 7.4) for 15 mins at room temperature. Finally, apoptosis quantification was carried out using a FACSCalibur flow cytometer and CellQuest software (BD Biosciences, Franklin Lakes, NY, USA).

Cell death inhibition assays were carried out by using the blocking anti-human TRAIL mAb (500 ng/mL, clone RIK2, BD Biosciences), with the pan-caspase inhibitor z-VAD-fmk (30 μM, Bachem, Bubendorf, Switzerland) or with the specific caspase-8 inhibitor IETD-fmk (30 μM, Bachem). Cells were pre-incubated with RIK2, z-VAD-fmk, or IET-fmk for 1 h prior to the addition of TRAIL (both sTRAIL and LUV-TRAIL).

4.5. Clonogenic Assay

Clonogenic survival was analyzed as previously described [22–24]. Briefly, 5000 cells per well were seeded into 6-well plates and left to be attached to the bottom overnight. The following day, cells were incubated with sTRAIL or LUV-TRAIL (300 ng/mL for A673 cells and 1000 ng/mL for HT-1080 and RD cells) for 24 h in presence or absence of FVP (50 nM for HT-1080 cells and 200 nM for A673 and RD cells). Then, culture medium was removed and replaced with fresh medium, and surviving cells were cultured for 10 days. After that, cells were washed twice with PBS, fixed with pure methanol for 30 min at 4 °C, and stained with crystal violet (1% in 50% ethanol). The measurement of the absorbance at 550 nm after dissolving crystal violet with DMSO was performed for quantifying the clonogenic assays. Data were expressed as the percentage of absorbance with respect to the respective control (untreated cells).

4.6. Western Blot Analysis

The study of the expression of the main proteins involved in the extrinsic apoptotic pathway was carried out using Western blot analysis as previously described [19–21]. Briefly, cells (5×10^6) were lysed at 4 °C with 100 µL of a buffer containing 1% Triton X-100 and protease and phosphatase inhibitors. Then, lysated cells were separated by 12% SDS-PAGE, transferred to PVDF membranes, and blocked with TBS-T buffer (10 mM Tris/HCl, pH 8.0, 0.12 M NaCl, 0.1% Tween-20, 0.05% sodium azide) containing 5% skimmed milk. PVDF membranes were incubated with mAbs against caspase-8 (BD Biosciences), caspase-3 (Cell Signaling, Danvers, MA; USA), Bid (BD Biosciences), PARP-1 (BD biosciences), caspase-9 (MBL, Woburn, MA; USA), cFLIP (clone NF6, Enzo, Farmingdale, NY, USA), Mcl-1 (Santa Cruz Biotech, Dallas, TX), Bcl-xL (Cell Signaling), or XIAP (BD Biosciences) in TBS-T containing 2% skimmed milk. Anti-β-actin mAb (Sigma, Saint Louis, MO; USA) was used as protein loading control. Pierce ECL Western Blotting Substrate (when used horseradish peroxidase-labeled secondary antibody, Life Technologies, Carlsbad, CA, USA) or the CDP-Star substrate (when used phosphatase alkaline-labeled secondary antibody, Merck, Darmstadt, Germany) were used to display the proteins.

4.7. Surface Expression of Death Receptors

The analysis of surface expression of death receptors was performed as follows: 1×10^5 cells were incubated with either anti-DR4, anti-DR5, anti-DcR1, anti-DcR2 monoclonal antibodies or isotype control, all of them PE-conjugated (eBioscience, San Diego, CA, USA) in PBS containing 5% FCS for 30 min at room temperature. Then, cells were analyzed by using a FACSCalibur flow cytometer and Cell Quest software (BD Biosciences).

4.8. Nuclear Staining

Hoechst 33342 staining was carried out for analyzing the nuclear morphological changes. Briefly, cells were seeded (10^5 cells) in 24-well plates in complete medium and incubated in presence or absence of FVP for 1 h (200 nM for A673 and RD cells and 50 nM for HT-1080 cells). After that, cells were treated with sTRAIL or LUV-TRAIL (1000 ng/mL) overnight. Then, cells were fixed with 4% PFA for 30 min, washed with PBS, and labeled with 1 µg/mL of Hoechst 33342 (Invitrogen, Dublin, Ireland). Finally, photographs were taken using a fluorescence microscope (E600/E400, Nikon, Tokio, Japan) equipped with digital photograph system (DXM 1200F, Nikon) at original magnification at 400×.

4.9. Statistical Analysis

GraphPad Prism 5 software was used to carry out computer-based statistical analysis. Results showed indicate the mean ± SD of at least three different experiments. Student's t test for non-paired variants was performed to evaluate statistical significance. A $p < 0.05$ value was considered to be significant.

Author Contributions: A.G.-L. performed almost all experiments. D.D.M. designed and performed several experiments and wrote the manuscript. A.A. supervised experiments and help to write the manuscript. L.M.-L. designed experimental work, supervised experiments, and wrote the manuscript.

Acknowledgments: This study has been funded by Instituto de Salud Carlos III through the project PI16/00526 (LML) (Co-funded by European Regional Development Fund/European Social Fund, "Investing in your future"). Ana Gallego-Lleida was supported by pre-doctoral fellowships from Gobierno de Aragón.

Conflicts of Interest: The authors have not conflict of interest. Alberto Anel and Luis Martinez-Lostao have filed a patent application (W02011020933) for the use of liposome-bound Apo2L/TRAIL.

References

1. Siegel, R.L.; Miller, K.D.; Jemal, A. Cancer statistics, 2018. *CA Cancer J. Clin.* **2018**, *68*, 7–30. [CrossRef] [PubMed]
2. Borden, E.C.; Baker, L.H.; Bell, R.S.; Bramwell, V.; Demetri, G.D.; Eisenberg, B.L.; Fletcher, C.D.; Fletcher, J.A.; Ladanyi, M.; Meltzer, P.; et al. Soft tissue sarcomas of adults: State of the translational science. *Clin. Cancer Res.* **2003**, *9*, 1941–1956. [PubMed]
3. Wilky, B.A.; Jones, R.L.; Keedy, V.L. The Current Landscape of Early Drug Development for Patients with Sarcoma. *Am. Soc. Clin. Oncol. Educ. Book* **2017**, *37*, 807–810. [CrossRef] [PubMed]
4. Kawai, A.; Yonemori, K.; Takahashi, S.; Araki, N.; Ueda, T. Systemic Therapy for Soft Tissue Sarcoma: Proposals for the Optimal Use of Pazopanib, Trabectedin, and Eribulin. *Adv. Ther.* **2017**, *34*, 1556–1571. [CrossRef] [PubMed]
5. Gamie, Z.; Kapriniotis, K.; Papanikolaou, D.; Haagensen, E.; Da Conceicao Ribeiro, R.; Dalgarno, K.; Krippner-Heidenreich, A.; Gerrand, C.; Tsiridis, E.; Rankin, K.S. TNF-related apoptosis-inducing ligand (TRAIL) for bone sarcoma treatment: Pre-clinical and clinical data. *Cancer Lett.* **2017**, *409*, 66–80. [CrossRef] [PubMed]
6. Pitti, R.M.; Marsters, S.A.; Ruppert, S.; Donahue, C.J.; Moore, A.; Ashkenazi, A. Induction of apoptosis by Apo-2 ligand, a new member of the tumor necrosis factor cytokine family. *J. Biol. Chem.* **1996**, *271*, 12687–12690. [CrossRef] [PubMed]
7. Wiley, S.R.; Schooley, K.; Smolak, P.J.; Din, W.S.; Huang, C.P.; Nicholl, J.K.; Sutherland, G.R.; Smith, T.D.; Rauch, C.; Smith, C.A.; et al. Identification and characterization of a new member of the TNF family that induces apoptosis. *Immunity* **1995**, *3*, 673–682. [CrossRef]
8. Holland, P.M. Death receptor agonist therapies for cancer, which is the right TRAIL? *Cytokine Growth Factor Rev.* **2014**, *25*, 185–193. [CrossRef] [PubMed]
9. Lemke, J.; von Karstedt, S.; Zinngrebe, J.; Walczak, H. Getting TRAIL back on track for cancer therapy. *Cell Death Differ.* **2014**, *21*, 1350–1364. [CrossRef] [PubMed]
10. Micheau, O.; Shirley, S.; Dufour, F. Death receptors as targets in cancer. *Br. J. Pharmacol.* **2013**, *169*, 1723–1744. [CrossRef] [PubMed]
11. Von Karstedt, S.; Montinaro, A.; Walczak, H. Exploring the TRAILs less travelled: TRAIL in cancer biology and therapy. *Nat. Rev. Cancer* **2017**, *17*, 352–366. [CrossRef] [PubMed]
12. Martinez-Lostao, L.; Marzo, I.; Anel, A.; Naval, J. Targeting the Apo2L/TRAIL system for the therapy of autoimmune diseases and cancer. *Biochem. Pharmacol.* **2012**, *83*, 1475–1483. [CrossRef] [PubMed]
13. Wajant, H.; Gerspach, J.; Pfizenmaier, K. Engineering death receptor ligands for cancer therapy. *Cancer Lett.* **2013**, *332*, 163–174. [CrossRef] [PubMed]
14. De Miguel, D.; Lemke, J.; Anel, A.; Walczak, H.; Martinez-Lostao, L. Onto better TRAILs for cancer treatment. *Cell Death Differ.* **2016**, *23*, 733. [CrossRef] [PubMed]
15. Bernardi, S.; Secchiero, P.; Zauli, G. State of art and recent developments of anti-cancer strategies based on TRAIL. *Recent Pat. Anticancer Drug Discov.* **2012**, *7*, 207–217. [CrossRef] [PubMed]
16. Lim, B.; Allen, J.E.; Prabhu, V.V.; Talekar, M.K.; Finnberg, N.K.; El-Deiry, W.S. Targeting TRAIL in the treatment of cancer: New developments. *Expert Opin. Ther. Targets* **2015**, *19*, 1171–1185. [CrossRef] [PubMed]
17. Martinez-Lorenzo, M.J.; Anel, A.; Gamen, S.; Monlen, I.; Lasierra, P.; Larrad, L.; Pineiro, A.; Alava, M.A.; Naval, J. Activated human T cells release bioactive Fas ligand and APO2 ligand in microvesicles. *J. Immunol.* **1999**, *163*, 1274–1281. [PubMed]
18. Monleon, I.; Martinez-Lorenzo, M.J.; Monteagudo, L.; Lasierra, P.; Taules, M.; Iturralde, M.; Pineiro, A.; Larrad, L.; Alava, M.A.; Naval, J.; et al. Differential secretion of Fas ligand- or APO2 ligand/TNF-related apoptosis-inducing ligand-carrying microvesicles during activation-induced death of human T cells. *J. Immunol.* **2001**, *167*, 6736–6744. [CrossRef] [PubMed]
19. De Miguel, D.; Basanez, G.; Sanchez, D.; Malo, P.G.; Marzo, I.; Larrad, L.; Naval, J.; Pardo, J.; Anel, A.; Martinez-Lostao, L. Liposomes decorated with Apo2L/TRAIL overcome chemoresistance of human hematologic tumor cells. *Mol. Pharm.* **2013**, *10*, 893–904. [CrossRef] [PubMed]
20. De Miguel, D.; Gallego-Lleyda, A.; Anel, A.; Martinez-Lostao, L. Liposome-bound TRAIL induces superior DR5 clustering and enhanced DISC recruitment in histiocytic lymphoma U937 cells. *Leuk. Res.* **2015**, *39*, 657–666. [CrossRef] [PubMed]

21. De Miguel, D.; Gallego-Lleyda, A.; Galan-Malo, P.; Rodriguez-Vigil, C.; Marzo, I.; Anel, A.; Martinez-Lostao, L. Immunotherapy with liposome-bound TRAIL overcome partial protection to soluble TRAIL-induced apoptosis offered by down-regulation of Bim in leukemic cells. *Clin. Transl. Oncol.* **2015**, *17*, 657–667. [CrossRef] [PubMed]
22. De Miguel, D.; Gallego-Lleyda, A.; Ayuso, J.M.; Erviti-Ardanaz, S.; Pazo-Cid, R.; del Agua, C.; Fernandez, L.J.; Ochoa, I.; Anel, A.; Martinez-Lostao, L. TRAIL-coated lipid-nanoparticles overcome resistance to soluble recombinant TRAIL in non-small cell lung cancer cells. *Nanotechnology* **2016**, *27*, 185101. [CrossRef] [PubMed]
23. De Miguel, D.; Gallego-Lleyda, A.; Ayuso, J.M.; Pawlak, A.; Conde, B.; Ochoa, I.; Fernandez, L.J.; Anel, A.; Martinez-Lostao, L. Improved Anti-Tumor Activity of Novel Highly Bioactive Liposome-Bound TRAIL in Breast Cancer Cells. *Recent Pat. Anticancer Drug Discov.* **2016**, *11*, 197–214. [CrossRef] [PubMed]
24. De Miguel, D.; Gallego-Lleyda, A.; Ayuso, J.M.; Pejenaute-Ochoa, D.; Jarauta, V.; Marzo, I.; Fernandez, L.J.; Ochoa, I.; Conde, B.; Anel, A.; et al. High-order TRAIL oligomer formation in TRAIL-coated lipid nanoparticles enhances DR5 cross-linking and increases antitumour effect against colon cancer. *Cancer Lett.* **2016**, *383*, 250–260. [CrossRef] [PubMed]
25. Fandy, T.E.; Ross, D.D.; Gore, S.D.; Srivastava, R.K. Flavopiridol synergizes TRAIL cytotoxicity by downregulation of FLIPL. *Cancer Chemother. Pharmacol.* **2007**, *60*, 313–319. [CrossRef] [PubMed]
26. Gamen, S.; Anel, A.; Pérez-Galán, P.; Lasierra, P.; Johnson, D.; Piñeiro, A.; Naval, J. Doxorubicin treatment activates a Z-VAD-sensitive caspase, which causes deltapsim loss, caspase-9 activity, and apoptosis in Jurkat cells. *Exp. Cell Res.* **2000**, *258*, 223–235. [CrossRef] [PubMed]
27. Wang, S.; Ren, W.; Liu, J.; Lahat, G.; Torres, K.; Lopez, G.; Lazar, A.J.; Hayes-Jordan, A.; Liu, K.; Bankson, J.; et al. TRAIL and doxorubicin combination induces proapoptotic and antiangiogenic effects in soft tissue sarcoma in vivo. *Clin. Cancer Res.* **2010**, *16*, 2591–2604. [CrossRef] [PubMed]
28. Harati, K.; Chromik, A.M.; Bulut, D.; Goertz, O.; Hahn, S.; Hirsch, T.; Klein-Hitpass, L.; Lehnhardt, M.; Uhl, W.; Daigeler, A. TRAIL and taurolidine enhance the anticancer activity of doxorubicin, trabectedin and mafosfamide in HT1080 human fibrosarcoma cells. *Anticancer Res.* **2012**, *32*, 2967–2984. [PubMed]
29. Balsas, P.; Lopez-Royuela, N.; Galan-Malo, P.; Anel, A.; Marzo, I.; Naval, J. Cooperation between Apo2L/TRAIL and bortezomib in multiple myeloma apoptosis. *Biochem. Pharmacol.* **2009**, *77*, 804–812. [CrossRef] [PubMed]
30. Shanker, A.; Brooks, A.D.; Tristan, C.A.; Wine, J.W.; Elliott, P.J.; Yagita, H.; Takeda, K.; Smyth, M.J.; Murphy, W.J.; Sayers, T.J. Treating metastatic solid tumors with bortezomib and a tumor necrosis factor-related apoptosis-inducing ligand receptor agonist antibody. *J. Natl. Cancer Inst.* **2008**, *100*, 649–662. [CrossRef] [PubMed]
31. Frezza, A.M.; Stacchiotti, S.; Gronchi, A. Systemic treatment in advanced soft tissue sarcoma: What is standard, what is new? *BMC Med.* **2017**, *15*, 109. [CrossRef] [PubMed]
32. Lazar, A.J.; Trent, J.C.; Lev, D. Sarcoma molecular testing: Diagnosis and prognosis. *Curr. Oncol. Rep.* **2007**, *9*, 309–315. [CrossRef] [PubMed]
33. Harati, K.; Emmelmann, S.; Behr, B.; Goertz, O.; Hirsch, T.; Kapalschinski, N.; Kolbenschlag, J.; Stricker, I.; Tannapfel, A.; Lehnhardt, M.; et al. Evaluation of the safety and efficacy of TRAIL and taurolidine use on human fibrosarcoma xenografts in vivo. *Oncol. Lett.* **2016**, *11*, 1955–1961. [CrossRef] [PubMed]
34. Kang, Z.; Sun, S.Y.; Cao, L. Activating Death Receptor DR5 as a Therapeutic Strategy for Rhabdomyosarcoma. *ISRN Oncol.* **2012**, *2012*, 395952. [CrossRef] [PubMed]
35. Hotta, T.; Suzuki, H.; Nagai, S.; Yamamoto, K.; Imakiire, A.; Takada, E.; Itoh, M.; Mizuguchi, J. Chemotherapeutic agents sensitize sarcoma cell lines to tumor necrosis factor-related apoptosis-inducing ligand-induced caspase-8 activation, apoptosis and loss of mitochondrial membrane potential. *J. Orthop. Res.* **2003**, *21*, 949–957. [CrossRef]
36. Karlisch, C.; Harati, K.; Chromik, A.M.; Bulut, D.; Klein-Hitpass, L.; Goertz, O.; Hirsch, T.; Lehnhardt, M.; Uhl, W.; Daigeler, A. Effects of TRAIL and taurolidine on apoptosis and proliferation in human rhabdomyosarcoma, leiomyosarcoma and epithelioid cell sarcoma. *Int. J. Oncol.* **2013**, *42*, 945–956. [CrossRef] [PubMed]
37. Li, X.; Huang, T.; Jiang, G.; Gong, W.; Qian, H.; Zou, C. Proteasome inhibitor MG132 enhances TRAIL-induced apoptosis and inhibits invasion of human osteosarcoma OS732 cells. *Biochem. Biophys. Res. Commun.* **2013**, *439*, 179–186. [CrossRef] [PubMed]

38. Lu, G.; Punj, V.; Chaudhary, P.M. Proteasome inhibitor Bortezomib induces cell cycle arrest and apoptosis in cell lines derived from Ewing's sarcoma family of tumors and synergizes with TRAIL. *Cancer Biol. Ther.* **2008**, *7*, 603–608. [CrossRef] [PubMed]
39. Cai, D.; Latham, V.M., Jr.; Zhang, X.; Shapiro, G.I. Combined depletion of cell cycle and transcriptional cyclin-dependent kinase activities induces apoptosis in cancer cells. *Cancer Res.* **2006**, *66*, 9270–9280. [CrossRef] [PubMed]
40. Jiang, J.; Matranga, C.B.; Cai, D.; Latham, V.M., Jr.; Zhang, X.; Lowell, A.M.; Martelli, F.; Shapiro, G.I. Flavopiridol-induced apoptosis during S phase requires E2F-1 and inhibition of cyclin A-dependent kinase activity. *Cancer Res.* **2003**, *63*, 7410–7422. [PubMed]
41. Li, Y.; Tanaka, K.; Li, X.; Okada, T.; Nakamura, T.; Takasaki, M.; Yamamoto, S.; Oda, Y.; Tsuneyoshi, M.; Iwamoto, Y. Cyclin-dependent kinase inhibitor, flavopiridol, induces apoptosis and inhibits tumor growth in drug-resistant osteosarcoma and Ewing's family tumor cells. *Int. J. Cancer* **2007**, *121*, 1212–1218. [CrossRef] [PubMed]
42. Morris, D.G.; Bramwell, V.H.; Turcotte, R.; Figueredo, A.T.; Blackstein, M.E.; Verma, S.; Matthews, S.; Eisenhauer, E.A. A Phase II Study of Flavopiridol in Patients with Previously Untreated Advanced Soft Tissue Sarcoma. *Sarcoma* **2006**, *2006*, 64374. [CrossRef] [PubMed]
43. Dickson, M.A.; Rathkopf, D.E.; Carvajal, R.D.; Grant, S.; Roberts, J.D.; Reid, J.M.; Ames, M.M.; McGovern, R.M.; Lefkowitz, R.A.; Gonen, M.; et al. A phase I pharmacokinetic study of pulse-dose vorinostat with flavopiridol in solid tumors. *Investig. New Drugs* **2011**, *29*, 1004–1012. [CrossRef] [PubMed]
44. Luke, J.J.; D'Adamo, D.R.; Dickson, M.A.; Keohan, M.L.; Carvajal, R.D.; Maki, R.G.; de Stanchina, E.; Musi, E.; Singer, S.; Schwartz, G.K. The cyclin-dependent kinase inhibitor flavopiridol potentiates doxorubicin efficacy in advanced sarcomas: Preclinical investigations and results of a phase I dose-escalation clinical trial. *Clin. Cancer Res.* **2012**, *18*, 2638–2647. [CrossRef] [PubMed]
45. Palacios, C.; Yerbes, R.; Lopez-Rivas, A. Flavopiridol induces cellular FLICE-inhibitory protein degradation by the proteasome and promotes TRAIL-induced early signaling and apoptosis in breast tumor cells. *Cancer Res.* **2006**, *66*, 8858–8869. [CrossRef] [PubMed]
46. Miyashita, K.; Shiraki, K.; Fuke, H.; Inoue, T.; Yamanaka, Y.; Yamaguchi, Y.; Yamamoto, N.; Ito, K.; Sugimoto, K.; Nakano, T. The cyclin-dependent kinase inhibitor flavopiridol sensitizes human hepatocellular carcinoma cells to TRAIL-induced apoptosis. *Int. J. Mol. Med.* **2006**, *18*, 249–256. [CrossRef] [PubMed]
47. Feoktistova, M.; Geserick, P.; Kellert, B.; Dimitrova, D.P.; Langlais, C.; Hupe, M.; Cain, K.; MacFarlane, M.; Hacker, G.; Leverkus, M. cIAPs block Ripoptosome formation, a RIP1/caspase-8 containing intracellular cell death complex differentially regulated by cFLIP isoforms. *Mol. Cell* **2011**, *43*, 449–463. [CrossRef] [PubMed]
48. Pop, C.; Oberst, A.; Drag, M.; Van Raam, B.J.; Riedl, S.J.; Green, D.R.; Salvesen, G.S. FLIP(L) induces caspase 8 activity in the absence of interdomain caspase 8 cleavage and alters substrate specificity. *Biochem. J.* **2011**, *433*, 447–457. [CrossRef] [PubMed]
49. Martinez-Lostao, L.; Garcia-Alvarez, F.; Basanez, G.; Alegre-Aguaron, E.; Desportes, P.; Larrad, L.; Naval, J.; Jose Martinez-Lorenzo, M.; Anel, A. Liposome-bound APO2L/TRAIL is an effective treatment in a rheumatoid arthritis model. *Arthritis Rheum.* **2010**, *62*, 2272–2282. [CrossRef] [PubMed]
50. MacFarlane, M.; Ahmad, M.; Srinivasula, S.M.; Fernandes-Alnemri, T.; Cohen, G.M.; Alnemri, E.S. Identification and molecular cloning of two novel receptors for the cytotoxic ligand TRAIL. *J. Biol. Chem.* **1997**, *272*, 25417–25420. [CrossRef] [PubMed]
51. Mosmann, T. Rapid colorimetric assay for cellular growth and survival: Application to proliferation and cytotoxicity assays. *J. Immunol. Methods* **1983**, *65*, 55–63. [CrossRef]

International Journal of
Molecular Sciences

MDPI

Review

Targeting Macrophages as a Potential Therapeutic Intervention: Impact on Inflammatory Diseases and Cancer

Mirco Ponzoni *, Fabio Pastorino, Daniela Di Paolo, Patrizia Perri and Chiara Brignole *

Laboratory of Experimental Therapy in Oncology, Istituto Giannina Gaslini, Via G. Gaslini 5, 16147 Genoa, Italy; fabiopastorino@gaslini.org (F.P.); danieladipaolo@gaslini.org (D.D.P.); patriziaperri@gaslini.org (P.P.)

* Correspondence: mircoponzoni@gaslini.org (M.P.); chiarabrignole@gaslini.org (C.B.); Tel.: +39-10-56363539 (M.P.); +39-10-56363533 (C.B.); Fax: +39-10-3779820 (M.P. & C.B.)

Received: 12 June 2018; Accepted: 30 June 2018; Published: 4 July 2018

Abstract: Macrophages, cells belonging to the innate immune system, present a high plasticity grade, being able to change their phenotype in response to environmental stimuli. They play central roles during development, homeostatic tissue processes, tissue repair, and immunity. Furthermore, it is recognized that macrophages are involved in chronic inflammation and that they play central roles in inflammatory diseases and cancer. Due to their large involvement in the pathogenesis of several types of human diseases, macrophages are considered to be relevant therapeutic targets. Nanotechnology-based systems have attracted a lot of attention in this field, gaining a pivotal role as useful moieties to target macrophages in diseased tissues. Among the different approaches that can target macrophages, the most radical is represented by their depletion, commonly obtained by means of clodronate-containing liposomal formulations and/or depleting antibodies. These strategies have produced encouraging results in experimental mouse models. In this review, we focus on macrophage targeting, based on the results so far obtained in preclinical models of inflammatory diseases and cancer. Pros and cons of these therapeutic interventions will be highlighted.

Keywords: macrophage targeting; macrophage depletion; nanomedicine; preclinical models; inflammatory diseases; cancer

---

1. Introduction

Macrophages, an essential component of the innate immune system, present heterogeneous functional phenotypes correlated to a high plasticity grade [1]. They play central roles in both physiological and pathological conditions, being involved in developmental and homeostatic tissue processes and in a plethora of human diseases. As a consequence, they have been deeply studied, and the idea to exploit them as therapeutic targets has focused the interest of experts of different research areas from inflammation to cancer.

At present, it is widely accepted that a variety of diseases (i.e., diabetes, atherosclerosis, rheumatoid arthritis, obesity, and cancer) are associated with chronic inflammation [2]. Furthermore, it is recognized that macrophages are the major players of chronic inflammation in most human diseases [1,3]. The knowledge about their biology, mechanisms of action, and activation phenotypes has been largely increased in the last few years. Macrophages have an innate and elevated propensity to adapt to the neighboring microenvironment and to rapidly change in response to environmental stimuli. Thus, it appears difficult to find out and design a unique therapeutic strategy based on macrophage modulation that is easily applicable to different kinds of human pathologies.

In the last decade, many studies have been addressed to understanding the mechanisms involved in macrophage activation and to relate them to macrophage function. The well-known bipolar model [4,5] distinguishes macrophages into two main different subsets: pro-inflammatory (M1, classically activated via interferon-γ) and anti-inflammatory (M2, alternatively activated via interleukin 4 and 13). However, a vast literature suggests that this is only a simplification, with M1 and M2 macrophages representing only the two extremes of a huge spectrum of activation states [6,7]. Indeed, this classification takes into consideration neither the fact that M1 and M2 stimuli do not exist as single entities in tissues, nor the source and the context of environmental stimuli. Moreover, the dichotomy of M1 and M2 macrophages derives from the pre-genomic era, when few markers were considered to establish differences and similarities in macrophage responses to stimuli [6]. Finally, it is important to underline that macrophages can also develop mixed M1/M2 phenotypes in different pathological conditions, where it is possible to find macrophages with intermediate activation status [8,9].

Due to their large involvement in several high-impact diseases, macrophages have acquired a lot of interest as potential therapeutic targets. With this focus, several approaches have been developed, ranging from depletion to re-programming/re-polarization [10]. In this field, nanotechnology-based systems (i.e., liposomes, dendrimers, gold nanoparticles, and polymeric nanoparticles) have attracted attention [11,12], also due to their potential to be modified according to the disease site to be targeted. Furthermore, they have been demonstrated to be more effective than conventional delivery systems and with limited side effects [13]. In line with this, the understanding of the biology and the activation states of macrophages in different disease conditions has been, and will be in the near future, very helpful to design therapeutic protocols of personalized medicine.

Here, we reexamine the data obtained in preclinical mouse models of inflammatory diseases and cancer. In particular, the targeting of macrophages led to encouraging results, highlighting the possibility of translating these therapeutic interventions to the clinical practice.

2. Macrophages in Health and Disease

Macrophages are innate immune cells characterized by evolutionarily conserved functions, such as host defense against infection and tissue homeostasis. They were first identified by Metchnikoff in the late 19th century as tissue-residing immune cells belonging to the reticulum endothelial system and endowed with the ability to detect and phagocytose dying cells, harmful molecules, and pathogens [14,15]. A new classification was made in the 1970s, when van Furth and colleagues proposed macrophages as being part of the mononuclear phagocytic system (MPS) [16].

Based on their differential localization within the body, macrophages were also classified in discrete subsets as tissue-resident macrophages (TRM) [17]. They present different transcriptional profiles, associated to heterogeneous phenotypes, as a consequence of the tissue-specific function [7]. Primarily described as immune sentinels able to protect the tissue in which they are located, TRM are characterized by several steady-state functions. Indeed, in healthy tissues, they are responsible for homeostasis maintenance through the clearance of senescent cells and regulation of tissue metabolism. Based on their host tissue, they were defined as Langerhans cells in the skin, Kupffer cells in the liver, microglia within the brain, alveolar macrophages in the lung, peritoneal macrophages in the peritoneum, and osteoclasts in the bone. Initially, it was supposed that TRM were derived from circulating blood monocytes. At present, a new concept suggests that they can derive either from embryonic progenitors, seeding the tissues during development and being able to maintain their population independently from blood-borne precursors in adulthood, or from hematopoietic stem cell-derived circulating monocytes [18].

For many years, it has been accepted as a paradigm model that macrophages can adopt opposite activation states in response to environmental stimuli sensed primarily by Toll-like receptors. Based on this assumption, they have been distinguished as classically (M1) and alternatively (M2) activated macrophages. M1 and M2 macrophages are pro-inflammatory and anti-inflammatory, respectively. The main activities associated to M1/M2 phenotypes can be summarized as "Fight" for M1 macrophages and "Fix" for M2 macrophages. These two diverse programs take place as a response to environmental stimuli through enzymatic pathways responsible for arginine catabolism [5]. Macrophages that follow the "Fight" program, thanks to inducible nitric oxide synthase (iNOS), convert arginine to citrulline and nitric oxide (NO), inhibiting cell proliferation and inducing cancer cell death. On the other hand, macrophages activating the "Fix" program are able to promote cell proliferation and repair, via the metabolism of arginine through arginase I (ArgI), that produces ornithine and urea.

Notably, it is clear that, at present, the extraordinary plasticity of macrophages is not limited to these two opposed phenotypes, representing only the extremes of a wide and continuous array of activation states [19,20].

The concept that macrophages take part as major players in a series of human diseases associated with chronic inflammation has been a matter of debate for a long time. At present, much evidence supports this theory and a vast literature is dedicated to this field. Furthermore, many efforts have been made to understand the role of macrophages in inflammation and cancer and to reveal the different phenotypes they can acquire in inflammation-driven pathological conditions. Although the dichotomy of M1 versus M2 does not entirely satisfy the reality, it is acknowledged that under pathological conditions, macrophages can be found in both the activation states.

In the last decade, researchers have tried to decipher which polarization/phenotype is associated to a specific inflammation-related disease. The findings obtained led to the understanding that, in most of the pathological conditions analyzed, macrophages can temporally acquire different phenotypes, due to the evolution of the disease.

2.1. *Inflammatory Diseases*

Several studies evidenced the importance of macrophages in diseases such as asthma, atherosclerosis, rheumatoid arthritis, osteoarthritis, endometriosis, diabetes type 1 and 2, and obesity.

In asthma, Zaslona and colleagues demonstrated the role of alveolar macrophages, concluding that resident alveolar macrophages play a protective role during the early phase of allergic lung inflammation. On the other hand, they stated that the recruited monocytes, putative precursors of alveolar macrophages, are involved in asthma pathogenesis by promoting lung inflammation [21]. Moreover, Jiang and colleagues recently updated the role of macrophages in asthma, concluding that both M1 and M2 subsets participate in asthma pathogenesis, being alternately predominant in the different phases of the disease [22]. The aforementioned situation, with M1 and M2 macrophages that alternate their predominance under pathological conditions, is found in several other human pathologies.

In atherosclerosis, a chronic inflammatory disease characterized by the narrowing and thickening of the arteries caused by buildup of plaque around the artery wall, macrophages participate in all the stages of the disease, being involved in pathogenesis and progression [23]. Summarily, pro-inflammatory macrophages are implicated in plaque initiation and progression, while the anti-inflammatory ones are involved in plaque stabilization. Furthermore, it has been reported that both macrophage subtypes (M1/M2) can be found in atherosclerotic plaques, where they are differently distributed [24].

In rheumatoid arthritis, a common autoimmune disorder characterized by chronic arthritis that leads to joint destruction, macrophages have a leading role in pathogenesis and disease progression. It has been reported that an increased number of sublining macrophages in the synovium is an early hallmark of active rheumatic disease. Furthermore, the degree of synovial macrophage infiltration is strictly related to joint erosion [25]. In the context of rheumatoid arthritis, the phenotype of the macrophages is heterogeneous and still needs to be completely elucidated. Of note, in synovial tissues and fluids, macrophages can present both pro-inflammatory and anti-inflammatory phenotypes [26].

In osteoarthritis, not commonly recognized as an inflammatory disease, macrophages directly participate in synovial inflammation through the production of TNF-α and IL-1β and, indirectly, through the activation of synovial fibroblasts [27].

Recently, the contribution of macrophages in metabolic diseases such as diabetes [28] and in obesity-related adipose inflammation, frequently associated with the development of insulin resistance and type 2 diabetes, has been also reported [29].

2.2. Cancer

Inflammatory cells have been recognized as a key component of the tumor microenvironment (TME), which can be represented as an ecological niche playing a critical role in cancer development, progression, and control. Within this niche, a complex interplay between different cell types (proliferating cancer cells, stromal cells, and infiltrating immune cells) leads to the establishment of a microenvironment, whose components influence each other to promote tumor progression. Macrophages are the dominant portion of infiltrating leukocytes in all tumors, where they are defined as tumor-associated macrophages (TAMs) and are mostly characterized by an M2-like phenotype [30]. They mainly derive from circulating monocytes, recruited at the tumor site by chemotactic factors.

Macrophages participate in both tumor initiation and development/progression. It is now accepted that non-malignant TME cells, being macrophages the most prominent ones, play a tumor-initiating role, generating a persistent inflammatory response against environmental stimuli [31,32]. Indeed, it has been reported that different inflammatory conditions (i.e., inflammatory bowel disease, Crohn's disease, asbestosis, etc.) are highly correlated with the increased risk of cancer development [33]. Noteworthy, macrophages involved in the phase of tumor initiation are characterized by a pro-inflammatory phenotype. Once the tumor is established, macrophages switch to an anti-inflammatory phenotype. It is customary to say that macrophages are educated by neoplastic cells to acquire a protumor phenotype [32,34,35].

TAMs support tumor progression through different mechanisms, promoting angiogenesis and tumor spreading at distant sites and creating an immunosuppressive environment [36]. Furthermore, it seems that different subsets of TAMs can be found at the tumor site, where they can be either temporally distinguished, depending on disease evolution, or geographically determined on the basis of their location. In this regard, macrophages found in hypoxic areas of the tumor are prevalently involved in tumor angiogenesis, while those residing at the tumor–stroma interface take part in invasion and metastasis [30,37].

TAMs are abundant in established tumors and their presence is often associated with increased tumor progression and invasion [32]. Indeed, preclinical and clinical evidence shows that a prevalence of TAMs within tumors is associated with a worse overall prognosis in a high proportion of solid tumors (i.e., breast cancer, ovarian cancer, melanoma, etc.; Table 1) [38–44]. Until now, to our knowledge, only in a few cases (i.e., colorectal cancer and non-small cell lung cancer) it has been reported that a high density of macrophages is associated with improved overall survival [45–47]. Doubtful is the role of TAMs in gastric cancer; a study by Ohno and colleagues proves, indeed, their positive correlation with improved survival [48], while more recent papers demonstrate that an high infiltration of TAMs is a predictor of poor prognosis for gastric cancer patients [49,50].

Table 1. Prognostic significance of macrophages in tumors. TAMs: tumor-associated macrophages, M1:pro-inflammatory, M2: anti-inflammatory.

Cancer Type	Prognosis Association	References
Melanoma	M2 TAMs accumulation in malignant melanoma was confirmed as biomarker of poor prognosis.	[43] Falleni, M. et al.
Breast	High density of TAMs was associated with malignant phenotype and poor survival in breast cancer patients, emerging as a novel prognostic factor.	[40] Zhao, X. et al.
Ovarian Cancer	High CD163⁺ TAMs infiltration was associated with poor prognosis, while a high M1/M2 ratio predicted better prognosis.	[41] Yuan, X. et al.
Bladder Cancer	A high TAMs count was associated with a lower 5-year survival rate with respect to low TAMs count, and it was proposed as a predictor of clinical outcome.	[44] Hanada, T. et al.
Colorectal Cancer	Macrophage infiltration was associated with improved patient survival. No difference in prognosis was found in patients with different ratios of M1/M2 infiltration.	[45] Edin, S. et al.
Non-small cell lung cancer	Independently of the density of CD68⁺ TAMs, the localization and M1/M2 polarization were suggested as potential prognostic markers.	[42] Mei, J. et al.

3. Macrophage Targeting

Despite macrophages being characterized by an elevated grade of plasticity, they have been the interest of researchers as potential therapeutic targets. They can acquire different phenotypes, frequently overlapping each others, in response to the continuous environmental stimuli they are subjected to. For these reasons, the design of therapeutic interventions has been very challenging, due most of all to the lack of unambiguous markers that can identify and distinguish pathological macrophages from their safe and useful counterparts. Several strategies have been pursued and adopted with the aim of manipulating macrophages, and their re-education and depletion can be considered the most important methods. In this scenario, nanotechnology-based systems have been greatly implicated and have led to important results in preclinical models [51–53].

3.1. Macrophage Re-Education

Macrophage re-education is one of the most promising strategies realized in the field of anticancer therapy, aimed at re-programming M2-like, protumor TAMs to M1-like macrophages endowed with antitumor functions [36,54]. In this field, nanotechnology-based systems have been explored, resulting as being applicable for successful preclinical achievements [12]. Recently, Song and colleagues [55] used bioconjugated manganese dioxide nanoparticles to efficiently prime TAMs to an antitumor M1-like phenotype, also resulting in higher sensitivity and response to chemotherapy, via the reduction of tumor hypoxia, in a murine breast cancer model.

Furthermore, Zanganeh and colleagues demonstrated that ferumoxytal nanoparticles significantly inhibited tumor growth in a breast cancer model, with an increased number of M1 macrophages at the tumor site [56]. In the same work, ferumoxital nanoparticles were able to inhibit liver and lung metastases in a mouse model of small cell lung cancer. Also in this case, metastases reduction was paralleled by the reversal of TAMs to the tumoricidal M1 phenotype.

An interesting study performed by Ortega and colleagues [57] demonstrated that mannosylated nanoparticles (Mn-NPs), encapsulating small interfering RNAs (siRNAs) against IκBα, were able to selectively target TAMs via the mannose receptor (CD206) and to restore the NF-κB signaling pathways, finally resulting in the induction of cytotoxic and immunostimulatory functions of TAMs in vitro. In a previous work, the same authors demonstrated that Mn-NPs are biocompatible and able to enhance the uptake of TAMs in vivo [58]. This last finding enables one to envisage that the proposed Mn-NPs could be translated to the in vivo setting.

Huang and colleagues developed a nucleic acid-encapsulating carrier endowed with two important features: (1) selectivity for TAMs (targeted to the macrophage galactose-type lectin I) and (2) microenvironment-responsiveness (pH-sensitive). These nanocarriers, encapsulating anti-IL-10 and anti-IL-10R oligonucleotides, were tested in a preclinical mouse model of liver cancer, and were demonstrated to be effective in suppressing the protumor functions of TAMs, stimulating their antitumor activities [59].

3.2. Macrophage Depletion

In pathological conditions, the more radical approach used to block macrophages is represented by their direct killing, named "macrophage depletion". This strategy is based on the use of either depleting antibodies, such as anti-colony-stimulating factor 1 receptor (anti-CSF1R), or molecules exerting a specific toxicity against macrophages, such as bisphosphonates (i.e., clodronate and zoledronic acid) and trabectedin.

3.2.1. Monoclonal Antibodies to the Colony-Stimulating Factor 1 Receptor

In a recent study performed in colorectal adenocarcinoma and fibrosarcoma mouse models, the depletion of TAMs was obtained by the use of a high-affinity humanized anti-colony-stimulating factor 1 receptor (CSF1R) monoclonal antibody (RG7155) [60]. The treatment with RG7155 induced a selective apoptosis of $CSF1R^+CD163^+$ M2-like macrophages, but not of $CD80^+$ M1-like macrophages, in vitro, and depleted TAMs in tumor-bearing mice. This was also paralleled by an increase of cytotoxic effector $CD8^+$ T cells, followed by delayed tumor growth and metastases formation.

In cervical and mammary carcinoma mouse models, the depletion of TAMs, obtained by means of a highly selective CSF1R inhibitor, resulted in the arrest and delay of tumor growth, respectively. Moreover, the depletion of macrophages was paralleled by the increase of $CD8^+$ T cells infiltrating cervical and breast carcinomas [61].

More recently, the evaluation of the safety, pharmacokinetics, pharmacodynamics, and antitumor activity of a fully human antibody against CSF1R (AMG 820) in a first-in-humans phase-I study for advanced solid tumors was published [62]. This study demonstrated that the anti-CSF1R antibody AMG 820 was tolerated with manageable toxicity up to 20 mg/kg every two weeks. Nevertheless, the treatment with AMG 820 as the single agent resulted in limited antitumor activity.

3.2.2. Liposomal Bisphosphonates

Liposomes, self-closed structures characterized by one or more phospholipid bilayers delimiting an inner aqueous space, were developed many years ago, and soon after demonstrated their great potential as vehicles for drug delivery [63–65]. Since their discovery more than 50 years ago, they have been a matter of extensive investigation, finding applicability in different medical areas. At present, due to both biological and technological features, they are considered to be the most successful drug delivery systems developed to date [66]. Several liposomal formulations are in clinical trials and some of them have been already released on the market for the cure of different kind of diseases [65,67].

The use of liposomes for drug delivery and their consequent therapeutic efficacy strictly depend on their physicochemical properties, with size and membrane charge being the most important ones [65]. Indeed, the fate of administered liposomes is closely associated to their physical features. For these reasons, researchers have been focusing their attention on the optimization of methods to synthesize liposomes suitable for in vivo use. Indeed, the inherent liposome composition renders them as extremely versatile and safe tools for medical purposes, due to their high biocompatibility and biodegradability, low immunogenicity, and drug protection.

Depending on the nature of the diseased tissues to be targeted, it is important to take into account the overall dimension of the liposome-based cargos [68,69]. Indeed, inflammatory or solid tumor tissues are characterized by enhanced vascular permeability. This feature can be exploited by nanocarriers of appropriate size (namely 50–100 nm), which can passively enter the leaky vessels of pathological tissues, accumulating at the target site. This phenomenon is known as the enhanced permeability and retention effect (EPR) and it is of particular interest for drug-loaded liposomes used for cancer treatment [70,71]. On the other hand, larger liposomes easily interact with plasma proteins undergoing opsonisation, a prerequisite for their recognition and capture by cells of the reticulum endothelial system (RES). The uptake of liposomes by phagocytes of the RES can be prevented by the surface modification of the liposomes themselves. Regarding this matter, the most used method is represented by the insertion of polyethylene glycol (PEG) chains at the liposome surface with the aid of cross-linker lipids [72–74]. Outer membrane modification with PEG (a method defined as "PEGylation") confers to liposomes stealth features, leading to increased blood circulation time, a great characteristic for drug-delivery systems suitable for application in cancer therapy [75].

As underlined before, liposomes present the inherent property to target mononuclear phagocytic cells. This feature can be exploited in the case of a macrophage-targeted therapy. In this regard, large multilamellar clodronate-contained liposomes were developed by van Rooijen and colleagues in the 1990s as specific tools for the transient depletion of macrophages [76]. Clodronate belongs to the drug family of bisphosphonates and is used for the treatment of osteolytic bone disease and post menopausal osteoporosis because of its ability to inhibit osteoclast function [77].

Large clodronate liposomes, due to their big size, are rapidly recognized and taken up by macrophages. Once captured and ingested by them, clodronate liposomes are exposed to lysosomal phospholipases that, after degradation of the phospholipid bilayer, lead to the release of the entrapped drug within the cytoplasmic compartment of the cells. This, in turn, causes apoptotic cell death via an ATP-dependent mechanism [76]. This technique is called "liposome-mediated macrophage suicide".

Clodronate liposomes can efficiently deplete tissue-resident macrophages of different body districts, preferentially depending on the route of administration [78]. Indeed, Kupffer cells, spleen macrophages, and bone marrow macrophages can be depleted by intravenous injection (*i.v.*) of clodronate liposomes, while intraperitoneal injection (*i.p.*) is needed for peritoneal macrophages. Moreover, the subcutaneous (*s.c.*) administration is able to deplete macrophages of the draining lymph nodes.

3.3. *Nanomedicine and Macrophage Functionality: Disease Models*

This section focuses on the application of macrophage depletion in a series of mouse disease models.

3.3.1. Inflammatory Diseases

Rheumatoid arthritis is one of the first inflammatory diseases in which liposomal clodronate was applied [79,80]. In experimental mouse models of arthritis, a single intraarticular injection of clodronate liposomes determined the reversible depletion of synovial phagocytic cells, accompanied by the reduction of cartilage destruction [79,80]. In an open study conducted on rheumatoid arthritis-affected patients scheduled for knee joint replacement, a single intraarticular injection of clodronate liposomes led to effective macrophage depletion in the synovial lining [81]. Moreover, the depletion of macrophages in the lining was also paralleled by a significant reduction of the expression of adhesion molecules, such as ICAM-1 and VCAM-1. These molecules are widely expressed in the synovium of rheumatoid arthritis patients. Of note, the procedure was well tolerated and resulted as being nontoxic. This last study represents, to our knowledge, the only example of macrophage depletion performed in humans. The results obtained are very important and encouraging for further investigation in this field and for designing protocols to be applied in the treatment of rheumatoid arthritis.

In a study conducted by Bu and colleagues, it was concluded that visceral adipose tissue macrophages (VATMs) can be regarded as a potential target for the development of drugs to be used for both the prevention and therapy of obesity and obesity-related complications [82]. Indeed, in a high-fat-diet animal model, the depletion of VATMs obtained by intraperitoneal injection of clodronate liposomes blocked the high-fat-induced weight gain and the development of insulin resistance. Gene expression analysis demonstrated that VATMs depletion was associated to the downregulation of genes involved in lipogenesis and gluconeogenesis.

Similar results were also obtained by Feng and colleagues, which demonstrated that the intraperitoneal injection of clodronate liposomes in diet-induced obese (DIO) mice reduced VATMs and improved systemic glucose homeostasis and insulin sensitivity. This was also paralleled by increased blood levels of the insulin-sensitizing hormone adiponectin [83].

Endometriosis, although its pathogenesis is still matter of debate, can be considered to be a chronic inflammatory disease that affects a large fraction of menstruating women. It is characterized by the growth of vascularized endometrial tissue in aberrant locations outside of the uterus, principally the pelvis, and it is associated with symptoms such as chronic pelvic pain, dysmenorrhea, and reduced fertility [84,85]. The endometriosis process is also accompanied by altered immune surveillance in the local peritoneal microenvironment. It has been reported that infiltrating macrophages are abundant in endometriotic lesions, where they contribute to the production of elevated amounts of pro-inflammatory and chemotactic cytokines.

In this context, the depletion of macrophages by the use of clodronate liposomes was very useful to understand their role in the establishment and development of the disease. In an experimental mouse model of endometriosis [86], clodronate liposomes were *i.p.* injected at different time points, with respect to the implantation of the endometrium. Two different treatment protocols were performed. In the first one, macrophages were depleted at early times; specifically, clodronate liposomes were injected at days 0, 4, and 8 from the implantation of the endometrial tissue in recipient mice. Clodronate liposomes significantly reduced the percentage of F4/80- and CD11b-positive cells in the peritoneum of sacrificed animals, with respect to mice treated with PBS-containing liposomes. This was also accompanied by a significant reduction in the weight of endometriotic lesions. These data indicate that, in the absence of macrophages, the syngeneic endometrium retains the ability to adhere to the peritoneal layer; nevertheless, the lesions fail to grow. In the second set of experiments, macrophages were depleted at later times (4 and 8 days after lesion engraftment), when endometriotic lesions had already been established and organized. In this case, macrophages depletion did not affect the number of endometriotic lesions, which was similar in treated and untreated animals. However, the total weight of lesions was significantly lower in mice subjected to ablative treatment. The results obtained indicated that the recruitment of the macrophages into the lesions is not only an early event sufficient for the initial development of the lesions, but it is also a necessary step for their subsequent successful establishment. These findings are straightforward because they suggest that it could be possible to treat human endometriosis by depleting local macrophages. Furthermore, it has been established that macrophages in the peritoneal fluid and/or those infiltrating the endometriotic lesions of patients express markers of alternative activation, such as CD163 (hemoglobin scavenger receptor) and CD206. In the forthcoming future, it can be envisaged to develop new formulations of clodronate liposomes directly targeted to these antigens. Due to the fact that macrophages have also a protective function and are involved in homeostatic tissue processes, the possibility of selectively targeting a specific subset of them is of particular interest for the development of patient-tailored treatment protocols.

Several studies have been also conducted on inflammatory diseases of the lung, such as granulomatous inflammation in response to *Mycobacterium tuberculosis* and chronic obstructive pulmonary disease (COPD). COPD is a life-threatening inflammatory disease of the lung characterized by chronic airway inflammation, mucus hypersecretion, and airway remodeling. In a cigarette smoke-induced COPD mouse model, Beckett and colleagues provided evidence for the pivotal role of macrophages in the pathogenesis of the disease [87]. Indeed, macrophage depletion in the lung, obtained by means of intranasal administration of clodronate liposomes given along an 8-week period of smoke exposure, resulted in a reduced smoke-induced epithelial thickening and emphysema development. Furthermore, macrophage depletion also determined protection against alteration of the lung function.

In a mouse model of pulmonary tuberculosis, it was demonstrated that the depletion of alveolar macrophages (AM), achieved by intranasal delivery of clodronate liposomes, was able to protect mice from lethality. AM depletion was associated with reduced outgrowth of mycobacteria in lungs and liver and to a polarized production of type-I cytokines in the lung tissue; moreover, AM-depleted mice displayed deficient granuloma formation. These results provide new insights for the design of novel therapeutic strategies against intracellular bacterial diseases [88].

3.3.2. Cancer

The direct killing of TAMs is a strategy aimed at eradicating macrophages at tumor sites, with the intention to abolish and/or interrupt the network of signals that favors tumor growth and progression.

Zhan and colleagues used a glucomannan polysaccharide with high affinity for the mannose receptor to specifically deliver alendronate (ALN) to TAMs. In both in vitro and in vivo experiments, this ALN–glucomannan bioconjugated formulation preferentially accumulated into macrophages, leading them to undergo apoptosis. Furthermore, in a subcutaneous sarcoma mouse model, the intratumor injection of ALN–glucomannan bioconjugated formulation determined the effective depletion of TAMs at the tumor site, reduced angiogenesis, led to the recovery of local immune surveillance, and significantly reduced tumor progression [89].

Hattori and colleagues tested the effectiveness of folate (FL)-decorated zoledronic acid (ZOL)-encapsulating liposomes in inhibiting tumor angiogenesis and tumor growth in a murine colon adenocarcinoma model. Here, although the FL–ZOL liposomes were able to induce selective cytotoxicity in vitro via the folate receptor, a severe toxicity limited their use in vivo [90].

Clodronate liposomes have been successfully applied in various mouse models of cancer with the aim of depleting macrophages at the tumor site. Banciu and colleagues demonstrated that the depletion of TAMs in a murine melanoma model was associated with a reduction of tumor volume [91]. Moreover, in a teratocarcinoma model, the depletion of TAMs resulted in the inhibition of tumor angiogenesis and in the inhibition of tumor growth [92]. In a chemically induced mouse lung adenocarcinoma model, the effective depletion of TAMs by means of clodronate liposomes was paralleled by a 50% reduction of tumor burden and tumor cell proliferation, as detected by the Ki67 marker, with respect to control animals treated with a vehicle only [93]. In this model, the treatment was more effective in slowing the growth of larger tumors, while the overall number of lung tumor nodules was not affected, being similar in treated and control mice.

Nevertheless, one of the major concerns of the use of clodronate liposomes to deplete TAMs relies on the fact that ablating strategies may also compromise the protective functions of macrophages themselves. Indeed, therapeutic approaches based on macrophage depletion have been associated with an increased risk of infection [94]. For this reason, our laboratory recently developed a new clodronate-containing liposomal formulation able to reach the tumor site without being massively detected and taken up by the cells of the MPS. The development of liposomes of appropriate physicochemical properties (i.e., size, Z-potential, and membrane modification) leads to the depletion of macrophages at different sites. Indeed, as already described before, large-sized liposomes are rapidly taken up by phagocytic cells [65], representing an excellent tool to preferentially target tissue macrophages rather than TAMs. Thus, to safely reach the tumor site, avoiding early recognition by the MPS, liposomes are subjected to PEGylation [74,95].

In a recent study, Piaggio and colleagues developed a new liposomal formulation of clodronate (Clo-Lipo-DOTAP) characterized by stealth features and small size, necessary to gain long circulating time and EPR-mediated tumor targeting properties, respectively. Moreover, the new formulation was not toxic (blood levels of kidney and liver markers were not altered), was cell-specific (immune cells other than macrophages were not affected), and resulted effective at clodronate dosage from 5 to 10 times lower than that used in previous studies [53].

Clo-Lipo-DOTAP were used in two xenograft mouse models of melanoma, the first one resembling the hallmarks of early stage primary tumor, and the second representing the lung metastatic behavior of advanced melanoma [53]. In both cases, the treatment with Clo-Lipo-DOTAP showed inhibition of tumor growth. In the primary tumor model, the depletion of TAMs determined a reduction of the volume of the subcutaneous tumors; this was further accompanied by a reduced tumor vascularization.

In the metastatic melanoma model, the antitumor effect driven by TAMs depletion was evident in the significant decrease of pulmonary tumor nodules, also paralleled by a reduction of tumor microvessel density.

Plasma levels of a number of cytokines, chemokines, and growth factors involved in and/or related to TAMs polarization, chemoattraction, and functions (IL-10, Mo KC, TNF-α, VEGF, and PDGF-bb) were statistically decreased in Clo-Lipo-DOTAP-treated mice with respect to controls. This result further highlights the antitumor efficacy of this new formulation.

Interestingly, tumor masses of treated animals, although significantly smaller with respect to controls, present a proliferating index comparable to that derived from tumors of untreated mice [53], as underlined by a similar percentage of Ki67-positive nuclei. These findings clearly indicate that the antitumor effectiveness, obtained after the depletion of TAMs, was derived from the inhibition of tumor angiogenesis and from the reduction of inflammatory-related signals, while it was not the result of direct tumor cell killing. All together, these data are encouraging and suggest that this approach could be used, in combination with first-line therapy, for the treatment of advanced-stage melanoma. Indeed, despite the recent application of promising protocols based on immunotherapy and targeted therapy [96], melanoma continues to be a very challenging tumor (Table 2).

Table 2. Nanomedicine and macrophage functionality: macrophage depletion in disease models. *i.t.*: intratracheal; *i.v.*: intravenous; *s.c.*: subcutaneous; *i.p.*: intraperitoneum; *i.a.*: intraarticular.

Disease Model	Clodronate Liposomal (Administration Schedule)	Achievements	References
Rheumatoid arthritis	(1) Single *i.a.* injection of clodronate liposomes in a mouse model of arthritis. (2) Single *i.a.* injection of clodronate liposomes in rheumatoid arthritis-affected patients	(1) The reversible depletion of synovial macrophages was accompanied by the reduction of cartilage destruction. (2) The effective depletion of macrophages in the synovial lining was paralleled by the reduction of adhesion molecules (ICAM-1 and VCAM-1 in the lining).	[79] van Lent P.L. et al. [80] van Lent P.L. et al. [81] Barrera P. et al.
Endometriosis	I treatment schedule: clodronate liposomes were *i.p.* injected at days 0, 4, and 8 with respect to the injection of endometrial tissues in recipient mice. II treatment schedule: clodronate liposomes were *i.p.* injected at days 4 and 8 with respect to endometriotic lesions implantation.	I schedule: significant reduction of F4/80$^+$ and CD11b$^+$ cells in the peritoneum of sacrificed mice was observed with respect to those treated with PBS-containing liposomes. Reduction in the weight of endometriotic lesions. II schedule: the delayed treatment did not affect the number of endometriotic lesions, while their total weight was statistically reduced.	[86] Bacci M. et al.
Lung adenocarcinoma (chemically-induced mouse lung adenocarcinoma)	Single *i.t.* instillation followed by *i.v.* injection, starting 2 days after *i.t.* administration and continuing once weekly for 5 consecutive weeks.	After 4–6 weeks of clodronate liposome treatment, a 50% decrease in the number of alveolar macrophages was observed. Tumour burden was reduced by 50% compared to vehicle-treated mice. The proliferation index of tumour cells (Ki67) was also attenuated.	[93] Fritz J.M. et al.
Melanoma (B16/F10 mouse melanoma model)	Clo-Lipo-DOTAP formulation was administered *i.v.*, starting 7 days after B16/F10 cells inoculum, and was repeated 3 times with 3-day interval between injections. The same schedule was used for both *s.c.* and metastatic model.	*s.c.* model: depletion of TAMs was associated with a reduction of tumor volumes and reduced tumor vascularization. Metastatic model: reduction of pulmonary tumor nodules and decreased tumor microvessel density was observed.	[53] Piaggio F. et al.

4. Conclusions and Future Perspectives

The nanotechnology-based systems described in this review and intended to manipulate macrophages in both inflammatory diseases and cancer have been demonstrated to be beneficial in improving the outcome in preclinical mouse models, holding promise for the future design of therapeutic interventions (Table 3). However, it appears evident that macrophage-targeted therapies are intrinsically characterized by limitations. Indeed, their effectiveness is closely related to an in-progress therapeutic window. In fact, the phenotypic re-reversal and re-population of macrophages in the case of re-education and depletion strategies, respectively, can occur after the discontinuance of the treatment.

Table 3. Nanotechnology-based systems and macrophage manipulation in preclinical models. ODNs: oligonucleotides.

Type of Nanocarriers		Pre-Clinical Models	Results	References
Passive targeting	Ferumoxytal nanoparticles	Breast cancer Small cell lung cancer	Inhibition of tumor growth (breast cancer) and liver and lung metastases (small cell lung cancer) mediated by reversal of TAMs to tumoricidal M1 macrophages.	[56] Zanganeh, S. et al.
	Clodronate Liposomes	Rheumatoid arthritis	Depletion of macrophages obtained by intra-articular injection of clodronate liposomes led to the reduction of cartilage destruction.	[79] van Lent, P.L. et al. [80] Van Lent, P.L. et al.
	Clodronate Liposomes	Obesity	Depletion of VATMs obtained by intraperitoneal injection of clodronate liposomes resulted in the block of high-fat-induced weight gain and insulin resistance	[82] Bu, L. et al.
	Clodronate Liposomes	Endometriosis	Depletion of macrophages obtained by intraperitoneal injection of clodronate liposomes resulted in weight reduction of endometriotic lesions	[86] Bacci, M. et al.
	Clodronate Liposomes	Melanoma	Depletion of TAMs was associated with a reduction of tumor volume	[89] Banciu M. et al.
	Clodronate Liposomes	Teratocarcinoma	Depletion of TAMs was associated with the inhibition of tumor angiogenesis and tumor growth	[92] Zeisberger, S.M. et al.
	Clo-Lipo-DOTAP	Melanoma	Depletion of TAMs was associated with a reduction of primary tumor growth and number of lung tumor nodules paralleled by the inhibition of tumor angiogenesis.	[53] Piaggio, F. et al.
Active targeting	Mannosylated nanoparticles encapsulating siRNAs against IκBα	Ex vivo macrophages	The restoration of NF-κB signaling pathways determined the induction of cytotoxic and immunostimulatory activities of TAMs	[57] Ortega, R.A. et al.
	Galactose-type lectin I-decorated nanocarriers encapsulating IL-10 and IL-10R ODNs	Liver cancer	Suppression of protumor functions of TAMs led to induction of their antitumor activities with final inhibition of tumor growth.	[59] Huang, Z. et al.
	ALN–glucomannan bioconjugates	Sarcoma	TAM depletion led to the local recovery of the immune surveillance at the tumor site and resulted in reduced tumor progression.	[89] Zhan, X. et al.
	Folate-decorated zoledronic acid-containing liposomes	Colon adenocarcinoma	Zoledronic liposomes targeted to the folate receptor were able to induce selective cytotoxicity in vitro; nevertheless, their use in vivo was limited due to the severe toxicity of zoledronic acid.	[90] Hattori, Y. et al.

Moreover, the overall challenge that nanotechnology-based systems have to face is represented by their poor selectivity. The continuous changes in phenotype/activation state that macrophages adopt in response to the constant environmental stimuli result in the unspecific targeting of tissue-resident macrophages, which play a pivotal role in tissue homeostasis and body defense. Indeed, the depletion of macrophages is frequently associated to increased risk of infection [97].

The aforementioned limitations determine the need to find novel treatments able to balance the cost–benefit ratio of this approach. More durable and specifically targeted interventions might be necessary to obtain long-lasting results, while sparing resident macrophages. In light of new insights on the activation states and related markers of macrophages in several human diseases, it is reasonable to hypothesize that it will be feasible to develop, in the near future, new nanocarriers targeted to specific subsets of macrophages involved in the pathogenesis and/or progression of a given pathology.

As a note of caution, we think that further investigations are needed to attempt translating the results so far obtained to the clinic. Indeed, despite the positive results obtained in preclinical studies and the proof-of-concept clinical study performed on rheumatoid arthritis [81], to our knowledge, no approved clinical trials are currently running and/or recruiting patients. The lack of translation to the clinical practice could be due to the intrinsic limitation related to the method of macrophage depletion and to the main obstacles usually faced by nanodelivery systems (i.e., pharmaceutical manufacturing, government regulations, intellectual properties), issues already reviewed elsewhere [75]. Moreover, the phase-I study performed with the use of the anti-CSF1R antibody also revealed its limits, supporting the hypothesis that this therapeutic approach might be applied only in a combination therapy [62]. In our opinion, a treatment strategy based on the combination of macrophage re-programming and macrophage depletion, which could result in a more deep and long-lasting microenvironment priming, should lead to improved therapeutic results in both inflammatory diseases and cancer. In this regard, a nanotechnology-based system could be an exceptionally suitable moiety. The development of nanocarriers decorated with macrophage-specific ligands and loaded with a combination of molecules for both macrophage killing and re-programming would, indeed, represents a true step ahead to design innovative protocols for macrophage-driven pathologies. This would pave the way for future patient-tailored therapies.

Author Contributions: C.B. participated to the conception and design of the study, collection of the data, and drafting of the manuscript. M.P., F.P., D.D.P., and P.P. participated in revising and discussing the accuracy and the intellectual content of the manuscript and to approving its final version.

Funding: This work was partially funded by a grant from the Associazione Italiana Ricerca Cancro (AIRC IG 18474 to M.P.).

Conflicts of Interest: The authors declare no conflict of interest.

References

1. Wynn, T.A.; Chawla, A.; Pollard, J.W. Macrophage biology in development, homeostasis and disease. *Nature* **2013**, *496*, 445–455. [CrossRef] [PubMed]
2. Schultze, J.L.; Schmieder, A.; Goerdt, S. Macrophage activation in human diseases. *Semin. Immunol.* **2015**, *27*, 249–256. [CrossRef] [PubMed]
3. Mantovani, A.; Biswas, S.K.; Galdiero, M.R.; Sica, A.; Locati, M. Macrophage plasticity and polarization in tissue repair and remodelling. *J. Pathol.* **2013**, *229*, 176–185. [CrossRef] [PubMed]
4. Sica, A.; Mantovani, A. Macrophage plasticity and polarization: In vivo veritas. *J. Clin. Investig.* **2012**, *122*, 787–795. [CrossRef] [PubMed]
5. Mills, C.D. M1 and M2 Macrophages: Oracles of Health and Disease. *Crit. Rev. Immunol.* **2012**, *32*, 463–488. [CrossRef] [PubMed]
6. Martinez, F.O.; Gordon, S. The M1 and M2 paradigm of macrophage activation: Time for reassessment. *F1000Prime Rep.* **2014**, *6*, 13. [CrossRef] [PubMed]
7. Gautier, E.L.; Shay, T.; Miller, J.; Greter, M.; Jakubzick, C.; Ivanov, S.; Helft, J.; Chow, A.; Elpek, K.G.; Gordonov, S.; et al. Gene-expression profiles and transcriptional regulatory pathways that underlie the identity and diversity of mouse tissue macrophages. *Nat. Immunol.* **2012**, *13*, 1118–1128. [CrossRef] [PubMed]
8. Vogel, D.Y.; Vereyken, E.J.; Glim, J.E.; Heijnen, P.D.; Moeton, M.; van der Valk, P.; Amor, S.; Teunissen, C.E.; van Horssen, J.; Dijkstra, C.D. Macrophages in inflammatory multiple sclerosis lesions have an intermediate activation status. *J. Neuroinflamm.* **2013**, *10*, 809. [CrossRef] [PubMed]
9. Pettersen, J.S.; Fuentes-Duculan, J.; Suarez-Farinas, M.; Pierson, K.C.; Pitts-Kiefer, A.; Fan, L.; Belkin, D.A.; Wang, C.Q.; Bhuvanendran, S.; Johnson-Huang, L.M.; et al. Tumor-associated macrophages in the cutaneous SCC microenvironment are heterogeneously activated. *J. Investig. Dermatol.* **2011**, *131*, 1322–1330. [CrossRef] [PubMed]
10. Schultze, J.L. Reprogramming of macrophages—New opportunities for therapeutic targeting. *Curr. Opin. Pharmacol.* **2016**, *26*, 10–15. [CrossRef] [PubMed]
11. Singh, Y.; Pawar, V.K.; Meher, J.G.; Raval, K.; Kumar, A.; Shrivastava, R.; Bhadauria, S.; Chourasia, M.K. Targeting tumor associated macrophages (TAMs) via nanocarriers. *J. Control. Release* **2017**, *254*, 92–106. [CrossRef] [PubMed]

12. Andon, F.T.; Digifico, E.; Maeda, A.; Erreni, M.; Mantovani, A.; Alonso, M.J.; Allavena, P. Targeting tumor associated macrophages: The new challenge for nanomedicine. *Semin. Immunol.* **2017**, *34*, 103–113. [CrossRef] [PubMed]
13. Shi, J.; Votruba, A.R.; Farokhzad, O.C.; Langer, R. Nanotechnology in drug delivery and tissue engineering: From discovery to applications. *Nano Lett.* **2010**, *10*, 3223–3230. [CrossRef] [PubMed]
14. Davies, L.C.; Jenkins, S.J.; Allen, J.E.; Taylor, P.R. Tissue-resident macrophages. *Nat. Immunol.* **2013**, *14*, 986–995. [CrossRef] [PubMed]
15. Tauber, A.I. Metchnikoff and the phagocytosis theory. *Nat. Rev. Mol. Cell Biol.* **2003**, *4*, 897–901. [CrossRef] [PubMed]
16. Van Furth, R.; Cohn, Z.A.; Hirsch, J.G.; Humphrey, J.H.; Spector, W.G.; Langevoort, H.L. Mononuclear phagocytic system: New classification of macrophages, monocytes and of their cell line. *Bull. World Health Organ.* **1972**, *47*, 651–658. [PubMed]
17. Epelman, S.; Lavine, K.J.; Randolph, G.J. Origin and functions of tissue macrophages. *Immunity* **2014**, *41*, 21–35. [CrossRef] [PubMed]
18. Dey, A.; Allen, J.; Hankey-Giblin, P.A. Ontogeny and polarization of macrophages in inflammation: Blood monocytes versus tissue macrophages. *Front. Immunol.* **2014**, *5*, 683. [CrossRef] [PubMed]
19. Murray, P.J.; Allen, J.E.; Biswas, S.K.; Fisher, E.A.; Gilroy, D.W.; Goerdt, S.; Gordon, S.; Hamilton, J.A.; Ivashkiv, L.B.; Lawrence, T.; et al. Macrophage activation and polarization: Nomenclature and experimental guidelines. *Immunity* **2014**, *41*, 14–20. [CrossRef] [PubMed]
20. Guilliams, M.; Ginhoux, F.; Jakubzick, C.; Naik, S.H.; Onai, N.; Schraml, B.U.; Segura, E.; Tussiwand, R.; Yona, S. Dendritic cells, monocytes and macrophages: A unified nomenclature based on ontogeny. *Nat. Rev. Immunol.* **2014**, *14*, 571–578. [CrossRef] [PubMed]
21. Zaslona, Z.; Przybranowski, S.; Wilke, C.; van Rooijen, N.; Teitz-Tennenbaum, S.; Osterholzer, J.J.; Wilkinson, J.E.; Moore, B.B.; Peters-Golden, M. Resident alveolar macrophages suppress, whereas recruited monocytes promote, allergic lung inflammation in murine models of asthma. *J. Immunol.* **2014**, *193*, 4245–4253. [CrossRef] [PubMed]
22. Jiang, Z.; Zhu, L. Update on the role of alternatively activated macrophages in asthma. *J. Asthma Allergy* **2016**, 101–107. [CrossRef] [PubMed]
23. Moore, K.J.; Sheedy, F.J.; Fisher, E.A. Macrophages in atherosclerosis: A dynamic balance. *Nat. Rev. Immunol.* **2013**, *13*, 709–721. [CrossRef] [PubMed]
24. Bobryshev, Y.V.; Ivanova, E.A.; Chistiakov, D.A.; Nikiforov, N.G.; Orekhov, A.N. Macrophages and Their Role in Atherosclerosis: Pathophysiology and Transcriptome Analysis. *Biomed. Res. Int.* **2016**, 9582430. [CrossRef] [PubMed]
25. Udalova, I.A.; Mantovani, A.; Feldmann, M. Macrophage heterogeneity in the context of rheumatoid arthritis. *Nat. Rev. Rheumatol.* **2016**, *12*, 472–485. [CrossRef] [PubMed]
26. Li, J.; Hsu, H.C.; Mountz, J.D. Managing macrophages in rheumatoid arthritis by reform or removal. *Curr. Rheumatol. Rep.* **2012**, *14*, 445–454. [CrossRef] [PubMed]
27. Bondeson, J.; Wainwright, S.D.; Lauder, S.; Amos, N.; Hughes, C.E. The role of synovial macrophages and macrophage-produced cytokines in driving aggrecanases, matrix metalloproteinases, and other destructive and inflammatory responses in osteoarthritis. *Arthritis Res. Ther.* **2006**, *8*, R187. [CrossRef] [PubMed]
28. Hill, A.A.; Reid Bolus, W.; Hasty, A.H. A decade of progress in adipose tissue macrophage biology. *Immunol. Rev.* **2014**, *262*, 134–152. [CrossRef] [PubMed]
29. Weisberg, S.P.; McCann, D.; Desai, M.; Rosenbaum, M.; Leibel, R.L.; Ferrante, A.W., Jr. Obesity is associated with macrophage accumulation in adipose tissue. *J. Clin. Investig.* **2003**, *112*, 1796–1808. [CrossRef] [PubMed]
30. Lewis, C.E.; Pollard, J.W. Distinct role of macrophages in different tumor microenvironments. *Cancer Res.* **2006**, *66*, 605–612. [CrossRef] [PubMed]
31. Balkwill, F.R.; Mantovani, A. Cancer-related inflammation: Common themes and therapeutic opportunities. *Semin. Cancer Biol.* **2012**, *22*, 33–40. [CrossRef] [PubMed]
32. Noy, R.; Pollard, J.W. Tumor-associated macrophages: From mechanisms to therapy. *Immunity* **2014**, *41*, 49–61. [CrossRef] [PubMed]
33. Coussens, L.M.; Werb, Z. Inflammation and cancer. *Nature* **2002**, *6917*, 860–867. [CrossRef] [PubMed]
34. Pollard, J.W. Tumour-educated macrophages promote tumour progression and metastasis. *Nat. Rev. Cancer* **2004**, *1*, 71–78. [CrossRef] [PubMed]

35. Stout, R.D.; Watkins, S.K.; Suttles, J. Functional plasticity of macrophages: In situ reprogramming of tumor-associated macrophages. *J. Leukoc. Biol.* **2009**, *86*, 1105–1109. [CrossRef] [PubMed]
36. Mantovani, A.; Marchesi, F.; Malesci, A.; Laghi, L.; Allavena, P. Tumour-associated macrophages as treatment targets in oncology. *Nat. Rev. Clin. Oncol.* **2017**, *14*, 399–416. [CrossRef] [PubMed]
37. Pollard, J.W. Macrophages define the invasive microenvironment in breast cancer. *J. Leukoc. Biol.* **2008**, *84*, 623–630. [CrossRef] [PubMed]
38. Zhang, Q.W.; Liu, L.; Gong, C.Y.; Shi, H.S.; Zeng, Y.H.; Wang, X.Z.; Zhao, Y.W.; Wei, Y.Q. Prognostic significance of tumor-associated macrophages in solid tumor: A meta-analysis of the literature. *PLoS ONE* **2012**, *7*, e50946. [CrossRef] [PubMed]
39. Bingle, L.; Brown, N.J.; Lewis, C.E. The role of tumour-associated macrophages in tumour progression: Implications for new anticancer therapies. *J. Pathol.* **2002**, *196*, 254–265. [CrossRef] [PubMed]
40. Zhao, X.; Qu, J.; Sun, Y.; Wang, J.; Liu, X.; Wang, F.; Zhang, H.; Wang, W.; Ma, X.; Gao, X.; et al. Prognostic significance of tumor-associated macrophages in breast cancer: A meta-analysis of the literature. *Oncotarget* **2017**, *8*, 30576–30586. [CrossRef] [PubMed]
41. Yuan, X.; Zhang, J.; Li, D.; Mao, Y.; Mo, F.; Du, W.; Ma, X. Prognostic significance of tumor-associated macrophages in ovarian cancer: A meta-analysis. *Gynecol. Oncol.* **2017**, *147*, 181–187. [CrossRef] [PubMed]
42. Mei, J.; Xiao, Z.; Guo, C.; Pu, Q.; Ma, L.; Liu, C.; Lin, F.; Liao, H.; You, Z.; Liu, L. Prognostic impact of tumor-associated macrophage infiltration in non-small cell lung cancer: A systemic review and meta-analysis. *Oncotarget* **2016**, *7*, 34217–34228. [CrossRef] [PubMed]
43. Falleni, M.; Savi, F.; Tosi, D.; Agape, E.; Cerri, A.; Moneghini, L.; Bulfamante, G.P. M1 and M2 macrophages' clinicopathological significance in cutaneous melanoma. *Melanoma Res.* **2017**, *27*, 200–210. [CrossRef] [PubMed]
44. Hanada, T.; Nakagawa, M.; Emoto, A.; Nomura, T.; Nasu, N.; Nomura, Y. Prognostic value of tumor-associated macrophage count in human bladder cancer. *Int. J. Urol.* **2000**, *7*, 263–269. [CrossRef] [PubMed]
45. Edin, S.; Wikberg, M.L.; Oldenborg, P.A.; Palmqvist, R. Macrophages: Good guys in colorectal cancer. *Oncoimmunology* **2013**, *2*, e23038. [CrossRef] [PubMed]
46. Forssell, J.; Oberg, A.; Henriksson, M.L.; Stenling, R.; Jung, A.; Palmqvist, R. High macrophage infiltration along the tumor front correlates with improved survival in colon cancer. *Clin. Cancer Res.* **2007**, *13*, 1472–1479. [CrossRef] [PubMed]
47. Welsh, T.J.; Green, R.H.; Richardson, D.; Waller, D.A.; O'Byrne, K.J.; Bradding, P. Macrophage and mast-cell invasion of tumor cell islets confers a marked survival advantage in non-small-cell lung cancer. *J. Clin. Oncol.* **2005**, *23*, 8959–8967. [CrossRef] [PubMed]
48. Ohno, S.; Inagawa, H.; Dhar, D.K.; Fujii, T.; Ueda, S.; Tachibana, M.; Suzuki, N.; Inoue, M.; Soma, G.; Nagasue, N. The degree of macrophage infiltration into the cancer cell nest is a significant predictor of survival in gastric cancer patients. *Anticancer Res.* **2003**, *23*, 5015–5022. [PubMed]
49. Zhang, J.; Yan, Y.; Yang, Y.; Wang, L.; Li, M.; Wang, J.; Liu, X.; Duan, X.; Wang, J. High Infiltration of Tumor-Associated Macrophages Influences Poor Prognosis in Human Gastric Cancer Patients, Associates with the Phenomenon of EMT. *Medicine* **2016**, *95*, e2636. [CrossRef] [PubMed]
50. Wang, X.L.; Jiang, J.T.; Wu, C.P. Prognostic significance of tumor-associated macrophage infiltration in gastric cancer: A meta-analysis. *Genet. Mol. Res.* **2016**, *15*. [CrossRef] [PubMed]
51. Ahsan, F.; Rivas, I.P.; Khan, M.A.; Torres Suarez, A.I. Targeting to macrophages: Role of physicochemical properties of particulate carriers—Liposomes and microspheres—On the phagocytosis by macrophages. *J. Control. Release* **2002**, *79*, 29–40. [CrossRef]
52. Lee, W.H.; Loo, C.Y.; Traini, D.; Young, P.M. Nano- and micro-based inhaled drug delivery systems for targeting alveolar macrophages. *Expert Opin. Drug Deliv.* **2015**, *12*, 1009–1026. [CrossRef] [PubMed]
53. Piaggio, F.; Kondylis, V.; Pastorino, F.; Di Paolo, D.; Perri, P.; Cossu, I.; Schorn, F.; Marinaccio, C.; Murgia, D.; Daga, A.; et al. A novel liposomal Clodronate depletes tumor-associated macrophages in primary and metastatic melanoma: Anti-angiogenic and anti-tumor effects. *J. Control. Release* **2016**, 165–177. [CrossRef] [PubMed]
54. Aras, S.; Zaidi, M.R. TAMeless traitors: Macrophages in cancer progression and metastasis. *Br. J. Cancer* **2017**, *117*, 1583–1591. [CrossRef] [PubMed]

55. Song, M.; Liu, T.; Shi, C.; Zhang, X.; Chen, X. Bioconjugated Manganese Dioxide Nanoparticles Enhance Chemotherapy Response by Priming Tumor-Associated Macrophages toward M1-like Phenotype and Attenuating Tumor Hypoxia. *ACS Nano* **2016**, *10*, 633–647. [CrossRef] [PubMed]
56. Zanganeh, S.; Hutter, G.; Spitler, R.; Lenkov, O.; Mahmoudi, M.; Shaw, A.; Pajarinen, J.S.; Nejadnik, H.; Goodman, S.; Moseley, M.; et al. Iron oxide nanoparticles inhibit tumour growth by inducing pro-inflammatory macrophage polarization in tumour tissues. *Nat. Nanotechnol.* **2016**, *11*, 986–994. [CrossRef] [PubMed]
57. Ortega, R.A.; Barham, W.; Sharman, K.; Tikhomirov, O.; Giorgio, T.D.; Yull, F.E. Manipulating the NF-kappaB pathway in macrophages using mannosylated, siRNA-delivering nanoparticles can induce immunostimulatory and tumor cytotoxic functions. *Int. J. Nanomed.* **2016**, 2163–2177. [CrossRef] [PubMed]
58. Ortega, R.A.; Barham, W.J.; Kumar, B.; Tikhomirov, O.; McFadden, I.D.; Yull, F.E.; Giorgio, T.D. Biocompatible mannosylated endosomal-escape nanoparticles enhance selective delivery of short nucleotide sequences to tumor associated macrophages. *Nanoscale* **2015**, *7*, 500–510. [CrossRef] [PubMed]
59. Huang, Z.; Zhang, Z.; Jiang, Y.; Zhang, D.; Chen, J.; Dong, L.; Zhang, J. Targeted delivery of oligonucleotides into tumor-associated macrophages for cancer immunotherapy. *J. Control. Release* **2012**, *158*, 286–292. [CrossRef] [PubMed]
60. Ries, C.H.; Cannarile, M.A.; Hoves, S.; Benz, J.; Wartha, K.; Runza, V.; Rey-Giraud, F.; Pradel, L.P.; Feuerhake, F.; Klaman, I.; et al. Targeting tumor-associated macrophages with anti-CSF-1R antibody reveals a strategy for cancer therapy. *Cancer Cell* **2014**, *25*, 846–859. [CrossRef] [PubMed]
61. Strachan, D.C.; Ruffell, B.; Oei, Y.; Bissell, M.J.; Coussens, L.M.; Pryer, N.; Daniel, D. CSF1R inhibition delays cervical and mammary tumor growth in murine models by attenuating the turnover of tumor-associated macrophages and enhancing infiltration by $CD8^+$ T cells. *Oncoimmunology* **2013**, *2*, e26968. [CrossRef] [PubMed]
62. Papadopoulos, K.P.; Gluck, L.; Martin, L.P.; Olszanski, A.J.; Tolcher, A.W.; Ngarmchamnanrith, G.; Rasmussen, E.; Amore, B.M.; Nagorsen, D.; Hill, J.S.; et al. First-in-Human Study of AMG 820, a Monoclonal Anti-Colony-Stimulating Factor 1 Receptor Antibody, in Patients with Advanced Solid Tumors. *Clin. Cancer Res.* **2017**, *23*, 5703–5710. [CrossRef] [PubMed]
63. Gregoriadis, G. Drug entrapment in liposomes. *FEBS Lett.* **1973**, *36*, 292–296. [CrossRef]
64. Gregoriadis, G.; Ryman, B.E. Liposomes as carriers of enzymes or drugs: A new approach to the treatment of storage diseases. *Biochem. J.* **1971**, *124*, 58P. [CrossRef] [PubMed]
65. Allen, T.M.; Cullis, P.R. Liposomal drug delivery systems: From concept to clinical applications. *Adv. Drug Deliv. Rev.* **2013**, *65*, 36–48. [CrossRef] [PubMed]
66. Felice, B.; Prabhakaran, M.P.; Rodriguetablez, A.P.; Ramakrishna, S. Drug delivery vehicles on a nano-engineering perspective. *Mater. Sci. Eng. C Mater. Biol. Appl.* **2014**, 178–195. [CrossRef] [PubMed]
67. Bozzuto, G.; Molinari, A. Liposomes as nanomedical devices. *Int. J. Nanomed.* **2015**, 975–999. [CrossRef] [PubMed]
68. Euliss, L.E.; DuPont, J.A.; Gratton, S.; DeSimone, J. Imparting size, shape, and composition control of materials for nanomedicine. *Chem. Soc. Rev.* **2006**, *35*, 1095–1104. [CrossRef] [PubMed]
69. Debbage, P. Targeted drugs and nanomedicine: Present and future. *Curr. Pharm. Des.* **2009**, *15*, 153–172. [CrossRef] [PubMed]
70. Gabizon, A.A. Selective tumor localization and improved therapeutic index of anthracyclines encapsulated in long-circulating liposomes. *Cancer Res.* **1992**, *52*, 891–896. [PubMed]
71. Peer, D.; Karp, J.M.; Hong, S.; Farokhzad, O.C.; Margalit, R.; Langer, R. Nanocarriers as an emerging platform for cancer therapy. *Nat. Nanotechnol.* **2007**, *2*, 751–760. [CrossRef] [PubMed]
72. Immordino, M.L.; Dosio, F.; Cattel, L. Stealth liposomes: Review of the basic science, rationale, and clinical applications, existing and potential. *Int. J. Nanomed.* **2006**, *1*, 297–315.
73. Ishida, T.; Harashima, H.; Kiwada, H. Interactions of liposomes with cells in vitro and in vivo: Opsonins and receptors. *Curr. Drug Metab.* **2001**, *2*, 397–409. [CrossRef] [PubMed]
74. Gabizon, A.A. Stealth liposomes and tumor targeting: One step further in the quest for the magic bullet. *Clin. Cancer Res.* **2001**, *7*, 223–225. [PubMed]
75. Sercombe, L.; Veerati, T.; Moheimani, F.; Wu, S.Y.; Sood, A.K.; Hua, S. Advances and Challenges of Liposome Assisted Drug Delivery. *Front.Pharmacol.* **2015**, *6*, 286. [CrossRef] [PubMed]

76. Van Rooijen, N.; Sanders, A. Liposome mediated depletion of macrophages: Mechanism of action, preparation of liposomes and applications. *J. Immunol Methods* **1994**, *174*, 83–93. [CrossRef]
77. Ghinoi, V.; Brandi, M.L. Clodronate: Mechanisms of action on bone remodelling and clinical use in osteometabolic disorders. *Expert Opin. Pharmacother.* **2002**, *3*, 1643–1656. [CrossRef] [PubMed]
78. van Rooijen, N.; van Kesteren-Hendrikx, E. Clodronate liposomes: Perspectives in research and therapeutics. *J. Liposome Res.* **2002**, *12*, 81–94. [CrossRef] [PubMed]
79. Van Lent, P.L.; van den Bersselaar, L.; van den Hoek, A.E.; van de Ende, M.; Dijkstra, C.D.; van Rooijen, N.; van de Putte, L.B.; van den Berg, W.B. Reversible depletion of synovial lining cells after intra-articular treatment with liposome-encapsulated dichloromethylene diphosphonate. *Rheumatol. Int.* **1993**, *13*, 21–30. [CrossRef] [PubMed]
80. Van Lent, P.L.; Holthuysen, A.E.; Van Rooijen, N.; Van De Putte, L.B.; Van Den Berg, W.B. Local removal of phagocytic synovial lining cells by clodronate-liposomes decreases cartilage destruction during collagen type II arthritis. *Ann. Rheum. Dis.* **1998**, *57*, 408–413. [CrossRef] [PubMed]
81. Barrera, P.; Blom, A.; van Lent, P.L.; van Bloois, L.; Beijnen, J.H.; van Rooijen, N.; de Waal Malefijt, M.C.; van de Putte, L.B.; Storm, G.; van den Berg, W.B. Synovial macrophage depletion with clodronate-containing liposomes in rheumatoid arthritis. *Arthritis Rheum.* **2000**, *43*, 1951–1959. [CrossRef]
82. Bu, L.; Gao, M.; Qu, S.; Liu, D. Intraperitoneal injection of clodronate liposomes eliminates visceral adipose macrophages and blocks high-fat diet-induced weight gain and development of insulin resistance. *AAPS J.* **2013**, *15*, 1001–1011. [CrossRef] [PubMed]
83. Feng, B.; Jiao, P.; Nie, Y.; Kim, T.; Jun, D.; van Rooijen, N.; Yang, Z.; Xu, H. Clodronate liposomes improve metabolic profile and reduce visceral adipose macrophage content in diet-induced obese mice. *PLoS ONE* **2011**, *6*, e24358. [CrossRef] [PubMed]
84. Capobianco, A.; Rovere-Querini, P. Endometriosis, a disease of the macrophage. *Front. Immunol.* **2013**, *4*, 9. [CrossRef] [PubMed]
85. Ahn, S.H.; Monsanto, S.P.; Miller, C.; Singh, S.S.; Thomas, R.; Tayade, C. Pathophysiology and Immune Dysfunction in Endometriosis. *Biomed. Res. Int.* **2015**, 795976. [CrossRef] [PubMed]
86. Bacci, M.; Capobianco, A.; Monno, A.; Cottone, L.; Di Puppo, F.; Camisa, B.; Mariani, M.; Brignole, C.; Ponzoni, M.; Ferrari, S.; et al. Macrophages are alternatively activated in patients with endometriosis and required for growth and vascularization of lesions in a mouse model of disease. *Am. J. Pathol.* **2009**, *175*, 547–556. [CrossRef] [PubMed]
87. Beckett, E.L.; Stevens, R.L.; Jarnicki, A.G.; Kim, R.Y.; Hanish, I.; Hansbro, N.G.; Deane, A.; Keely, S.; Horvat, J.C.; Yang, M.; et al. A new short-term mouse model of chronic obstructive pulmonary disease identifies a role for mast cell tryptase in pathogenesis. *J. Allergy Clin. Immunol.* **2013**, *131*, 752–762. [CrossRef] [PubMed]
88. Leemans, J.C.; Juffermans, N.P.; Florquin, S.; van Rooijen, N.; Vervoordeldonk, M.J.; Verbon, A.; van Deventer, S.J.; van der Poll, T. Depletion of alveolar macrophages exerts protective effects in pulmonary tuberculosis in mice. *J. Immunol.* **2001**, *166*, 4604–4611. [CrossRef] [PubMed]
89. Zhan, X.; Jia, L.; Niu, Y.; Qi, H.; Chen, X.; Zhang, Q.; Zhang, J.; Wang, Y.; Dong, L.; Wang, C. Targeted depletion of tumour-associated macrophages by an alendronate-glucomannan conjugate for cancer immunotherapy. *Biomaterials* **2014**, *35*, 10046–10057. [CrossRef] [PubMed]
90. Hattori, Y.; Yamashita, J.; Sakaida, C.; Kawano, K.; Yonemochi, E. Evaluation of antitumor effect of zoledronic acid entrapped in folate-linked liposome for targeting to tumor-associated macrophages. *J. Liposome Res.* **2015**, *25*, 131–140. [CrossRef] [PubMed]
91. Banciu, M.; Metselaar, J.M.; Schiffelers, R.M.; Storm, G. Antitumor activity of liposomal prednisolone phosphate depends on the presence of functional tumor-associated macrophages in tumor tissue. *Neoplasia* **2008**, *10*, 108–117. [CrossRef] [PubMed]
92. Zeisberger, S.M.; Odermatt, B.; Marty, C.; Zehnder-Fjallman, A.H.; Ballmer-Hofer, K.; Schwendener, R.A. Clodronate-liposome-mediated depletion of tumour-associated macrophages: A new and highly effective antiangiogenic therapy approach. *Br. J. Cancer.* **2006**, *95*, 272–281. [CrossRef] [PubMed]
93. Fritz, J.M.; Tennis, M.A.; Orlicky, D.J.; Lin, H.; Ju, C.; Redente, E.F.; Choo, K.S.; Staab, T.A.; Bouchard, R.J.; Merrick, D.T.; et al. Depletion of tumor-associated macrophages slows the growth of chemically induced mouse lung adenocarcinomas. *Front Immunol.* **2014**, *5*, 587. [CrossRef] [PubMed]

94. Purnama, C.; Ng, S.L.; Tetlak, P.; Setiagani, Y.A.; Kandasamy, M.; Baalasubramanian, S.; Karjalainen, K.; Ruedl, C. Transient ablation of alveolar macrophages leads to massive pathology of influenza infection without affecting cellular adaptive immunity. *Eur. J. Immunol.* **2014**, *44*, 2003–2012. [CrossRef] [PubMed]
95. Li, S.D.; Huang, L. Stealth nanoparticles: High density but sheddable PEG is a key for tumor targeting. *J. Control. Release* **2010**, *145*, 178–181. [CrossRef] [PubMed]
96. Postow, M.A.; Callahan, M.K.; Wolchok, J.D. Immune Checkpoint Blockade in Cancer Therapy. *J. Clin. Oncol.* **2015**, *33*, 1974–1982. [CrossRef] [PubMed]
97. Schneider, C.; Nobs, S.P.; Heer, A.K.; Kurrer, M.; Klinke, G.; van Rooijen, N.; Vogel, J.; Kopf, M. Alveolar macrophages are essential for protection from respiratory failure and associated morbidity following influenza virus infection. *PLoS Pathog.* **2014**, *10*, e1004053. [CrossRef] [PubMed]

MDPI
St. Alban-Anlage 66
4052 Basel
Switzerland
Tel. +41 61 683 77 34
Fax +41 61 302 89 18
www.mdpi.com

International Journal of Molecular Sciences Editorial Office
E-mail: ijms@mdpi.com
www.mdpi.com/journal/ijms

www.ingramcontent.com/pod-product-compliance
Lightning Source LLC
Chambersburg PA
CBHW040316270926
51850CB00007BC/1500
9783039363742